C. Blagden, George Gilpin

Tables for Reducing the Quantities by Weight, in Any Mixture of Pure Spirit and Water,

to Those by Measure; And for Determining the Proportion, by Measure, of

Each of the Two Substances in Such Mixtures.

C. Blagden, George Gilpin

Tables for Reducing the Quantities by Weight, in Any Mixture of Pure Spirit and Water,
to Those by Measure; And for Determining the Proportion, by Measure, of Each of the Two Substances in Such Mixtures.

ISBN/EAN: 9783337346089

Printed in Europe, USA, Canada, Australia, Japan

Cover: Foto ©berggeist007 / pixelio.de

More available books at **www.hansebooks.com**

XX. *Tables for reducing the Quantities by Weight, in any Mixture of pure Spirit and Water, to those by Measure; and for determining the Proportion, by Measure, of each of the two Substances in such Mixtures.* By Mr. George Gilpin, *Clerk to the Royal Society.* Communicated by Sir Charles Blagden, *Knt. Sec. R. S.*

Read June 19th, 1794.

THESE tables are founded on the experiments of which the results were given in the Report and Supplementary Report on the best method of proportioning the excise on spirituous liquors. They are computed for every degree of heat from 30° to 80°, and for the addition or subtraction of every one part in a hundred of water or spirit; but as the experiments themselves were made only to every fifth degree of heat, and every five in the hundred of water or spirit, the intermediate places are filled up by interpolation in the usual manner, with allowance for second differences.

Every table consists of eight columns, and there are two tables for every degree of heat. In the first column of the *first* of the two tables, are given the proportions of spirit and water by weight, 100 parts of spirit being taken as the constant number, to which additions are made successively of one part of water from 1 to 99 inclusively. The first column in the *second* table has 100 parts of water for the constant number, with the parts of spirit decreasing successively by unity, from

100 to 1 inclusively. It must be observed, that each of these tables occupying one page, is divided in the middle for adapting it more conveniently to the size of the paper; but the whole of each page is to be considered as one continued table. The second column of all the tables gives the specific gravities of the corresponding mixtures of spirit and water in the first column, taken from the table of specific gravities in the Supplementary Report, the intermediate spaces being filled up by interpolation. In the third column 100 parts by measure of pure spirit, at the temperature marked on the top of every separate table, is assumed as the constant standard number, to which the respective quantities of water by measure, at the same temperature, are to be proportioned in the next column. The fourth column, therefore, contains the proportion of water by measure, to 100 measures of spirit, answering to the proportions by weight in the same horizontal line of the first column. The fifth column shews the number of parts which the quantities of spirit and water contained in the third and fourth columns would measure when the mixture has been completed; that is, the bulk of the whole mixture after the concentration, or mutual penetration, has fully taken place. The sixth column, deduced from the three preceding ones, gives the effect of that concentration, or how much smaller the volume of the whole mixture is, than it would be if there was no such principle as the mutual penetration. The seventh column shews the quantity of pure spirit by measure, at the temperature in the table, contained in 100 measures of the mixture laid down in the fifth column. Lastly, the eighth column gives the decimal multiplier, by means of which the quantity by measure of standard pure spirit, of ,825 specific gravity at

60° of heat, may at once be ascertained, the temperature and specific gravity of the liquor being given ; pursuant to the idea suggested in the Report, that " the simplest and most equitable " method of levying the duty on spirituous liquors would be, " to consider rectified spirit as the true and only excisable " matter."

It may be proper to add a short account of the method pursued in computing some of the columns of these tables. Columns I. II. and III. require no other explanation than has been already given. Col. IV. is obtained thus : divide the specific gravity of the pure spirit, at the temperature in the table, by the specific gravity of water at the same temperature: then, for the *first* of the two tables for each degree of heat, the proportion is, as 100 is to the quantity of water by weight in the first column, so is the quotient of the abovementioned division to the quantity of water by measure sought; for the *second* of the two tables the proportion is, as the quantity of spirit by weight in the first column is to 100, so is that same quotient to the quantity of water by measure sought.

Col. V. requires more calculation. The first step is to compute what the specific gravity of the mixture in question would be if no concentration took place ; to obtain which, the constant number 100 (indicating the quantity by measure) of pure spirit, is to be multiplied by the specific gravity of pure spirit at the temperature in the table, and the corresponding measure of water in the fourth column is also to be multiplied by its specific gravity at the given temperature ; these two products being added together, their sum is to be divided by the sum of the absolute quantities of spirit and water by measure in the same horizontal line of the third and fourth columns:

then the proportion is, as this quotient (or what the specific gravity would be without concentration) is to the real specific gravity as found in the same horizontal line of the second column of the table, so is the sum of the quantities of spirit and water in the third and fourth columns inversely to the bulk of the mixture.

Col. VI. is obtained by subtracting the real bulk of the mixture in col. V. from the sum of the quantities of spirit and water in col. III. and IV. the difference between them being the diminution occasioned by the concentration on that whole quantity.   Col. VII. is obviously to be computed by the following proportion : as the bulk of the whole quantity of the mixture in col. V. is to 100 (the constant quantity), so is 100 to the quantity of pure spirit *per cent.* at the temperature of the table. Col. VIII. is formed by reducing the volume of the spirit *per cent.* at the temperature of the table, to its volume at 60°, by the following proportion : as ,825 (the specific gravity of pure spirit at 60°) is to its specific gravity at the given temperature, so is the number in the seventh column to the volume of pure spirit, at 60° of heat, contained in 100 parts by measure of the mixture at the temperature of the table : this divided by 100 is the decimal multiplier sought ; the product of which into any measure of a spirituous liquor of the corresponding specific gravity and temperature, will be the true quantity of standard pure spirit, at 60° of heat, contained in that liquor.

It may very probably be thought right, for the future use of the revenue, to compute another set of tables, in which the degree of heat standing at the head of each table, the first column of it shall be even numbers of specific gravity.   This would be proper for looking out at once the quantities of spirit and water in a mixture, from its heat and specific gravity, as immediately determined by experiment. For scientific purposes also, tables should be constructed to shew the regular increments and decrements of the concentration, by equal variations in the proportions of spirit and water: but these, and others of a similar nature, which might be suggested, do not belong to the present subject. C. B.

TABLE I.   HEAT 30°.

I. Spirit and water by weight.	II. Specific gravity.	III. Spirit by measure.	IV. Water by measure.	V. Bulk of mixture.	VI. Diminution of bulk.	VII. Quantity of Spirit per cent.	VIII. Decimal multipliers.
Sp. + W.							
100 + 0	,83896	100	—	100,00	—	100,00	1,0169
1	,84129	—	0,84	100,72	0,12	99,28	1,0096
2	,84355	—	1,68	101,45	0,23	98,57	1,0024
3	,84575	—	2,51	102,18	0,33	97,87	,9953
4	,84788	—	3,35	102,91	0,44	97,18	,9882
100 + 5	,84995	—	4,19	103,64	0,55	96,49	,9812
6	,85197	—	5,03	104,38	0,65	95,80	,9743
7	,85394	—	5,87	105,12	0,75	95,13	,9674
8	,85586	—	6,71	105,87	0,84	94,46	,9606
9	,85773	—	7,54	106,61	0,93	93,80	,9539
100 + 10	,85957	—	8,38	107,36	1,02	93,14	,9472
11	,86138	—	9,22	108,11	1,11	92,50	,9406
12	,86316	—	10,06	108,86	1,20	91,86	,9341
13	,86490	—	10,90	109,61	1,29	91,23	,9277
14	,86660	—	11,73	110,37	1,36	90,61	,9214
100 + 15	,86825	—	12,57	111,12	1,45	89,99	,9151
16	,86984	—	13,41	111,87	1,54	89,38	,9089
17	,87139	—	14,25	112,63	1,62	88,77	,9028
18	,87291	—	15,09	113,39	1,70	88,17	,8967
19	,87440	—	15,93	114,15	1,78	87,58	,8907
100 + 20	,87585	—	16,77	114,91	1,86	87,00	,8847
21	,87729	—	17,60	115,67	1,93	86,42	,8788
22	,87871	—	18,44	116,44	2,00	85,85	,8730
23	,88010	—	19,28	117,21	2,07	85,29	,8673
24	,88147	—	20,12	117,98	2,14	84,73	,8616
100 + 25	,88282	—	20,96	118,75	2,21	84,18	,8560
26	,88414	—	21,80	119,53	2,27	83,64	,8505
27	,88544	—	22,63	120,31	2,32	83,10	,8451
28	,88672	—	23,47	121,09	2,38	82,57	,8397
29	,88797	—	24,31	121,87	2,44	82,05	,8344
100 + 30	,88921	—	25,15	122,66	2,50	81,53	,8291
31	,89043	—	25,99	123,43	2,56	81,02	,8239
32	,89163	—	26,83	124,21	2,62	80,51	,8188
33	,89281	—	27,66	124,98	2,68	80,01	,8137
34	,89397	—	28,50	125,76	2,74	79,52	,8087
100 + 35	,89511	—	29,34	126,53	2,81	79,03	,8037
36	,89623	—	30,18	127,31	2,87	78,55	,7988
37	,89733	—	31,02	128,09	2,93	78,07	,7939
38	,89842	—	31,86	128,87	2,99	77,60	,7891
39	,89949	—	32,69	129,65	3,04	77,13	,7844
100 + 40	,90054	—	33,54	130,43	3,11	76,77	,7797
41	,90158	—	34,38	131,21	3,17	76,21	,7751
42	,90260	—	35,22	131,99	3,23	75,76	,7705
43	,90361	—	36,05	132,77	3,28	75,32	,7660
44	,90460	—	36,89	133,55	3,34	74,88	,7615
100 + 45	,90558	—	37,72	134,33	3,39	74,44	,7570
46	,90654	—	38,56	135,12	3,44	74,01	,7526
47	,90749	—	39,40	135,90	3,50	73,58	,7483
48	,90842	—	40,24	136,69	3,55	73,16	,7440
49	,90933	—	41,08	137,47	3,61	72,74	,7397
100 + 50	,91023	100	41,92	138,26	3,66	72,33	,7355
51	,91111	—	42,75	139,04	3,71	71,92	,7314
52	,91197	—	43,59	139,83	3,76	71,51	,7273
53	,91282	—	44,43	140,62	3,81	71,11	,7232
54	,91366	—	45,27	141,41	3,86	70,72	,7191
100 + 55	,91449	—	46,11	142,20	3,91	70,33	,7151
56	,91531	—	46,95	142,99	3,96	69,94	,7112
57	,91612	—	47,78	143,78	4,00	69,55	,7073
58	,91691	—	48,62	144,57	4,05	69,17	,7034
59	,91770	—	49,46	145,36	4,10	68,80	,6996
100 + 60	,91847	—	50,30	146,15	4,15	68,42	,6958
61	,91923	—	51,14	146,94	4,20	68,05	,6920
62	,91998	—	51,98	147,73	4,25	67,69	,6883
63	,92072	—	52,81	148,52	4,29	67,33	,6846
64	,92145	—	53,65	149,32	4,33	66,97	,6810
100 + 65	,92217	—	54,49	150,11	4,38	66,62	,6774
66	,92288	—	55,33	150,90	4,43	66,27	,6739
67	,92358	—	56,17	151,70	4,47	65,92	,6704
68	,92427	—	57,01	152,49	4,52	65,58	,6669
69	,92496	—	57,84	153,29	4,55	65,24	,6634
100 + 70	,92563	—	58,68	154,08	4,60	64,90	,6600
71	,92629	—	59,52	154,88	4,64	64,57	,6566
72	,92694	—	60,36	155,67	4,69	64,24	,6532
73	,92758	—	61,20	156,47	4,73	63,91	,6499
74	,92821	—	62,04	157,26	4,78	63,59	,6466
100 + 75	,92889	—	62,87	158,06	4,81	63,27	,6434
76	,92951	—	63,71	158,85	4,86	62,95	,6402
77	,93012	—	64,55	159,65	4,90	62,63	,6370
78	,93078	—	65,39	160,45	4,94	62,32	,6338
79	,93132	—	66,23	161,25	4,98	62,02	,6306
100 + 80	,93191	—	67,07	162,05	5,02	61,71	,6275
81	,93249	—	67,90	162,84	5,06	61,41	,6244
82	,93306	—	68,74	163,64	5,10	61,11	,6214
83	,93363	—	69,58	164,44	5,14	60,81	,6184
84	,93419	—	70,42	165,24	5,18	60,52	,6154
100 + 85	,93474	—	71,26	166,04	5,22	60,22	,6124
86	,93529	—	72,10	166,84	5,26	59,94	,6095
87	,93583	—	72,93	167,64	5,29	59,65	,6066
88	,93636	—	73,77	168,44	5,33	59,37	,6037
89	,93689	—	74,61	169,24	5,37	59,09	,6008
100 + 90	,93741	—	75,45	170,04	5,41	58,81	,5980
91	,93792	—	76,29	170,85	5,44	58,53	,5952
92	,93843	—	77,13	171,65	5,48	58,26	,5924
93	,93893	—	77,96	172,45	5,51	57,99	,5897
94	,93942	—	78,80	173,25	5,55	57,72	,5870
100 + 95	,93991	—	79,64	174,05	5,59	57,45	,5843
96	,94038	—	80,47	174,85	5,62	57,19	,5816
97	,94085	—	81,30	175,65	5,65	56,93	,5790
98	,94131	—	82,14	176,46	5,68	56,67	,5763
99	,94177	—	82,98	177,27	5,71	56,41	,5736

TABLE II.   HEAT 30°.

I. Water and spirit by weight.	II. Specific gravity.	III. Spirit by measure.	IV. Water by measure.	V. Bulk of mixture.	VI. Diminution of bulk.	VII. Quantity of spirit per cent.	VIII. Decimal multipliers.
W. + Sp.							
100+100	,94222	100	83,83	178,08	5,75	56,15	,5710
99	,94267	—	84,69	178,90	5,79	55,89	,5684
98	,94312	—	85,55	179,74	5,81	55,63	,5658
97	,94357	—	86,43	180,59	5,84	55,37	,5631
96	,94402	—	87,33	181,45	5,88	55,11	,5604
100 + 95	,94447	—	88,25	182,33	5,92	54,84	,5577
94	,94492	—	89,18	183,24	5,94	54,57	,5550
93	,94537	—	90,14	184,17	5,97	54,30	,5522
92	,94583	—	91,12	185,12	6,00	54,02	,5494
91	,94629	—	92,13	186,09	6,04	53,74	,5465
100 + 90	,94675	—	93,15	187,08	6,07	53,45	,5436
89	,94723	—	94,19	188,09	6,10	53,17	,5407
88	,94771	—	95,26	189,12	6,14	52,88	,5377
87	,94821	—	96,36	190,18	6,18	52,58	,5347
86	,94870	—	97,48	191,26	6,22	52,28	,5317
100 + 85	,94920	—	98,63	192,37	6,26	51,98	,5286
84	,94970	—	99,80	193,51	6,29	51,68	,5255
83	,95021	—	101,00	194,67	6,33	51,37	,5224
82	,95071	—	102,24	195,86	6,38	51,06	,5192
81	,95122	—	103,50	197,09	6,41	50,74	,5160
100 + 80	,95173	—	104,79	198,34	6,45	50,42	,5127
79	,95225	—	106,12	199,62	6,50	50,09	,5094
78	,95275	—	107,48	200,95	6,53	49,76	,5060
77	,95327	—	108,88	202,31	6,57	49,43	,5026
76	,95378	—	110,31	203,70	6,61	49,09	,4992
100 + 75	,95429	—	111,78	205,13	6,65	48,75	,4957
74	,95479	—	113,29	206,61	6,68	48,40	,4922
73	,95529	—	114,84	208,13	6,71	48,05	,4886
72	,95580	—	116,44	209,69	6,75	47,69	,4850
71	,95630	—	118,08	211,29	6,79	47,33	,4813
100 + 70	,95681	—	119,76	212,94	6,82	46,96	,4775
69	,95733	—	121,49	214,64	6,85	46,59	,4738
68	,95786	—	123,28	216,39	6,89	46,21	,4700
67	,95838	—	126,12	218,19	6,93	45,83	,4661
66	,95891	—	127,02	220,05	6,97	45,44	,4621
100 + 65	,95944	—	128,98	221,97	7,01	45,05	,4581
64	,95520	—	130,99	223,95	7,04	44,65	,4541
63	,96050	—	133,06	225,99	7,07	44,25	,4500
62	,96103	—	135,21	228,10	7,11	43,84	,4458
61	,96156	—	137,43	230,28	7,15	43,43	,4416
100 + 60	,96209	—	139,72	232,54	7,18	43,00	,4373
59	,96262	—	142,08	234,87	7,21	42,58	,4330
58	,96314	—	144,53	237,29	7,24	42,14	,4286
57	,96367	—	147,07	239,80	7,27	41,70	,4241
56	,96418	—	149,70	242,40	7,30	41,25	,4195
100 + 55	,96470	—	152,42	245,09	7,33	40,80	,4149
54	,96520	—	155,22	247,87	7,35	40,34	,4102
53	,96570	—	158,14	250,77	7,37	39,87	,4054
52	,96620	—	161,19	253,79	7,40	39,40	,4006
51	,96670	—	164,36	256,92	7,44	38,92	,3957

I. Water and spirit by weight.	II. Specific gravity.	III. Spirit by measure.	IV. Water by measure.	V. Bulk of mixture.	VI. Diminution of bulk.	VII. Quantity of spirit per cent.	VIII. Decimal multipliers.
W. + Sp.							
100 + 50	,96719	100	167,67	260,19	7,48	38,43	,3908
49	,96769	—	171,09	263,63	7,46	37,93	,3858
48	,96819	—	174,65	267,18	7,47	37,43	,3806
47	,96869	—	178,37	270,89	7,48	36,92	,3754
46	,96919	—	182,25	274,75	7,50	36,40	,3701
100 + 45	,96967	—	186,30	278,79	7,51	35,87	,3648
44	,97015	—	190,53	283,02	7,51	35,33	,3593
43	,97062	—	194,96	287,46	7,50	34,79	,3538
42	,97109	—	199,60	292,11	7,49	34,23	,3482
41	,97155	—	204,47	296,98	7,49	33,67	,3424
100 + 40	,97200	—	209,58	302,12	7,46	33,10	,3366
39	,97244	—	214,96	307,49	7,47	32,52	,3307
38	,97288	—	220,62	313,17	7,45	31,94	,3247
37	,97332	—	226,58	319,17	7,41	31,34	,3186
36	,97375	—	232,87	325,51	7,36	30,73	,3124
100 + 35	,97418	—	239,53	332,18	7,35	30,11	,3061
34	,97461	—	246,57	339,26	7,31	29,48	,2997
33	,97504	—	254,04	346,79	7,25	28,84	,2932
32	,97548	—	261,98	354,79	7,19	28,19	,2866
31	,97591	—	270,43	363,29	7,14	27,53	,2799
100 + 30	,97635	—	279,44	372,36	7,08	26,86	,2731
29	,97679	—	289,08	382,04	7,04	26,19	,2662
28	,97723	—	299,41	392,43	6,98	25,48	,2591
27	,97768	—	310,50	403,60	6,90	24,78	,2519
26	,97814	—	322,44	415,63	6,81	24,06	,2446
100 + 25	,97860	—	335,34	428,65	6,69	23,33	,2372
24	,97906	—	349,31	442,73	6,58	22,59	,2297
23	,97954	—	364,49	458,03	6,46	21,83	,2220
22	,98004	—	381,06	474,72	6,34	21,06	,2142
21	,98055	—	399,21	492,99	6,22	20,28	,2063
100 + 20	,98108	—	419,17	513,08	6,09	19,49	,1982
19	,98163	—	441,23	535,29	5,94	18,68	,1900
18	,98221	—	465,74	559,95	5,79	17,86	,1816
17	,98282	—	493,14	587,50	5,64	17,02	,1731
16	,98345	—	523,96	618,48	5,48	16,17	,1644
100 + 15	,98412	—	558,89	653,58	5,31	15,30	,1556
14	,98481	—	598,81	693,69	5,12	14,41	,1466
13	,98555	—	644,88	739,94	4,94	13,51	,1375
12	,98633	—	698,62	793,88	4,74	12,59	,1281
11	,98716	—	762,13	857,59	4,54	11,66	,1186
100 + 10	,988c4	—	838,34	934,04	4,30	10,71	,1089
9	,98896	—	931,49	1027,42	4,07	9,73	,0990
8	,98995	—	1047,92	1144,10	3,82	8,74	,0889
7	,99101	—	1197,63	1294,05	3,58	7,73	,0786
6	,99214	—	1397,23	1493,91	3,32	6,69	,0681
100 + 5	,99334	—	1670,68	1773,63	3,05	5,64	,0573
4	,99460	—	2095,85	2193,08	2,77	4,56	,0464
3	,99602	—	2794,46	2891,94	2,52	3,46	,0352
2	,99751	—	4191,71	4289,38	2,33	2,33	,0237
1	,99910	—	8383,40	8481,13	2,27	1,18	,0120

TABLE I.   HEAT 31°.

I. Spirit and water by weight.	II. Specific gravity.	III. Spirit by measure.	IV. Water by measure.	V. Bulk of mixture.	VI. Diminution of bulk.	VII. Quantity of spirit per cent.	VIII. Decimal multipliers.
Sp. + W.							
100 + 0	,83852	100	—	100,00	—	100,00	1,0164
1	,84084	—	0,84	100,72	0,12	99,28	1,0091
2	,84310	—	1,68	101,45	0,23	98,57	1,0019
3	,84530	—	2,51	102,18	0,33	97,87	,9948
4	,84743	—	3,35	102,91	0,44	97,18	,9877
100 + 5	,84950	—	4,19	103,64	0,55	96,49	,9807
6	,85152	—	5,03	104,38	0,65	95,80	,9738
7	,85349	—	5,87	105,12	0,75	95,13	,9669
8	,85541	—	6,71	105,87	0,84	94,46	,9601
9	,85728	—	7,54	106,61	0,93	93,80	,9534
100 + 10	,85911	—	8,38	107,36	1,02	93,14	,9467
11	,86092	—	9,22	108,11	1,11	92,50	,9401
12	,86270	—	10,05	108,86	1,19	91,86	,9336
13	,86443	—	10,89	109,61	1,28	91,23	,9272
14	,86613	—	11,72	110,37	1,35	90,61	,9209
100 + 15	,86778	—	12,56	111,12	1,44	89,99	,9146
16	,86937	—	13,41	111,88	1,53	89,38	,9084
17	,87092	—	14,24	112,63	1,61	88,77	,9023
18	,87244	—	15,08	113,39	1,69	88,17	,8963
19	,87394	—	15,92	114,15	1,77	87,58	,8903
100 + 20	,87539	—	16,76	114,91	1,85	87,00	,8842
21	,87683	—	17,59	115,68	1,91	86,42	,8784
22	,87826	—	18,43	116,45	1,98	85,85	,8726
23	,87966	—	19,27	117,21	2,06	85,29	,8669
24	,88103	—	20,11	117,98	2,13	84,73	,8612
100 + 25	,88238	—	20,95	118,75	2,20	84,18	,8556
26	,88370	—	21,79	119,53	2,26	83,64	,8501
27	,88500	—	22,62	120,31	2,31	83,10	,8447
28	,88628	—	23,46	121,09	2,37	82,57	,8393
29	,88753	—	24,30	121,87	2,43	82,05	,8340
100 + 30	,88877	—	25,14	122,65	2,49	81,53	,8287
31	,88999	—	25,97	123,42	2,55	81,02	,8235
32	,89120	—	26,81	124,20	2,61	80,51	,8184
33	,89238	—	27,65	124,97	2,68	80,01	,8133
34	,89354	—	28,48	125,75	2,73	79,52	,8083
100 + 35	,89468	—	29,32	126,52	2,80	79,03	,8033
36	,89580	—	30,16	127,30	2,86	78,55	,7984
37	,89690	—	31,00	128,08	2,92	78,07	,7936
38	,89799	—	31,84	128,86	2,98	77,60	,7887
39	,89906	—	32,67	129,64	3,03	77,13	,7841
100 + 40	,90011	—	33,52	130,42	3,10	76,67	,7793
41	,90115	—	34,36	131,20	3,16	76,21	,7748
42	,90218	—	35,19	131,98	3,21	75,76	,7702
43	,90319	—	36,03	132,76	3,27	75,32	,7657
44	,90418	—	36,86	133,54	3,32	74,88	,7612
100 + 45	,90510	—	37,70	134,32	3,38	74,45	,7567
46	,90612	—	38,53	135,11	3,42	74,01	,7523
47	,90707	—	39,37	135,89	3,48	73,59	,7480
48	,90800	—	40,21	136,68	3,53	73,17	,7437
49	,90891	—	41,06	137,46	3,60	72,74	,7394
100 + 50	,90980	100	41,90	138,25	3,65	72,33	,7352
51	,91069	—	42,72	139,03	3,69	71,92	,7311
52	,91155	—	43,57	139,82	3,75	71,51	,7270
53	,91241	—	44,40	140,61	3,79	71,12	,7229
54	,91325	—	45,24	141,40	3,84	70,72	,7188
100 + 55	,91408	—	46,08	142,18	3,90	70,33	,7148
56	,91490	—	46,92	142,97	3,95	69,94	,7109
57	,91571	—	47,75	143,76	3,99	69,56	,7070
58	,91650	—	48,59	144,55	4,04	69,18	,7031
59	,91729	—	49,43	145,34	4,09	68,80	,6993
100 + 60	,91806	—	50,27	146,13	4,14	68,43	,6955
61	,91882	—	51,11	146,92	4,19	68,06	,6917
62	,91957	—	51,95	147,71	4,24	67,69	,6880
63	,92031	—	52,78	148,50	4,28	67,34	,6843
64	,92104	—	53,62	149,30	4,32	66,98	,6807
100 + 65	,92176	—	54,46	150,09	4,37	66,63	,6771
66	,92247	—	55,29	150,88	4,41	66,28	,6736
67	,92317	—	56,14	151,68	4,46	65,92	,6701
68	,92386	—	56,97	152,47	4,50	65,58	,6666
69	,92455	—	57,80	153,27	4,53	65,24	,6631
100 + 70	,92522	—	58,64	154,06	4,58	64,91	,6597
71	,92588	—	59,48	154,86	4,62	64,57	,6563
72	,92653	—	60,32	155,66	4,66	64,24	,6530
73	,92717	—	61,16	156,46	4,70	63,91	,6497
74	,92780	—	62,00	157,25	4,75	63,59	,6464
100 + 75	,92848	—	62,83	158,04	4,79	63,27	,6431
76	,92910	—	63,67	158,83	4,84	62,96	,6399
77	,92971	—	64,51	159,64	4,87	62,63	,6367
78	,93036	—	65,35	160,43	4,92	62,33	,6335
79	,93091	—	66,19	161,23	4,96	62,02	,6303
100 + 80	,93150	—	67,03	162,03	5,00	61,72	,6273
81	,93199	—	67,86	162,82	5,04	61,41	,6242
82	,93266	—	68,70	163,62	5,08	61,12	,6212
83	,93323	—	69,54	164,42	5,12	60,81	,6182
84	,93379	—	70,38	165,22	5,16	60,52	,6151
100 + 85	,93434	—	71,22	166,03	5,19	60,23	,6122
86	,93489	—	72,05	166,83	5,22	59,95	,6093
87	,93543	—	72,89	167,63	5,26	59,66	,6064
88	,93596	—	73,72	168,43	5,29	59,37	,6035
89	,93649	—	74,56	169,23	5,33	59,09	,6006
100 + 90	,93701	—	75,40	170,03	5,37	58,82	,5978
91	,93752	—	76,24	170,83	5,41	58,53	,5950
92	,93802	—	77,08	171,63	5,45	58,27	,5922
93	,93852	—	77,91	172,43	5,48	57,99	,5895
94	,93901	—	78,75	173,24	5,51	57,73	,5868
100 + 95	,93950	—	79,59	174,03	5,56	57,46	,5841
96	,93998	—	80,42	174,84	5,58	57,20	,5814
97	,94045	—	81,25	175,64	5,61	56,93	,5788
98	,94092	—	82,09	176,44	5,65	56,67	,5761
99	,94138	—	82,93	177,25	5,68	56,42	,5734

TABLE II.   HEAT 31°.

I. Water and spirit by weight.	II. Specific gravity.	III. Spirit by measure.	IV. Water by measure.	V. Bulk of mixture.	VI. Diminution of bulk.	VII. Quantity of spirit per cent.	VIII. Decimal multipliers.
W. + Sp.							
100+100	,94183	100	83,78	178,06	5,72	56,15	,5708
99	,94228	—	84,64	178,88	5,76	55,90	,5682
98	,94273	—	85,51	179,72	5,79	55,64	,5656
97	,94318	—	86,38	180,57	5,81	55,37	,5629
96	,94363	—	87,29	181,43	5,86	55,12	,5602
100 + 95	,94408	—	88,20	182,31	5,89	54,84	,5575
94	,94453	—	89,13	183,22	5,91	54,57	,5548
93	,94498	—	90,09	184,15	5,94	54,30	,5520
92	,94544	—	91,07	185,09	5,98	54,02	,5492
91	,94591	—	92,08	186,07	6,01	53,75	,5463
100 + 90	,94636	—	93,10	187,06	6,04	53,46	,5434
89	,94685	—	94,14	188,06	6,08	53,17	,5405
88	,94734	—	95,21	189,10	6,11	52,89	,5375
87	,94784	—	96,31	190,16	6,15	52,59	,5345
86	,94833	—	97,42	191,24	6,18	52,29	,5315
100 + 85	,94883	—	98,58	192,35	6,23	51,99	,5284
84	,91933	—	99,75	193,49	6,26	51,69	,5253
83	,94983	—	100,94	194,65	6,29	51,37	,5222
82	,95034	—	102,18	195,83	6,35	51,06	,5190
81	,95085	—	103,44	197,06	6,38	50,75	,5158
100 + 80	,95136	—	104,73	198,32	6,41	50,43	,5125
79	,95188	—	106,06	199,60	6,46	50,10	,5092
78	,95239	—	107,42	200,92	6,50	49,76	,5058
77	,95291	—	108,82	202,28	6,54	49,44	,5024
76	,95342	—	110,25	203,67	6,58	49,09	,4990
100 + 75	,95392	—	111,72	205,10	6,62	48,75	,4955
74	,95443	—	113,22	206,58	6,64	48,41	,4921
73	,95443	—	114,77	208,10	6,67	48,05	,4885
72	,95544	—	116,38	209,66	6,72	47,69	,4849
71	,95594	—	118,01	211,25	6,76	47,34	,4812
100 + 70	,95646	—	119,69	212,90	6,79	46,97	,4774
69	,95698	—	121,42	214,60	6,82	46,59	,4736
68	,95751	—	123,21	216,35	6,86	46,21	,4699
67	,95803	—	126,05	218,16	6,89	45,83	,4660
66	,95857	—	126,94	220,01	6,93	45,45	,4660
100 + 65	,95910	—	128,91	221,93	6,98	45,06	,4580
64	,95963	—	130,92	223,91	7,01	44,66	,4540
63	,96017	—	132,98	225,95	7,03	44,26	,4499
62	,96070	—	135,13	228,06	7,07	43,85	,4457
61	,96124	—	137,35	230,23	7,12	43,44	,4415
100 + 60	,96177	—	139,64	232,50	7,14	43,01	,4372
59	,96230	—	142,00	234,83	7,17	42,59	,4329
58	,96283	—	144,45	237,24	7,21	42,15	,4285
57	,96336	—	146,99	239,75	7,24	41,71	,4240
56	,96363	—	149,62	242,34	7,28	41,26	,4195
100 + 55	,96439	—	152,33	245,03	7,30	40,81	,4148
54	,96490	—	155,13	247,81	7,32	40,35	,4101
53	,96541	—	158,05	250,71	7,34	39,88	,4053
52	,96591	—	161,10	253,73	7,37	39,41	,4005
51	,96642	—	164,26	256,87	7,39	38,93	,3956
100 + 50	,96691	100	167,57	260,13	7,44	38,44	,3907
49	,96742	—	170,99	263,56	7,43	37,94	,3857
48	,96792	—	174,55	267,11	7,44	37,44	,3805
47	,96843	—	178,26	270,81	7,45	36,93	,3754
46	,96893	—	182,14	274,67	7,47	36,41	,3701
100 + 45	,96941	—	186,19	278,72	7,47	35,88	,3647
44	,96990	—	190,42	282,94	7,48	35,34	,3592
43	,97038	—	194,85	287,37	7,48	34,79	,3537
42	,97085	—	199,48	292,02	7,46	34,24	,3481
41	,97132	—	204,35	296,89	7,46	33,68	,3424
100 + 40	,97178	—	209,46	302,02	7,44	33,11	,3365
39	,97222	—	214,83	307,39	7,44	32,53	,3307
38	,97267	—	220,49	313,07	7,42	31,94	,3247
37	,97311	—	226,45	319,07	7,38	31,34	,3186
36	,97355	—	232,73	325,40	7,33	30,73	,3124
100 + 35	,97399	—	239,39	332,07	7,32	30,12	,3061
34	,97442	—	246,43	339,15	7,28	29,49	,2997
33	,97486	—	253,89	346,67	7,22	28,85	,2932
32	,97531	—	261,83	354,66	7,17	28,20	,2866
31	,97575	—	270,27	363,16	7,11	27,54	,2799
100 + 30	,97620	—	279,28	372,22	7,06	26,87	,2731
29	,97664	—	288,91	381,90	7,01	26,18	,2662
28	,97709	—	299,24	392,29	6,95	25,49	,2591
27	,97755	—	310,32	403,44	6,88	24,78	,2519
26	,97802	—	322,25	415,47	6,78	24,07	,2446
100 + 25	,97849	—	335,15	428,47	6,68	23,34	,2372
24	,97896	—	349,11	443,54	6,57	22,59	,2297
23	,97945	—	364,28	457,83	6,45	21,84	,2220
22	,97996	—	381,84	474,50	6,34	21,07	,2142
21	,98048	—	398,98	492,76	6,22	20,29	,2063
100 + 20	,98102	—	418,93	512,84	6,09	19,50	,1982
19	,98158	—	440,98	535,03	5,95	18,69	,1900
18	,98216	—	465,47	559,68	5,79	17,87	,1816
17	,98278	—	492,86	587,21	5,65	17,03	,1731
16	,98342	—	523,66	618,17	5,49	16,17	,1644
100 + 15	,98409	—	558,57	653,25	5,32	15,31	,1556
14	,98479	—	598,47	693,33	5,14	14,42	,1466
13	,98554	—	644,51	739,55	4,98	13,52	,1375
12	,98632	—	698,22	793,46	4,76	12,60	,1281
11	,98716	—	761,69	857,13	4,56	11,67	,1186
100 + 10	,98804	—	837,86	933,54	4,32	10,71	,1089
9	,98896	—	930,96	1026,86	4,10	9,74	,0990
8	,98992	—	1047,32	1143,47	3,85	8,74	,0889
7	,99103	—	1196,95	1293,34	3,61	7,73	,0786
6	,99216	—	1396,44	1493,08	3,36	6,70	,0681
100 + 5	,99336	—	1675,73	1772,64	3,09	5,64	,0573
4	,99466	—	2094,65	2191,85	2,80	4,56	,0464
3	,99605	—	2792,86	2890,32	2,54	3,46	,0352
2	,99754	—	4189,30	4886,96	2,34	2,33	,0237
1	,99913	—	8378,58	8476,39	2,19	1,18	,0120

Mr. GILPIN's Tables

TABLE I.   HEAT 32°.

I. Spirit and water by weight.	II. Specific gravity.	III. Spirit by measure.	IV. Water by measure.	V. Bulk of mixture.	VI. Diminution of bulk.	VII. Quantity of spirit per cent.	VIII. Decimal multipliers.
Sp. + W.							
100 + 0	,83807	100	—	100,00	—	100,00	1,0158
1	,84039	—	0,84	100,72	0,12	99,28	1,0086
2	,84265	—	1,68	101,45	0,23	98,57	1,0014
3	,84485	—	2,51	102,18	0,33	97,87	,9943
4	,84698	—	3,35	102,91	0,44	97,18	,9872
100 + 5	,84905	—	4,19	103,64	0,55	96,49	,9801
6	,85107	—	5,03	104,38	0,65	95,80	,9733
7	,85304	—	5,87	105,12	0,75	95,13	,9664
8	,85496	—	6,70	105,87	0,83	94,46	,9596
9	,85683	—	7,54	106,61	0,93	93,80	,9529
100 + 10	,85866	—	8,37	107,36	1,01	93,14	,9462
11	,86046	—	9,21	108,11	1,10	92,50	,9396
12	,86224	—	10,05	108,86	1,19	91,86	,9331
13	,86396	—	10,88	109,61	1,27	91,23	,9267
14	,86565	—	11,72	110,37	1,35	90,61	,9204
100 + 15	,86730	—	12,56	111,12	1,44	89,99	,9141
16	,86890	—	13,40	111,88	1,52	89,38	,9079
17	,87045	—	14,23	112,64	1,59	88,77	,9018
18	,87198	—	15,07	113,40	1,67	88,17	,8958
19	,87348	—	15,91	114,16	1,75	87,58	,8898
100 + 20	,87494	—	16,75	114,92	1,83	87,00	,8838
21	,87638	—	17,58	115,68	1,90	86,42	,8779
22	,87780	—	18,42	116,45	1,97	85,85	,8721
23	,87921	—	19,26	117,22	2,04	85,29	,8664
24	,88059	—	20,10	117,99	2,11	84,73	,8607
100 + 25	,88193	—	20,94	118,76	2,18	84,18	,8552
26	,88326	—	21,77	119,53	2,24	83,64	,8497
27	,88456	—	22,61	120,31	2,30	83,11	,8443
28	,88584	—	23,45	121,09	2,36	82,58	,8389
29	,88709	—	24,28	121,86	2,42	82,05	,8336
100 + 30	,88833	—	25,12	122,64	2,48	81,53	,8283
31	,88955	—	25,96	123,42	2,54	81,02	,8231
32	,89076	—	26,80	124,19	2,61	80,52	,8180
33	,89195	—	27,64	124,97	2,67	80,02	,8129
34	,89311	—	28,47	125,75	2,72	79,53	,8079
100 + 35	,89424	—	29,31	126,52	2,79	79,03	,8029
36	,89537	—	30,15	127,30	2,85	78,50	,7980
37	,89647	—	30,99	128,07	2,92	78,08	,7932
38	,89756	—	31,83	128,85	2,98	77,61	,7884
39	,89863	—	32,65	129,63	3,02	77,14	,7837
100 + 40	,89968	—	33,50	130,41	3,09	76,67	,7789
41	,90072	—	34,34	131,19	3,15	76,22	,7744
42	,90175	—	35,17	131,97	3,20	75,77	,7698
43	,90276	—	36,01	132,75	3,26	75,33	,7653
44	,90376	—	36,84	133,53	3,31	74,89	,7608
100 + 45	,90474	—	37,68	134,31	3,37	74,45	,7563
46	,90570	—	38,51	135,10	3,41	74,02	,7520
47	,90665	—	39,35	135,88	3,47	73,59	,7477
48	,90758	—	40,19	136,67	3,52	73,17	,7434
49	,90849	—	41,03	137,45	3,58	72,75	,7391
Sp. + W.							
100 + 50	,90937	100	41,87	138,24	3,63	72,34	,7348
51	,91027	—	42,70	139,02	3,68	71,93	,7308
52	,91113	—	43,54	139,81	3,73	71,52	,7266
53	,91199	—	44,38	140,60	3,78	71,12	,7226
54	,91284	—	45,22	141,39	3,83	70,73	,7185
100 + 55	,91367	—	46,06	142,17	3,89	70,34	,7145
56	,91448	—	46,90	142,96	3,94	69,95	,7106
57	,91530	—	47,73	143,75	3,98	69,56	,7067
58	,91609	—	48,56	144,54	4,02	69,18	,7028
59	,91687	—	49,40	145,33	4,07	68,81	,6990
100 + 60	,91764	—	50,24	146,12	4,12	68,43	,6952
61	,91840	—	51,08	146,91	4,17	68,06	,6914
62	,91916	—	51,92	147,70	4,22	67,70	,6877
63	,91990	—	52,75	148,49	4,26	67,34	,6840
64	,92063	—	53,59	149,29	4,30	66,98	,6804
100 + 65	,92135	—	54,43	150,08	4,35	66,63	,6769
66	,92206	—	55,26	150,87	4,39	66,28	,6733
67	,92276	—	56,11	151,67	4,44	65,93	,6698
68	,92345	—	56,94	152,46	4,48	65,59	,6663
69	,92413	—	57,77	153,26	4,51	65,25	,6628
100 + 70	,92481	—	58,61	154,05	4,56	64,91	,6594
71	,92547	—	59,45	154,85	4,60	64,58	,6560
72	,92612	—	60,28	155,64	4,64	64,25	,6527
73	,92676	—	61,12	156,44	4,68	63,92	,6494
74	,92739	—	61,97	157,24	4,73	63,60	,6461
100 + 75	,92806	—	62,80	158,03	4,77	63,28	,6428
76	,92869	—	63,64	158,82	4,82	62,96	,6396
77	,92930	—	64,48	159,63	4,85	62,64	,6364
78	,92994	—	65,31	160,42	4,89	62,33	,6333
79	,93050	—	66,15	161,22	4,93	62,03	,6300
100 + 80	,93109	—	67,00	162,02	4,98	61,72	,6270
81	,93168	—	67,82	162,81	5,01	61,42	,6240
82	,93226	—	68,66	163,61	5,05	61,12	,6210
83	,93283	—	69,50	164,41	5,09	60,82	,6179
84	,93339	—	70,34	165,21	5,13	60,53	,6149
100 + 85	,93394	—	71,18	166,01	5,17	60,23	,6119
86	,93449	—	72,01	166,81	5,20	59,95	,6091
87	,93503	—	72,85	167,61	5,24	59,66	,6062
88	,93556	—	73,68	168,41	5,27	59,38	,6032
89	,93609	—	74,52	169,21	5,31	59,10	,6004
100 + 90	,93661	—	75,36	170,01	5,35	58,82	,5975
91	,93712	—	76,20	170,82	5,38	58,54	,5948
92	,93762	—	77,04	171,62	5,42	58,27	,5920
93	,93812	—	77,87	172,42	5,45	58,00	,5892
94	,93861	—	78,71	173,22	5,49	57,73	,5866
100 + 95	,93910	—	79,55	174,02	5,53	57,46	,5838
96	,93958	—	80,38	174,82	5,56	57,21	,5812
97	,94005	—	81,21	175,62	5,59	56,94	,5785
98	,94052	—	82,04	176,42	5,62	56,68	,5758
99	,94099	—	82,88	177,23	5,65	56,42	,5732

TABLE II.   HEAT 32°.

I. Water and spirit by weight.	II. Specific gravity.	III. Spirit by measure.	IV. Water by measure.	V. Bulk of mixture.	VI. Diminution of bulk.	VII. Quantity of spirit per cent.	VIII. Decimal multipliers.
W. + Sp.							
100+100	,94144	100	83,74	178,04	5,70	56,16	,5706
99	,94189	—	84,59	178,86	5,73	55,91	,5679
98	,94233	—	85,46	179,70	5,76	55,65	,5653
97	,94278	—	86,33	180,55	5,78	55,38	,5626
96	,94333	—	87,24	181,41	5,83	55,12	,5600
100+95	,94369	—	88,15	182,29	5,86	54,85	,5573
94	,94413	—	89,08	183,20	5,88	54,58	,5545
93	,94459	,1	90,04	184,12	5,92	54,31	,5517
92	,94505	—	91,02	185,07	5,95	54,03	,5489
91	,94552	—	92,02	186,04	5,98	53,75	,5461
100+90	,94598	—	93,04	187,03	6,01	53,47	,5431
89	,94647	—	94,08	188,04	6,04	53,18	,5403
88	,94696	—	95,15	189,07	6,08	52,89	,5373
87	,94746	—	96,25	190,13	6,12	52,59	,5343
86	,94795	—	97,37	191,21	6,16	52,30	,5313
100+85	,94845	—	98,52	192,32	6,20	51,99	,5282
84	,94896	—	99,69	193,46	6,23	51,69	,5251
83	,94947	—	100,89	194,62	6,27	51,38	,5220
82	,94997	—	102,12	195,81	6,31	51,07	,5188
81	,95048	—	103,38	197,03	6,35	50,76	,5156
100+80	,95099	—	104,67	198,29	6,38	50,44	,5123
79	,95151	—	106,00	199,57	6,43	50,11	,5090
78	,95202	—	107,36	200,89	6,47	49,77	,5046
77	,95254	—	108,76	202,25	6,51	49,45	,5022
76	,95305	—	110,19	203,64	6,55	49,10	,4988
100+75	,95355	—	111,66	205,07	6,59	48,76	,4953
74	,95406	—	113,15	206,55	6,64	48,42	,4919
73	,95457	—	114,70	208,06	6,64	48,06	,4883
72	,95508	—	116,31	209,63	6,68	47,70	,4847
71	,95558	—	117,94	211,22	6,72	47,34	,4810
100+70	,95610	—	119,62	212,87	6,75	46,98	,4772
69	,95662	—	121,35	214,57	6,78	46,60	,4734
68	,95715	—	123,14	216,31	6,83	46,22	,4697
67	,95768	—	124,98	218,12	6,86	45,84	,4658
66	,95822	—	126,87	219,98	6,89	45,46	,4618
100+65	,95875	—	128,83	221,89	6,94	45,07	,4578
64	,95929	—	130,84	223,87	6,97	44,67	,4533
63	,95983	—	132,90	225,91	6,99	44,26	,4497
62	,96037	—	135,05	228,01	7,04	43,85	,4455
61	,96091	—	137,27	230,19	7,08	43,44	,4413
100+60	,96145	—	139,56	232,45	7,11	43,02	,4370
59	,96198	—	141,92	234,78	7,14	42,60	,4327
58	,96251	—	144,36	237,19	7,17	42,16	,4283
57	,96304	—	146,90	239,70	7,20	41,72	,4238
56	,96356	—	149,53	242,29	7,24	41,27	,4193
100+55	,96408	—	152,24	244,97	7,27	40,82	,4146
54	,96459	—	155,04	247,75	7,29	40,36	,4100
53	,96511	—	157,96	250,66	7,30	39,89	,4052
52	,96562	—	161,01	253,68	7,33	39,42	,4004
51	,96613	—	164,17	256,81	7,36	38,93	,3955

I. Water and spirit by weight.	II. Specific gravity.	III. Spirit by measure.	IV. Water by measure.	V. Bulk of mixture.	VI. Diminution of bulk.	VII. Quantity of spirit per cent.	VIII. Decimal multipliers.
W. + Sp.							
100+50	,96663	100	167,48	260,08	7,40	38,44	,3905
49	,96714	—	170,89	263,50	7,39	37,95	,3856
48	,96765	—	174,45	267,04	7,41	37,45	,3804
47	,96816	—	178,16	270,73	7,43	36,94	,3752
46	,96867	—	182,03	274,59	7,44	36,42	,3700
100+45	,96915	—	186,08	278,64	7,44	35,89	,3646
44	,96965	—	190,31	282,86	7,45	35,35	,3591
43	,97013	—	194,74	287,29	7,45	34,80	,3536
42	,97061	—	199,37	291,94	7,43	34,25	,3480
41	,97108	—	204,23	296,80	7,43	33,69	,3423
100+40	,97155	—	209,34	301,92	7,42	33,12	,3365
39	,97200	—	214,71	307,29	7,42	32,54	,3306
38	,97245	—	220,36	312,97	7,39	31,95	,3246
37	,97290	—	226,32	318,97	7,35	31,35	,3185
36	,97335	—	232,60	325,30	7,30	30,74	,3123
100+35	,97379	—	239,25	331,96	7,29	30,12	,3060
34	,97423	—	246,29	339,03	7,26	29,50	,2996
33	,97468	—	253,74	346,55	7,19	28,86	,2931
32	,97513	—	261,68	354,54	7,14	28,21	,2865
31	,97558	—	270,12	363,03	7,09	27,55	,2798
100+30	,97604	—	279,12	372,08	7,04	26,88	,2730
29	,97649	—	288,74	381,76	6,98	26,19	,2661
28	,97695	—	299,07	392,15	6,92	25,50	,2591
27	,97741	—	310,14	403,29	6,85	24,79	,2519
26	,97789	—	322,06	415,31	6,75	24,08	,2446
100+25	,97837	—	334,96	428,30	6,66	23,35	,2372
24	,97885	—	348,91	442,35	6,56	22,60	,2297
23	,97935	—	364,07	457,63	6,44	21,85	,2220
22	,97987	—	381,62	474,29	6,33	21,08	,2142
21	,98040	—	398,75	492,53	6,22	20,30	,2063
100+20	,98096	—	418,69	512,60	6,09	19,51	,1981
19	,98152	—	440,73	534,57	5,96	18,70	,1900
18	,98211	—	465,20	559,41	5,79	17,87	,1816
17	,98272	—	492,58	586,92	5,66	17,04	,1731
16	,98338	—	523,36	617,86	5,50	16,18	,1644
100+15	,98400	—	558,25	652,92	5,33	15,32	,1556
14	,98477	—	598,13	692,97	5,16	14,42	,1466
13	,98552	—	644,14	739,17	4,97	13,53	,1375
12	,98631	—	697,82	793,04	4,78	12,61	,1281
11	,98715	—	761,26	856,68	4,58	11,67	,1186
100+10	,98804	—	837,38	933,04	4,34	10,72	,1089
9	,98897	—	930,43	1026,30	4,13	9,74	,0990
8	,98998	—	1046,72	1142,85	3,87	8,75	,0889
7	,99105	—	1196,27	1292,63	3,64	7,73	,0786
6	,99218	—	1395,65	1492,25	3,40	6,70	,0681
100+5	,99338	—	1674,78	1771,65	3,13	5,64	,0573
4	,99469	—	2093,45	2190,63	2,82	4,56	,0464
3	,99606	—	2791,27	2888,71	2,56	3,46	,0352
2	,99757	—	4186,91	4284,58	2,33	2,33	,0237
1	,99916	—	8373,79	8471,67	2,12	1,18	,0120

TABLE I. HEAT 33°.

Left panel

I. Spirit and water by weight.	II. Specific gravity.	III. Spirit by measure.	IV. Water by measure.	V. Bulk of mixture.	VI. Diminution of bulk.	VII. Quantity of Spirit per cent.	VIII. Decimal multipliers.
Sp. + W.							
100 + 0	,83762	100	—	100,00	—	100,00	1,0153
1	,83994	—	0,84	100,72	0,12	99,28	1,0081
2	,84220	—	1,68	101,45	0,23	98,57	1,0009
3	,84440	—	2,51	102,18	0,33	97,87	,9938
4	,84653	—	3,35	102,91	0,44	97,18	,9867
100 + 5	,84860	—	4,19	103,64	0,55	96,49	,9796
6	,85062	—	5,03	104,38	0,65	95,80	,9728
7	,85259	—	5,86	105,12	0,74	95,13	,9659
8	,85451	—	6,70	105,87	0,83	94,46	,9591
9	,85638	—	7,53	106,61	0,92	93,80	,9524
100 + 10	,85820	—	8,37	107,36	1,01	93,14	,9457
11	,86000	—	9,21	108,11	1,10	92,50	,9391
12	,86178	—	10,04	108,86	1,18	91,86	,9326
13	,86349	—	10,88	109,61	1,27	91.23	,9262
14	,86518	—	11,71	110,37	1,34	90,61	,9199
100 + 15	,86683	—	12,55	111,12	1,43	89,99	,9136
16	,86842	—	13,39	111,88	1,51	89,38	,9074
17	,86998	—	14,23	112,64	1,59	88,77	,9013
18	,87151	—	15,07	113,40	1,67	88,17	,8953
19	,87302	—	15,90	114,16	1,74	87,58	,8893
100 + 20	,87448	—	16,74	114,92	1,82	87,00	,8833
21	,87592	—	17,57	115,69	1,88	86,42	,8774
22	,87735	—	18,41	116,46	1,95	85,85	,8717
23	,87876	—	19,25	117,22	2,03	85,29	,8660
24	,88014	—	20,09	117,99	2,10	84,73	,8603
100 + 25	,88149	—	20,93	118,76	2,17	84,18	,8548
26	,88281	—	21,76	119,53	2,23	83,64	,8493
27	,88412	—	22,59	120,31	2,28	83,11	,8439
28	,88540	—	23,44	121,08	2,36	82,58	,8385
29	,88665	—	24,27	121,86	2,41	82 05	,8332
100 + 30	,88789	—	25,11	122,64	2,47	81,53	,8279
31	,88911	—	25,94	123,41	2,53	81,03	,8227
32	,89032	—	26,78	124,19	2,59	80,52	,8176
33	,89151	—	27,62	124,96	2,66	80,02	,8125
34	,89267	—	28,45	125,74	2,71	79,53	,8075
100 + 35	,89380	—	29,29	126,51	2,78	79,04	,8025
36	,89494	—	30,13	127,29	2,84	78,56	,7976
37	,89604	—	30,97	128,06	2,91	78,08	,7928
38	,89713	—	31,81	128,84	2,97	77,61	,7880
39	,88820	—	32,63	129,63	3,00	77,14	,7833
100 + 40	,89925	—	33,48	130,41	3,07	76,68	,7785
41	,90029	—	34,32	131,19	3,13	76,22	,7740
42	,90132	—	35,15	131,97	3,18	75,77	,7694
43	,90233	—	35,99	132,75	3,24	75,33	,7649
44	,90333	—	36,82	133,53	3,29	74,89	,7604
100 + 45	,90431	—	37,66	134,31	3,35	74,46	,7559
46	,90527	—	38,49	135,09	3,40	74,02	,7517
47	,90622	—	39,33	135,87	3,46	73,60	,7473
48	,90715	—	40,17	136,66	3,51	73,18	,7431
49	,90807	—	41,01	137,44	3,57	72,75	,7388

Right panel

I. Spirit and Water by weight.	II. Specific gravity.	III. Spirit by measure.	IV. Water by measure.	V. Bulk of mixture.	VI. Diminution of bulk.	VII. Quantity of Spirit per cent.	VIII. Decimal multipliers.
Sp. + W.							
100 + 50	,90895	100	41,85	138,23	3,62	72,34	,7345
51	,90985	—	42,67	139,01	3,66	71,93	,7304
52	,91071	—	43,52	139,80	3,72	71,53	,7263
53	,91157	—	44,35	140,59	3,76	71,13	,7223
54	,91242	—	45,19	141,38	3,81	70,73	,7182
100 + 55	,91325	—	46,03	142,16	3,87	70,34	,7142
56	,91406	--	46,87	142,95	3,92	69,95	,7103
57	,91488	—	47,70	143,74	3,96	69,57	,7064
58	,91567	—	48,53	144,53	4,00	69,19	,7025
59	,91645	—	49,37	145,32	4,05	68,81	,6986
100 + 60	,91722	—	50,21	146,11	4,10	68,44	,6949
61	,91799	—	51,05	146,90	4,15	68,07	,6911
62	,91875	—	51,89	147,69	4,20	67,71	,6874
63	,91949	—	52,72	148,48	4,24	67,35	,6837
64	,92021	—	53,56	149,28	4,28	66,99	,6801
100 + 65	,92093	—	54,40	150,07	4,33	66,64	,6766
66	,92164	—	55,23	150,86	4,37	66,29	,6730
67	,92234	—	56,07	151,66	4,41	65,94	,6695
68	,92303	—	56,91	152,45	4,46	65,59	,6660
69	,92371	—	57,74	153,24	4,50	65,25	,6625
100 + 70	,92439	—	58,58	154,04	4,54	64,92	,6591
71	,92505	—	59,41	154,83	4,58	64,59	,6557
72	,92571	—	60,25	155,63	4,62	64,26	,6524
73	,92634	—	61,09	156,43	4,66	63,93	,6491
74	,92698	—	61,93	157,22	4,71	63,61	,6458
100 + 75	,92764	—	62,76	158,02	4,74	63,29	,6425
76	,92827	—	63,60	158,81	4,79	62,97	,6393
77	,92888	—	64,44	159,61	4,83	62,65	,6361
78	,92951	—	65,27	160,40	4,87	62,34	,6330
79	,93000	—	66,11	161,20	4,91	62,04	,6298
100 + 80	,93008	—	66,96	162,00	4,96	61,73	,6267
81	,93127	—	67,78	162,79	4,99	61,43	,6237
82	,93185	—	68,62	163,59	5,03	61,13	,6207
83	,93243	—	69,46	164,39	5,07	60,83	,6176
84	,93299	—	70,30	165,19	5,11	60,54	,6146
100 + 85	,93354	—	71,14	165,99	5,15	60,24	,6116
86	,93409	—	71,97	166,79	5,18	59,96	,6088
87	,93463	—	72,81	167,59	5,22	59,67	,6059
88	,93516	—	73,64	168,39	5,25	59,39	,6030
89	,93569	—	74,48	169,19	5,29	59,11	,6001
100 + 90	,93621	—	75,32	169,99	5,33	58,83	,5972
91	,93672	—	76,15	170,80	5,35	58,55	,5945
92	,93722	—	76,99	171,60	5,39	58,28	,5917
93	,93772	—	77,82	172,40	5,42	58.01	,5890
94	,93821	—	78,66	173,20	5,46	57,74	,5863
100 + 95	,93870	—	79,50	174,00	5,50	57,47	,5835
96	,93918	—	80,33	174,80	5,53	57,21	,5809
97	,93965	—	81,16	175,60	5,56	56,95	,5783
98	,94012	—	82,00	176,40	5,60	56,69	,5756
99	,94059	—	82,84	177,21	5,63	56,43	,5730

TABLE II. HEAT 33°.

I. Water and spirit by weight.	II. Specific gravity.	III. Spirit by measure.	IV. Water by measure.	V. Bulk of mixture.	VI. Diminution of bulk.	VII. Quantity of spirit per cent.	VIII. Decimal multipliers.	I. Water and spirit by weight.	II. Specific gravity.	III. Spirit by measure.	IV. Water by measure.	V. Bulk of mixture.	VI. Diminution of bulk.	VII. Quantity of spirit per cent.	VIII. Decimal multipliers.
W. + Sp								W. + Sp.							
100+100	,94105	100	83,69	178,02	5,67	56,16	,5703	100 + 50	,96635	100	167,38	260,02	7,36	38,45	,3904
99	,94149	—	84,54	178,84	5,70	55,91	,5677	49	,96687	—	170,79	263,43	7,36	37,96	,3855
98	,94194	—	85,41	179,68	5,73	55,65	,5651	48	,96738	—	174,35	266,97	7,38	37,46	,3803
97	,9¡239	—	86,28	180,53	5,75	55,39	,5624	47	,96790	—	178,06	270,66	7,40	36,95	,3752
96	,94284	—	87,19	181,39	5,80	55,13	,5598	46	,96841	—	181,93	274,51	7,42	36,43	,3699
100 + 95	,94329	—	88,10	182,27	5,83	54,85	,5570	100 + 45	,96890	—	185,97	278,56	7,41	35,90	,3645
94	,94374	—	89,03	183,18	5,85	54,59	,5543	44	,96940	—	190,20	282,78	7,42	35,36	,3590
93	,94420	—	89,99	184,10	5,89	54,32	,5515	43	,96989	—	194,63	287,20	7,43	34,81	,3535
92	,94466	—	90,97	185,04	5,93	54,04	,5487	42	,97037	—	199,26	291,85	7,41	34,26	,3479
91	,94513	—	91,97	186,02	5,95	53,76	,5459	41	,97085	—	204,11	296,71	7,40	33,70	,3422
100 + 90	,94560	—	92,99	187,00	5,99	53,48	,5429	100 + 40	,97132	—	209,22	301,82	7,40	33,13	,3364
89	,94609	—	94,03	188,01	6,02	53,19	,5401	39	,97178	—	214,59	307,20	7,39	32,55	,3305
88	,94658	—	95,10	189,05	6,05	52,90	,5371	38	,97223	—	220,23	312,87	7,36	31,96	,3245
87	,94708	—	96,20	190,10	6,10	52,60	,5341	37	,97269	—	226,19	318,87	7,32	31,36	,3184
86	,94758	—	97,31	191,19	6,12	52,31	,5311	36	,97315	—	232,47	325,19	7,28	30,75	,3122
100 + 85	,94808	—	98,47	192,29	6,18	52,00	,5280	100 + 35	,97359	—	239,11	331,85	7,26	30,13	,3059
84	,94859	—	99,64	193,43	6,21	51,70	,5249	34	,97404	—	246,15	338,92	7,23	29,51	,2996
83	,94910	—	100,83	194,59	6,24	51,39	,5218	33	,97450	—	253,60	346,43	7,17	28,87	,2931
82	,94960	—	102,06	195,78	6,28	51,08	,5186	32	,97495	—	261 53	354,41	7,12	28,22	,2865
81	,95011	—	103,32	197,00	6,32	50,77	,5154	31	,97541	—	269,97	362,90	7,07	27,56	,2798
100 + 80	,95062	—	104,61	198,26	6,35	50,45	,5121	100 + 30	,97588	—	278,95	371,94	7,02	26,89	,2730
79	,95114	—	105,94	199,54	6,40	50,11	,5088	29	,97634	—	288,58	381,61	6,97	26,20	,2661
78	,95165	—	107,30	200,86	6,44	49,78	,5044	28	,97681	—	298,90	392,00	6,90	25,51	,2590
77	,95217	—	108,69	202,22	6,47	49,46	,5020	27	,97728	—	309,96	403,14	6,82	24,80	,2519
76	,95268	—	110,13	203,61	6,52	49,11	,4986	26	,97776	—	321,88	415,15	6,73	24,09	,2446
100 + 75	,95320	—	111,59	205,04	6,55	48,77	,4952	100 + 25	,97825	—	334,77	428,12	6,65	23,36	,2371
74	,95370	—	113,09	206,52	6,57	48,42	,4917	24	,97874	—	348,71	442,16	6,55	22,61	,2296
73	,95421	—	114,64	208,03	6,61	48,07	,4881	23	,97925	—	363,86	457,43	6,43	21,86	,2220
72	,95472	—	116,24	209,59	6,65	47,71	,4845	22	,97978	—	381,40	474,08	6,32	21,09	,2142
71	,95522	—	117,87	211,19	6,68	47,35	,4808	21	,98033	—	398,52	492,30	6,22	20,31	,2062
100 + 70	,95574	—	119,55	212,83	6,72	46,99	,4770	100 + 20	,98090	—	418,45	512,36	6,09	19,52	,1981
69	,95626	—	121,28	214,53	6,75	46,61	,4732	19	,98146	—	440,48	534,51	5,97	18,71	,1899
68	,95680	—	123,07	216,28	6,79	46,23	,4693	18	,98206	—	464,94	559,14	5,80	17,88	,1816
67	,95733	—	124,91	218,09	6,82	45,85	,4656	17	,98269	—	492,30	586,63	5,67	17,05	,1731
66	,95787	—	126,80	219,94	6,86	45,47	,4616	16	,98334	—	523,06	617,55	5,51	16,19	,1644
100 + 65	,95843	—	128,75	221,85	6,90	45,08	,4576	100 + 15	,98403	—	557,93	652,59	5,34	15,33	,1556
64	,95895	—	130,77	223,83	6,94	44,68	,4536	14	,98475	—	597,79	692,62	5,17	14,43	,1466
63	,95949	—	132,83	225,86	6,97	44,27	,4495	13	,98550	—	643,77	738,79	4,98	13,54	,1375
62	,96004	—	134,97	227,97	7,00	43,86	,4453	12	,98630	—	697,42	792,62	4,80	12,62	,1281
61	,96058	—	137,19	230,14	7,05	43,45	,4411	11	,98715	—	760,83	856,23	4,60	11,68	,1186
100 + 60	,96112	—	139,48	232,41	7,07	43,03	,4369	100 + 10	,98804	—	836,90	932,54	4,36	10,72	,1089
59	,96166	—	41,84	234,73	7,11	42,60	,4325	9	,98898	—	929,90	1025,75	4,15	9,75	,0990
58	,96219	—	144,28	237,14	7,14	42,17	,4281	8	,98999	—	1046,13	1142,23	3,90	8,75	,0889
57	,96272	—	146,82	239,65	7,17	41,73	,4236	7	,99100	—	1195,59	1291,92	3,67	7,74	,0786
56	,96325	—	149,45	242,24	7,21	41,28	,4192	6	,99220	—	1394,86	1491,43	3,43	6 70	,0681
100 + 55	,96377	—	152,16	244,92	7,24	40,83	,4145	100 + 5	,99340	—	1673 83	1770,07	3,16	5,64	,0573
54	,96429	—	154,95	247,70	7,25	40,37	,4099	4	,99471	—	2092,26	2189,41	2,85	4,56	,0464
53	,96481	—	157,87	250,60	7,27	39,90	,4051	3	,99610	—	2789,69	2887,10	2,59	3 46	,0352
52	,96533	—	160,92	253,62	7,30	39,43	,4003	2	,99750	—	4184,53	4282,19	2,34	2,34	,0237
51	,96584	—	164,08	256,76	7,32	38,94	,3954	1	,99918	—	8369 04	8466,95	2,09	1,18	,0120

TABLE I.   HEAT 34°.

I. Spirit and water by weight.	II. Specific gravity.	III. Spirit by measure.	IV. Water by measure.	V. Bulk of mixture.	VI. Diminution of bulk.	VII. Quantity of spirit per cent.	VIII. Decimal multipliers.	I. Spirit and water by weight.	II. Specific gravity.	III. Spirit by measure.	IV. Water by measure.	V. Bulk of mixture.	VI. Diminution of Bulk.	VII. Quantity of spirit per cent.	VIII. Decimal multipliers.
Sp. + W.								Sp. + W.							
100 + 0	,83717	100	—	100,00	—	100,00	1,0147	100 + 50	,90853	100	41,82	138,22	3,60	72,35	,7342
1	,83949	—	0,84	100,72	0,12	99,28	1,0076	51	,90943	—	42,65	139,00	3,65	71,94	,7301
2	,84174	—	1,68	101,45	0,23	98,57	1,0004	52	,91029	—	43,49	139,79	3,70	71,53	,7259
3	,84395	—	2,51	102,18	0,33	97,87	,9932	53	,91115	—	44,33	140,58	3,75	71,13	,7219
4	,84608	—	3,35	102,91	0,44	97,18	,9861	54	,91200	—	45,17	141,37	3,80	70,74	,7179
100 + 5	,84815	—	4,19	103,64	0,55	96,49	,9791	100 + 55	,91283	—	46,01	142,15	3,86	70,35	,7138
6	,85017	—	5,03	104,38	0,65	95,80	,9722	56	,91365	—	46,84	142,94	3,90	69,96	,7100
7	,85213	—	5,86	105,12	0,74	95,13	,9654	57	,91446	—	47,67	143,73	3,94	69,57	,7061
8	,85405	—	6,70	105,87	0,83	94,46	,9586	58	,91525	—	48,51	144,52	3,99	69,19	,7021
9	,85593	—	7,53	106,61	0,92	93,80	,9519	59	,91603	—	49,34	145,31	4,03	68,82	,6983
100 + 10	,85774	—	8,37	107,36	1,01	93,14	,9452	100 + 60	,91681	—	50,18	146,10	4,08	68,44	,6945
11	,85954	—	9,20	108,11	1,09	92,50	,9386	61	,91758	—	51,02	146,89	4,13	68,07	,6908
12	,86131	—	10,04	108,86	1,18	91,86	,9321	62	,91833	—	51,86	147,68	4,18	67,71	,6871
13	,86302	—	10,88	109,61	1,27	91,23	,9257	63	,91907	—	52,69	148,47	4,22	67,35	,6834
14	,86470	—	11,71	110,37	1,34	90,61	,9194	64	,91979	—	53,53	149,27	4,26	66,99	,6798
100 + 15	,86635	—	12,55	111,13	1,42	89,99	,9131	100 + 65	,92051	—	54,37	150,06	4,31	66,64	,6763
16	,86794	—	13,39	111,89	1,50	89,38	,9069	66	,92122	—	55,20	150,85	4,35	66,29	,6727
17	,86951	—	14,22	112,64	1,58	88,77	,9008	67	,92192	—	56,04	151,65	4,39	65,94	,6692
18	,87105	—	15,06	113,40	1,66	88,17	,8948	68	,92261	—	56,88	152,44	4,44	65,60	,6657
19	,87255	—	15,89	114,16	1,73	87,58	,8888	69	,92329	—	57,71	153,23	4,48	65,26	,6622
100 + 20	,87402	—	16,73	114,93	1,80	87,00	,8829	100 + 70	,92397	—	58,55	154,03	4,52	64,92	,6588
21	,87547	—	17,56	115,69	1,87	86,42	,8770	71	,92464	—	59,38	154,82	4,56	64,59	,6554
22	,87690	—	18,40	116,46	1,94	85,85	,8712	72	,92530	—	60,22	155,62	4,60	64,26	,6521
23	,87831	—	19,24	117,23	2,01	85,30	,8656	73	,92593	—	61,05	156,41	4,64	63,93	,6488
24	,87969	—	20,08	118,00	2,08	84,74	,8599	74	,92657	—	61,90	157,21	4,69	63,61	,6455
100 + 25	,88104	—	20,92	118,76	2,16	84,19	,8543	100 + 75	,92722	—	62,73	158,00	4,73	63,29	,6422
26	,88237	—	21,74	119,54	2,20	83,65	,8489	76	,92785	—	63,56	158,80	4,76	62,97	,6390
27	,88368	—	22,58	120,31	2,27	83,11	,8435	77	,92846	—	64,41	159,60	4,81	62,65	,6358
28	,88495	—	23,42	121,08	2,34	82,58	,8381	78	,92908	—	65,23	160,39	4,84	62,34	,6327
29	,88621	—	24,26	121,85	2,41	82,06	,8328	79	,92968	—	66,07	161,19	4,88	62,04	,6295
100 + 30	,88745	—	25,09	122,64	2,45	81,54	,8275	100 + 80	,93027	—	66,92	161,98	4,94	61,73	,6265
31	,88807	—	25,93	123,41	2,52	81,03	,8223	81	,93086	—	67,74	162,77	4,97	61,43	,6234
32	,88988	—	26,77	124,18	2,59	80,53	,8172	82	,93144	—	68,58	163,57	5,01	61,13	,6204
33	,89107	—	27,61	124,96	2,65	80,03	,8121	83	,93202	—	69,42	164,37	5,05	60,83	,6173
34	,89223	—	28,44	125,73	2,71	79,53	,8071	84	,93259	—	70,26	165,17	5,09	60,54	,6143
100 + 35	,89337	—	29,28	126,51	2,77	79,04	,8021	100 + 85	,93314	—	71,10	165,97	5,13	60,25	,6114
36	,89451	—	30,12	127,29	2,83	78,56	,7992	86	,93369	—	71,93	166,77	5,16	59,96	,6085
37	,89561	—	30,95	128 06	2,89	78,08	,7924	87	,93423	—	72,77	167,57	5,20	59,68	,6057
38	,89670	—	31,79	128,83	2,96	77,61	,7876	88	,93476	—	73,60	168,37	5,23	59,39	,6027
39	,88777	—	32,62	129,62	3,00	77,14	,7829	89	,93529	—	74,44	169,17	5,27	59,11	,5998
100 + 40	,89882	—	33,46	130,40	3,06	76,68	,7782	100 + 90	,93581	—	75,28	169,97	5,31	58,83	,5970
41	,89986	—	34,30	131,18	3,12	76,23	,7736	91	,93632	—	76,11	170,78	5,33	58,55	,5942
42	,90089	—	35,13	131,96	3,17	75,78	,7690	92	,93682	—	76,95	171,58	5,37	58,28	,5914
43	,90190	—	35,97	132,74	3,23	75,34	,7645	93	,93732	—	77,78	172,38	5,40	58,01	,5887
44	,90290	—	36,80	133,52	3,28	74,89	,7600	94	,93781	—	78,62	173,18	5,44	57,74	,5860
100 + 45	,90388	—	37,64	134,30	3,34	74,46	,7556	100 + 95	,93830	—	79,46	173,98	5,48	57,47	,5833
46	,90484	—	38,47	135,08	3,39	74,03	,7513	96	,93878	—	80,28	174,78	5,50	57,22	,5806
47	,90579	—	39,31	135,86	3,45	73,60	,7469	97	,93925	—	81,12	175,58	5,54	56,95	,5780
48	,90672	—	40,15	136,65	3,50	73,18	,7427	98	,93972	—	81,95	176,38	5,57	56,69	,5754
49	,90764	—	40,98	137,43	3,55	72,76	,7384	99	,94019	—	82,80	177,19	5,61	56,44	,5728

TABLE II. HEAT 34°.

Water and spirit by weight.	Specific gravity.	Spirit by measure.	Water by measure.	Bulk of mixture.	Diminution of bulk.	Quantity of spirit per cent.	Decimal multipliers.
W. + Sp.							
100+100	,94065	100	83,64	178,00	5,64	56,17	,5701
99	,94110	—	84.49	178,82	5,67	55,92	,5074
98	,94154	—	85,36	179,66	5,70	55,06	,5648
97	,94199	—	86,23	180,51	5,72	55,39	,5621
96	,94244	—	87.14	181,37	5.77	55,13	,5595
100+95	,94209	—	88,05	182,25	5,80	54 06	,5508
94	,94335	—	88,98	183.15	5,82	54,59	,5540
93	,94361	—	89,84	184.07	5,87	54.33	,5512
92	,94427	—	90,92	185,02	5,90	54,05	,5484
91	,94475	—	91,92	186,00	5,92	53.77	,5456
100+90	,94522	—	92,94	186,98	5,90	53,40	,5427
89	,94571	—	93,98	187,99	5,99	53.19	,5399
88	,94021	—	95,04	189,02	6,02	52,91	,5369
87	,94671	—	96,14	190,08	6.06	52,61	,5339
86	,94721	—	97,20	191,16	6.10	52,32	,5309
100+85	,94771	—	98,41	192,26	6,15	52,01	,5278
84	,94822	—	99,58	193,40	6.18	51,71	,5247
83	,94873	—	100.77	194,56	6,21	51,40	,5216
82	,94923	—	102,00	195,75	6,25	51,08	,5184
81	,94974	—	103.26	196,97	6.29	50,77	,5152
100+80	,95025	—	104,55	198,23	6,32	50,45	,5119
79	,95077	—	105,88	199,51	6.37	50,12	,5086
78	,95128	—	107,24	200,83	6,41	49,79	,5052
77	,95180	—	108,63	202,19	6,44	49,46	,5016
76	,95232	—	110,07	203,58	6,49	49,12	,4984
100+75	,95203	—	111,53	205,01	6,52	48,78	,4950
74	,95334	—	113,02	206,49	6,53	48,43	,4915
73	,95485	—	114,57	207,99	6,58	48,08	,4879
72	,95436	—	116,17	209,55	6,62	47,72	,4843
71	,95480	—	117,80	211,10	6,64	47,36	,4806
100+70	,95538	—	119,48	212,80	6,68	46,99	,4766
69	,95591	—	121.21	214,50	6,71	46,62	,4730
68	,95645	—	123,00	210,24	6.76	46,24	,4693
67	,95698	—	124.84	218,05	6,79	45,86	,4654
66	,95757	—	125,73	219,90	6.83	45,47	,4615
100+65	,95800	—	128 08	221,81	6,87	45.08	,4575
64	,95801	—	130,69	223,79	6.90	44,69	,4535
63	,95916	—	132.76	225,82	6.94	44,28	,4494
62	,95971	—	134,90	227,92	6,98	43 87	,4451
61	,96026	—	137,11	230,10	7,01	43,42	,4409
100+60	,96080	—	139,40	432,30	7,04	43,04	,4367
59	,96134	—	141,70	234,08	7,08	42,61	,4324
58	,96187	—	144,20	237,09	7,11	42,18	,4280
57	,96241	—	146,74	239,60	7,14	41,74	,4235
56	,96295	—	149.30	242,19	7,17	41,29	,4191
100+55	,96340	—	152,07	244 87	7,20	40,8	,4144
54	,96399	—	154,86	247,65	7,21	40.38	,4098
53	,96453	—	157,78	250 55	7,23	39,91	,4050
52	,96504	—	160,82	253,57	7,25	39,44	,4002
51	,96555	—	163,99	256,70	7,29	38,95	,3953
100+50	,96607	100	107,28	259,96	7,32	38.46	,3903
49	,96660	—	170,70	263.37	7,33	37 97	,3853
48	,96711	—	174,25	266,90	7,35	37,47	,3802
47	,96764	—	177,96	270,59	7,37	36,96	,3751
46	,96815	—	181,83	274,44	7,39	36,44	,3698
100+45	,96865	—	185,87	278,48	7.39	35,91	,3644
44	,96915	—	190,09	282,70	7,39	35,37	,3589
43	,96965	—	194,52	287,12	7,40	34,82	,3534
42	,97013	—	199.15	291,76	7,39	34,27	,3478
41	,97062	—	204,00	296,62	7,38	33,71	,3421
100+40	,97110	—	209,10	301.73	7,37	33,14	,3303
39	,97156	—	214,47	307,11	7,36	32,56	,3304
38	,97202	—	220,11	312,78	7.33	31,97	,3245
37	,97248	—	226,06	318,77	7,29	31,37	,3184
36	,97294	—	232,34	325 09	7,25	30,76	,3122
100+35	,97339	—	238,98	331,74	7,24	30.14	,3059
34	,97385	—	246,01	338,80	7,21	29,52	,2995
33	,97432	—	253,46	346,31	7,15	28,88	,2930
32	,97478	—	261,38	354,28	7,10	28,23	,2864
31	,97525	—	269,82	362,77	7,05	27,57	,2797
100+30	,97,72	—	278,80	371,80	7,00	26,90	,2729
29	,97619	—	288,42	381,47	6,95	26,21	,2660
28	,97667	—	298,73	391,85	6,88	25,52	,2590
27	,97715	—	309,79	402,99	6,80	24,81	,2518
26	,97763	—	321,70	414,99	6,71	24,10	,2445
100+25	,97813	—	334,58	427,95	6,63	23,37	,2371
24	,97864	—	348,51	441.97	6,54	22,62	,2296
23	,97916	—	363,66	457,23	6,43	21,87	,2219
22	,97970	—	381,18	473,87	6,31	21,10	,2141
21	,98026	—	398,30	492,07	6,23	20,32	,2062
100+20	,98083	—	418,21	512,12	6,09	19,53	,1981
19	,98141	—	440,23	534,26	5,97	18,72	,1899
18	,98201	—	464,68	558,87	5,81	17,89	,1816
17	,98265	—	492,02	586,34	5,68	17,06	,1731
16	,98331	—	522,77	617,24	5,53	16,20	,1644
100+15	,98400	—	557,62	652,26	5,36	15,33	,1556
14	,98473	—	597,45	692,27	5,18	14,44	,1466
13	,98549	—	643,41	738,41	5,00	13,54	,1374
12	,98629	—	697,03	792,21	4,82	12,62	,1281
11	,98714	—	700,40	855,78	4,62	11,69	,1186
100+10	,98804	—	836,43	932,04	4,39	10,73	,1089
9	,98898	—	949,37	1025,20	4,17	9,75	,0990
8	,99001	—	1054,54	1141,61	3,93	8,76	,0889
7	,99107	—	1194,91	1291,21	3,70	7,74	,0786
6	,99215	—	1304,07	1490,61	3,42	6,71	,0681
100+5	,99342	—	1672,88	1769,69	3,19	5 65	,0573
4	,99473	—	2091,08	2188,19	2,89	4,57	,0464
3	,99621	—	2788 10	2885,49	2,61	3,46	,0352
2	,99761	—	4182,17	4279,81	2,30	2 34	,0237
1	,99920	—	8364,32	8464,24	2,08	1.18	,0120

TABLE I.   HEAT 35°.

I. Spirit and water by weight.	II. Specific gravity.	III. Spirit by measure.	IV. Water by measure.	V. Bulk of mixture.	VI. Diminution of bulk.	VII. Quantity of spirit per cent.	VIII. Decimal multipliers.	I. Spirit and water by weight.	II. Specific gravity.	III. Spirit by measure.	IV. Water by measure.	V. Bulk of mixture.	VI. Diminution of bulk.	VII. Quantity of spirit per cent.	VIII. Decimal multipliers.
Sp. + W.								Sp. + W.							
100 + 0	,83672	100	0,84	100,00	—	100,00	1,0142	100 + 50	,90811	100	41,80	138,21	3,59	72,35	,7338
1	,83904	—	0,84	100,72	0,12	99,28	1,0070	51	,90900	—	42,63	138,99	3,64	71,94	,7297
2	,84129	—	1,67	101,45	0,22	98,57	,9998	52	,90987	—	43,47	139,78	3,69	71,54	,7256
3	,84349	—	2,51	102,18	0,33	97,87	,9926	53	,91073	—	44,30	140,57	3,73	71,14	,7215
4	,84562	—	3,34	102,91	0,43	97,18	,9855	54	,91158	—	45,14	141,36	3,78	70,74	,7175
100 + 5	,84769	—	4,18	103,64	0,54	96,49	,9786	100 + 55	,91241	—	45,98	142,14	3,84	70,35	,7135
6	,84971	—	5,02	104,38	0,64	95,80	,9716	56	,91323	—	46,81	142,93	3,88	69,96	,7096
7	,85167	—	5,85	105,12	0,73	95,13	,9648	57	,91404	—	47,65	143,72	3,93	69,58	,7057
8	,85359	—	6,69	105,87	0,82	94,46	,9580	58	,91483	—	48,49	144,51	3,98	69,20	,7018
9	,85547	—	7,52	106,61	0,91	93,80	,9513	59	,91561	—	49 32	145,30	4,02	68,82	,6980
100 + 10	,85729	—	8,36	107,36	1,00	93,14	,9447	100 + 60	,91640	—	50,16	146,09	4,07	68,45	,6942
11	,85908	—	9,20	108,11	1,09	92,50	,9381	61	,91716	—	50,99	146,88	4,11	68,08	,6905
12	,86083	—	10,03	108,86	1,17	91,86	,9316	62	,91791	—	51,83	147,67	4,16	67,72	,6868
13	,86255	—	10,87	109,61	1,26	91,23	,9252	63	,91805	—	52 67	148,46	4,21	67,36	,6831
14	,86423	—	11,70	110,37	1,33	90,61	,9189	64	,91937	—	53,50	149,26	4,24	67,00	,6795
100 + 15	,86587	—	12,54	111,13	1,41	89,99	,9126	100 + 65	,92009	—	54,34	150,05	4,29	66,65	,0760
16	,86747	—	13,38	111,89	1,49	89,38	,9064	66	,92080	—	55,17	150,84	4,33	66,30	,6724
17	,86904	—	14,21	112,65	1,56	88,77	,9003	67	,92150	—	56,01	151,64	4,37	65,95	,6689
18	,87058	—	15,05	113,41	1,64	88,17	,8943	68	,92219	—	56,85	152,43	4,42	65,60	,6654
19	,87209	—	15,88	114,17	1,71	87,58	,8883	69	,92287	—	57,68	153,22	4,40	65,26	,6619
200 + 20	,87357	—	16,72	114,94	1,78	87,00	,8824	100 + 70	,92355	—	58,52	154,02	4,50	64,93	,6585
21	,87502	—	17,55	115,70	1,85	86,43	,8766	71	,92422	—	59,35	154,81	4,54	64,60	,6551
22	,87645	—	18,39	116,46	1,93	85,86	,8708	72	,92488	—	60,19	155,61	4,58	64,27	,6518
23	,87786	—	19,23	117,23	2,00	85,30	,8651	73	,92552	—	61,02	156,40	4,62	63,94	,6485
24	,87924	—	20,06	118,00	2,06	84 74	,8595	74	,92616	—	61,86	157,20	4,66	63,62	,6452
100 + 25	,88059	—	20,90	118,77	2,13	84,19	,8539	100 + 75	,92680	—	62,70	157,99	4,71	63,30	,6419
26	,88192	—	21,73	119,54	2,19	83,65	,8484	76	,92743	—	63,53	158,79	4,75	62,98	,6387
27	,88323	—	22,57	120,31	2,26	83,12	,8430	77	,92805	—	64,37	159,59	4,78	62,66	,6355
28	,88451	—	23,41	121,08	2,33	82,59	,8376	78	,92866	—	65,20	160,38	4,82	62,35	,6324
29	,88577	—	24,24	121,85	2,39	82,06	,8323	79	,92926	—	66,04	161,18	4,86	62,05	,6293
100 + 30	,88701	—	25,08	122,63	2,45	81,54	,8270	100 + 80	,92986	—	66,88	161,97	4,91	61,74	,0262
31	,88823	—	25,91	123,40	2,51	81,04	,8218	81	,93045	—	67,71	162,76	4,95	61,44	,6231
32	,88944	—	26,75	124,18	2,57	80,53	,8167	82	,93103	—	68,55	163,56	4,99	61,14	,6201
33	,89063	—	27,59	124,95	2,64	80,03	,8117	83	,93161	—	69,38	164,36	5,02	60,84	,6170
34	,89179	—	28,42	125,73	2,69	79,54	,8067	84	,93218	—	70,22	165,16	5,06	60,55	,6140
100 + 35	,89294	—	29,26	126,50	2,76	79,05	,8017	100 + 85	,93274	—	71,06	165,95	5,11	60,20	,6111
36	,89407	—	30,10	127,28	2,82	78,57	,7968	86	,93329	—	71,89	166,75	5,14	59,97	,6082
37	,89517	—	30,93	128,05	2,88	78,09	,7920	87	,93385	—	72,73	167,55	5,18	59,68	,6053
38	,89626	—	31,77	128,83	2,94	77,62	,7872	88	,93436	—	73,56	168,35	5,21	59,40	,6024
39	,89733	—	32,60	129,61	2,99	77,15	,7825	89	,93486	—	74 40	169,15	5,25	59 12	,5995
100 + 40	,89839	—	33,44	130,39	3,05	76,69	,7778	100 + 90	,93541	—	75,24	169 95	5,29	58,84	,5967
41	,89943	—	34,27	131,17	3,10	76,24	,7732	91	,93592	—	76,07	170,76	5,31	58,56	,5939
42	,90046	—	35,11	131,95	3 16	75,79	,7686	92	,93642	—	76,91	171,56	5,35	58,29	,5911
43	,90147	—	35,95	132,73	3,22	75,34	,7641	93	,93692	—	77,74	172,36	5,38	58,02	,5884
44	,90247	—	36,78	133,51	3,27	74,90	,7596	94	,93741	—	78,58	173,16	5,42	57,75	,5857
100 + 45	,90345	—	37,62	134,29	3,33	74,47	,7552	100 + 95	,93790	—	79 42	173,96	5,46	57,48	,5830
46	,90441	—	38,45	135,07	3,38	74,04	,7509	96	,93838	—	80,24	174,76	5,48	57,22	,5803
47	,90536	—	39,29	135,86	3,43	73,61	,7466	97	,93885	—	81,07	175,56	5,51	56,96	,5777
48	,90629	—	40,13	136,64	3,49	73,19	,7423	98	,93932	—	81,91	176,36	5,55	56,70	,5751
49	,90721	—	40,96	137,42	3,54	72,77	,7380	99	,93979	—	82,75	177,17	5,58	56,44	,5725

TABLE II.   HEAT 35°.

Left half

I. Water and spirit by weight.	II. Specific gravity.	III. Spirit by measure.	IV. Water by measure.	V. Bulk of mixture.	VI. Diminution of bulk.	VII. Quantity of spirit per cent.	VIII. Decimal multipliers.
W. + Sp.							
100+100	,94025	100	83,60	177,98	5,62	56,18	,5698
99	,94070	—	84.45	178,80	5,65	55,93	,5672
98	,94115	—	85,31	179.64	5,67	55,67	,5646
97	,94159	—	86,19	180,49	5,70	55,40	,5619
96	,94204	—	87,09	181.35	5,74	55,14	,5593
100+95	,94249	—	88,00	182,23	5,77	54,87	,5566
94	,94295	—	88.93	183,13	5,80	54,60	,5538
93	,94342	—	89,89	184,05	5,84	54,33	,5510
92	,94389	—	90,87	185,00	5,87	54,05	,5482
91	,94436	—	91,87	185.97	5,90	53,77	,5454
100+90	,94484	—	92,89	186,95	5,91	53,49	,5425
89	,94533	—	93,93	187,96	5,97	53.20	,5390
88	,94583	—	94.99	188,99	6,00	52,91	,5397
87	,94633	—	96,09	190,05	6,04	52,62	,5337
86	,94683	—	97,21	191.13	6,08	52,32	,5307
100+85	,94734	—	98,35	192,23	6,12	52,02	,5276
84	,94785	—	99,52	193,37	6,15	51,71	,5245
83	,94835	—	100.72	194,53	6,19	51,40	,5214
82	,94886	—	101,95	195,72	6,23	51,09	,5182
81	,9,937	—	103.20	196 94	6,26	50,78	,5150
100+80	,94988	—	104,49	198.20	6,29	50,46	,5117
79	,95040	—	105,82	199.48	6,34	50,13	,5084
78	,95091	—	107,18	200,80	6,38	49,80	,5051
77	,95143	—	108,57	202,16	6,41	49.47	,5017
76	,95194	—	110,00	203.55	6,45	49,13	,4983
100+75	,95246	—	111,46	204.98	6.48	48,79	,4948
74	,95297	—	112,96	206,45	6,51	48,44	,4913
73	,95348	—	114,51	207.90	6,55	48,09	,4877
72	,95399	—	116,10	209,52	6,58	47,73	,4841
71	,95450	—	117,74	211,13	6,61	47,37	,4804
100+70	,95502	—	119,42	212,77	6,65	47,00	,4767
69	,95555	—	121,15	214,46	6,69	46,63	,4729
68	,95609	—	122,93	216,21	6,72	46,25	,4691
67	,95663	—	124,77	218,01	6,76	45,87	,4652
66	,95717	—	126,66	219,86	6,80	45,48	,4613
100+65	,95772	—	128,61	221,77	6,84	45,09	,4573
64	,95827	—	130,62	223,74	6,88	44,70	,4533
63	,95882	—	132,69	225,77	6,92	44,29	,4492
62	,95938	—	134,83	227,88	6,95	43,88	,4450
61	,95993	—	137,04	230,06	6,98	43,47	,4408
100+60	,96048	—	139,32	232,31	7,01	43,05	,4366
59	,96102	—	141,68	234,63	7,05	42,62	,4322
58	,96155	—	144,12	237,04	7,08	42,19	,4278
57	,96209	—	146,66	239,55	7,11	41,75	,4234
56	,96262	—	149,28	242,14	7,14	41,30	,4189
100+55	96315	—	151,99	244,82	7,17	40,85	,4143
54	,96368	—	154,78	247,60	7,18	40,39	,4096
53	,96421	—	157,69	250,50	7,19	39,92	,4049
52	,96474	—	160,73	253,51	7,22	39,45	,4001
51	,96526	—	163,90	256,65	7,25	38,96	,3952

Right half

I. Water and spirit by weight.	II. Specific gravity.	III. Spirit by measure.	IV. Water by measure.	V. Bulk of mixture.	VI. Diminution of bulk.	VII. Quantity of spirit per cent.	VIII. Decimal multipliers.
W. + Sp.							
100 + 50	,96579	100	167,19	259,91	7,28	38,47	,3902
49	,96632	—	170,61	263,30	7,31	37,98	,3852
48	,96684	—	174,16	266,83	7,33	37,48	,3801
47	,96737	—	177,86	270,52	7,34	36,97	,3750
46	,96789	—	181,73	274,37	7,36	36,45	,3697
100 + 45	,96840	—	185,77	278,40	7,37	35,92	,3643
44	96890	—	189,99	282,62	7,37	35,38	,3588
43	,96940	—	194,41	287,04	7,37	34,83	,3533
42	,96989	—	199,04	291,67	7,37	34,28	,3477
41	,97038	—	203,89	296,53	7,36	33,72	,3420
100 + 40	,97086	—	208,99	301,64	7,35	33,15	,3362
39	,97133	—	214,35	307,02	7,33	32,57	,3303
38	,97180	—	219,99	312,69	7,30	31,98	,3244
37	,97226	—	225,94	318,67	7,27	31,38	,3183
36	,97273	—	232,21	324,98	7,23	30,77	,3121
100 + 35	97319	—	238,85	331,63	7,22	30,15	,3058
34	,97366	—	245,87	338,68	7,19	29,53	,2995
33	,97413	—	253,32	346,19	7,13	28,89	,2930
32	,97460	—	261,24	354,16	7,08	28,24	,2864
31	,97508	—	269,67	362,63	7,04	27,58	,2797
100 + 30	,97556	—	278,65	371,66	6,99	26,91	,2729
29	,97603	—	288,26	381,33	6,93	26,22	,2660
28	,97652	—	298,56	391,70	6,88	25,53	,2590
27	,97701	—	309,62	402,84	6,78	24,82	,2518
26	,97750	—	321,52	414.83	6,69	24,11	,2445
100 + 25	,97801	—	334,39	427,77	6,62	23,38	,2371
24	,97853	—	348,32	441,79	6,53	22,63	,2296
23	,97906	—	363,46	457,03	6,43	21,88	,2219
22	,97961	—	379,97	473,66	6,31	21,11	,2141
21	,98018	—	398,08	491,85	6,23	20,33	,2062
100 + 20	,98076	—	417,98	511,88	6,10	19,54	,1981
19	,98135	—	439,98	534,01	5,97	18,73	,1899
18	,98196	—	464,42	558,59	5,83	17,90	,1816
17	,98260	—	491,74	586,06	5,68	17,06	,1731
16	,98327	—	522,48	616,94	5,54	16,21	,1644
100 + 15	,98397	—	557,31	651,94	5,37	15,34	,1556
14	,98470	—	597,12	691,92	5,20	14,45	,1466
13	,98547	—	643,05	738,03	5,02	13,55	,1374
12	,98628	—	696,64	791,80	4,84	12,63	,1281
11	,98714	—	759,97	855,33	4,64	11,69	,1186
100 + 10	,98804	—	835,96	931,55	4,41	10,73	,1089
9	,98899	—	928,85	1024,65	4,20	9,76	,0990
8	,99000	—	1044,95	1140,99	3,96	8,77	,0889
7	,99108	—	1194,23	1290,51	3,72	7,75	,0786
6	,99222	—	1393,28	1489,79	3,49	6,71	,0681
100 + 5	,99344	—	1671,93	1768,71	3,22	5,65	,0573
4	,99474	—	2089,91	2186,98	2,93	4,57	,0464
3	,99013	—	2786,55	2883,90	2,65	3,47	,0352
2	,99762	—	4179,82	4277,44	2,38	2,34	,0237
1	,99921	—	8359,64	8457,55	2,09	1,18	,0120

Mr. GILPIN's Tables

TABLE I. HEAT 36°.

I. Spirit and water by weight.	II. Specific gravity.	III. Spirit by measure.	IV. Water by measure.	V. Bulk of mixture.	VI. Diminution of bulk.	VII. Quantity of spirit per cent.	VIII. Decimal multipliers.
Sp. + W.							
100 + 0	,85627	100	—	100,00	—	100,00	1,0137
1	,83858	—	0,84	100,72	0,12	99,28	1,0005
2	,84082	—	1,67	101,45	0,22	98,57	,9993
3	,84303	—	2,51	102,18	0,33	97,87	,9921
4	,84516	—	3,34	102,91	0,43	97,18	,9850
100 + 5	,84723	—	4,18	103,64	0,54	96,49	,9780
6	,84926	—	5,02	104,38	0,64	95,86	,9711
7	,85122	—	5,85	105,12	0,73	95,13	,9643
8	,85314	—	6,69	105,87	0,82	94,46	,9575
9	,85502	—	7,52	100,61	0,91	93,80	,9508
100 + 10	,85684	—	8,36	107,36	1,00	93,14	,9442
11	,85863	—	9,19	108,11	1,08	92,50	,9377
12	,86038	—	10,02	108,86	1,16	91,86	,9312
13	,86210	—	10,86	109,61	1,25	91,23	,9248
14	,86378	—	11,69	110,37	1,32	90 61	,9185
100 + 15	,86541	—	12,53	111,13	1,40	89,99	,9122
16	,86701	—	13,37	111,89	1,48	89,38	,9060
17	,86859	—	14,20	112,64	1,56	88,77	,8999
18	,87013	—	15,04	113,41	1,63	88,17	,8939
19	,87164	—	15,87	114,17	1,70	87,59	,8879
100 + 20	,87313	—	16,71	114,94	1,77	87,01	,8819
21	,87458	—	17,54	115,70	1,84	86,45	,8761
22	,87601	—	18,38	116,46	1,92	85,80	,8703
23	,87742	—	19,22	117,23	1,99	85,30	,8647
24	,87880	—	20,05	117,99	2,06	84,74	,8591
100 + 25	,88015	—	20,89	118,77	2,13	84,19	,8535
26	,88148	—	21,72	119,53	2,19	83,65	,8480
27	,88279	—	22,56	120,30	2,26	83,12	,8426
28	,88407	—	23,40	121,08	2,32	82,59	,8372
29	,88533	—	24,23	121,85	2,38	82 06	,8319
100 + 30	,88657	—	25,07	122,63	2,44	81,55	,8266
31	,88779	—	25,90	123,40	2,50	81,04	,8214
32	,88900	—	26,74	124,18	2,56	80,53	,8163
33	,89019	—	27,58	124,95	2,03	80,03	,8113
34	,89135	—	28,41	125,72	2,69	79 54	,8063
100 + 35	,89250	—	29,24	126,50	2,74	79,05	,8013
36	,89363	—	30,08	127,27	2,81	78,57	,7965
37	,89473	—	30,91	128,05	2,86	78,09	,7917
38	,89582	—	31,75	128,82	2,93	77,62	,7869
39	,89689	—	32,58	129,60	2,98	77,15	,7821
100 + 40	,89795	—	33,42	130,38	3,04	76,69	,7775
41	,89899	—	34,25	131,16	3,09	76,24	,7728
42	,90002	—	35,09	131,94	3,15	75,79	,7683
43	,90103	—	35,93	132,72	3,21	75,34	,7638
44	,90204	—	36,70	133,50	3,26	74,90	,7593
100 + 45	,90302	—	37,60	134,28	3,32	74,47	,7549
46	,90397	—	38,43	135,06	3,37	74,04	,7505
47	,90493	—	39,27	135,85	3,42	73,61	,7463
48	,90586	—	40,10	136,63	3,47	73,19	,7420
49	,90678	—	40,94	137,41	3,53	72,77	,7377
100 + 50	,90768	100	41,78	138,20	3,58	72,36	,7335
51	,90857	..	42,61	138,98	3,63	71,95	,7294
52	,90944	—	43,44	139,77	3,67	71,55	,7253
53	,91030	—	44,28	140,56	3,72	71,15	,7212
54	,91115	—	45,12	141,34	3,78	70,75	,7172
100 + 55	,91198	—	45,96	142,13	3,83	70,36	,7132
56	,91280	—	46,78	142,92	3,86	69,97	,7093
57	,91362	—	47 62	143,71	3,91	69,59	,7054
58	,91441	—	48 45	144 49	3,97	69,21	,7015
59	,91517	—	49 29	145,28	4,01	68,83	,6977
100 + 60	,91593	—	50,13	146,07	4,06	68,46	,6939
61	,91674	—	50,96	146,86	4,10	68,09	,6902
62	,91748	—	51,80	147,66	4,14	67,72	,6865
63	,91823	—	52 64	148,44	4,20	67,37	,6828
64	,91895	—	53,47	149 24	4,23	67,00	,6792
100 + 65	,91967	—	54,31	150,03	4,28	66,65	,6756
66	,92039	—	55,14	150,82	4,32	66,30	,6721
67	,92109	—	55,98	151,62	4,36	65,96	,6686
68	,92178	—	56,81	152,41	4,40	65,61	,6651
69	,92224	—	57,64	153,20	4,44	65,27	,6616
100 + 70	,92315	—	58,48	154,00	4,48	64,93	,6582
71	,92381	—	59,31	154,80	4,51	64,61	,6549
72	,92447	—	60,15	155,59	4,56	64,27	,6515
73	,92511	—	60 98	156,38	4,60	63,94	,6482
74	,92576	—	61,82	157,18	4,64	63,62	,6449
100 + 75	,92640	—	62 66	157,97	4,69	63,31	,6416
76	,92702	—	63,50	158,77	4,73	62,08	,6384
77	,92764	—	64,33	159,57	4 76	62,67	,6352
78	,92825	—	65,16	160,36	4,80	62,30	,6321
79	,92885	—	66,00	161,16	4,84	62,05	,6291
100 + 80	,92949	—	66,84	161,95	4,89	61,75	,6259
81	,93005	—	67,67	162,74	4,93	61,45	,6229
82	,93063	—	68,51	163,54	4,97	61,14	,6199
83	,93121	—	69,34	164 35	4,99	60,84	,6168
84	,93178	—	70,18	165,14	5,04	60,56	,6138
100 + 85	,93234	—	71,02	165,94	5,08	60,20	,6109
86	,93289	—	71,85	166,74	5,11	59,97	,6080
87	,93343	—	72,69	167,54	5,15	59,69	,6051
88	,93397	—	73,52	168 34	5,18	59,40	,6022
89	,93450	—	74,36	169,14	5,22	59,12	,5993
100 + 90	,93501	—	75,20	169,94	5,26	58,85	,5965
91	,93553	—	76,02	170,74	5,28	58,57	,5937
92	,93603	—	76,86	171,54	5,32	58 29	,5909
93	,93653	—	77,70	172,35	5,35	58,02	,5882
94	,93702	—	78,53	173,15	5,38	57,75	,5855
100 + 95	,93751	—	79,37	173,94	5,43	57,49	,5827
96	,93799	—	80,19	174,74	5,45	57,23	,5801
97	,93846	—	81,02	175,54	5,48	56,97	,5775
98	,93893	—	81,86	176,34	5,52	56,70	,5749
99	,93939	—	82,70	177,14	5,56	56,45	,5722

TABLE II.   HEAT 36°.

I. Water and Spirit by weight.	II. Specific gravity.	III. Spirit by measure.	IV. Water by measure.	V. Bulk of mixture.	VI. Diminution of bulk.	VII. Quantity of spirit per cent.	VIII. Decimal multipliers.
W. + Sp.							
100+100	,93986	100	83,55	177,95	5,60	56,19	,5696
99	,94030	—	84,40	178,77	5,63	55,93	,5669
98	,94076	—	85,26	179,61	5,65	55,67	,5643
97	,94120	—	86,14	180,46	5,68	55,41	,5617
96	,94165	—	87,04	181,32	5,72	55,15	,5591
100+95	,94210	—	87,95	182,20	5,75	54,88	,5563
94	,94257	—	88,88	183,11	5,77	54,61	,5536
93	,94304	—	89,84	184,02	5,82	54,34	,5508
92	,94351	—	90,82	184,97	5,85	54,06	,5480
91	,94398	—	91,81	185,94	5,87	53,78	,5452
100+90	,94446	—	92,84	186,92	5,92	53,50	,5423
89	,94496	—	93,87	187,94	5,93	53,21	,5394
88	,94546	—	94,94	188,96	5,98	52,92	,5365
87	,94596	—	96,04	190,03	6,01	52,63	,5335
86	,94646	—	97,15	191,11	6,04	52,33	,5305
100+85	,94696	—	98,30	192,21	6,09	52,03	,5274
84	,94748	—	99,46	193,34	6,12	51,72	,5243
83	,94798	—	100,66	194,50	6,16	51,41	,5212
82	,94849	—	101,89	195,69	6,20	51,09	,5180
81	,94900	—	103,14	196,91	6,23	50,79	,5148
100+80	,94950	—	104,43	198,17	6,26	50,47	,5115
79	,95003	—	105,76	199,45	6,31	50,13	,5082
78	,95054	—	107,12	200,77	6,35	49,80	,5049
77	,95106	—	108,51	202,13	6,38	49,47	,5015
76	,95157	—	109,94	203,52	6,42	49,13	,4981
100+75	,95208	—	111,39	204,94	6,45	48,79	,4946
74	,95261	—	112,89	206,41	6,48	48,45	,4911
73	,95312	—	114,44	207,92	6,52	48,09	,4875
72	,95364	—	116,03	209,48	6,55	47,73	,4839
71	,95415	—	117,67	211,09	6,58	47,38	,4803
100+70	,95466	—	119,35	212,73	6,62	47,01	,4765
69	,95521	—	121,08	214,42	6,66	46,64	,4728
68	,95575	—	122,86	216,17	6,69	46,26	,4690
67	,95629	—	124,7.	217,97	6,73	45,88	,4651
66	,95683	—	126,69	219,82	6,77	45,49	,4612
100+65	,95738	—	128,54	221,73	6,81	45,10	,4572
64	,95793	—	130,54	223,70	6,84	44,70	,4532
63	,95848	—	132,61	225,73	6,88	44,29	,4491
62	,95904	—	134,75	227,84	6,91	43,89	,4449
61	,95959	—	136,96	230,01	6,95	43,48	,4407
100+60	,96014	—	139,24	232,26	6,98	43,05	,4364
59	,96069	—	141,60	234,59	7,01	42,63	,4321
58	,96123	—	144,04	236,99	7,05	42,20	,4277
57	,96177	—	146,58	239,50	7,08	41,76	,4233
56	,96231	—	149,19	242,09	7,10	41,31	,4188
100+55	,96283	—	151,90	244,76	7,14	40,85	,4141
54	,96335	—	154,69	247,55	7,14	40,40	,4095
53	,96391	—	157,60	250,45	7,15	39,93	,4048
52	,96445	—	160,64	253,46	7,18	39,45	,4000
51	,96497	—	163,81	256,59	7,22	38,97	,3951
W. + Sp.							
100+50	,96550	100	167,09	259,85	7,24	38,48	,3901
49	,96604	—	170,51	263,23	7,28	37,99	,3851
48	,96656	—	174,06	266,76	7,30	37,48	,3800
47	,96710	—	177,76	270,45	7,31	36,98	,3749
46	,96762	—	181,63	274,30	7,33	36,46	,3696
100+45	,96814	—	185,66	278,33	7,33	35,92	,3642
44	,96864	—	189,88	282,54	7,34	35,39	,3588
43	,96915	—	194,30	286,96	7,34	34,84	,3533
42	,96964	—	198,93	291,59	7,34	34,29	,3477
41	,97014	—	203,77	296,44	7,33	33,73	,3420
100+40	,97063	—	208,87	301,55	7,32	33,16	,3361
39	,97110	—	214,23	306,92	7,31	32,58	,3303
38	,97158	—	219,86	312,59	7,27	31,99	,3243
37	,97205	—	225,81	318,56	7,25	31,39	,3182
36	,97253	—	232,08	324,86	7,22	30,78	,3121
100+35	,97300	—	238,71	331,52	7,19	30,16	,3058
34	,97347	—	245,73	338,56	7,17	29,54	,2994
33	,97395	—	253,18	346,06	7,12	28,90	,2929
32	,97442	—	261,09	354,03	7,06	28,25	,2863
31	,97491	—	269,52	362,50	7,02	27,59	,2796
100+30	,97540	—	278,49	371,52	6,97	26,92	,2728
29	,97587	—	288,10	381,18	6,92	26,23	,2659
28	,97637	—	298,39	391,55	6,84	25,54	,2589
27	,97687	—	309,44	402,67	6,77	24,83	,2518
26	,97737	—	321,34	414,66	6,68	24,11	,2445
100+25	,97789	—	334,20	427,59	6,61	23,39	,2371
24	,97841	—	348,12	441,60	6,52	22,64	,2296
23	,97895	—	363,26	456,83	6,43	21,89	,2219
22	,97951	—	379,76	473,45	6,31	21,12	,2141
21	,98009	—	397,86	491,63	6,23	20,34	,2062
100+20	,98068	—	417,75	511,64	6,11	19,54	,1981
19	,98128	—	439,73	533,76	5,97	18,73	,1899
18	,98189	—	464,16	558,32	5,84	17,91	,1816
17	,98254	—	491,47	585,77	5,70	17,07	,1731
16	,98322	—	522,19	616,64	5,55	16,21	,1644
100+15	,98393	—	557,00	651,61	5,39	15,34	,1556
14	,98466	—	596,79	691,57	5,22	14,46	,1466
13	,98544	—	642,69	737,65	5,04	13,56	,1374
12	,98625	—	696,25	791,39	4,86	12,63	,1281
11	,98712	—	759,55	854,88	4,67	11,70	,1186
100+10	,98803	—	835,50	931,05	4,45	10,74	,1089
9	,98912	—	928,33	1024,10	4,23	9,76	,0990
8	,98999	—	1044,37	1140,38	3,99	8,77	,0888
7	,99108	—	1193,57	1289,81	3,76	7,75	,0786
6	,99222	—	1392,51	1488,98	3,53	6,72	,0681
100+5	,99344	—	1671,01	1767,74	3,27	5,66	,0573
4	,99475	—	2088,75	2185,78	2,97	4,57	,0464
3	,99614	—	2785,00	2882,31	2,69	3,47	,0352
2	,99764	—	4177,50	4275,07	2,43	2,34	,0237
1	,99923	—	8354,99	8452,87	2,12	1,18	,0120

Mr. GILPIN's Tables

I. Spirit and water by weight.	II. Specific gravity.	III. Spirit by measure.	IV. Water by measure.	V. Bulk of mixture.	VI. Diminution of bulk.	VII. Quantity of spirit per cent.	VIII. Decimal multipliers.	I. Spirit and water by weight.	II. Specific gravity.	III. Spirit by measure.	IV. Water by measure.	V. Bulk of mixture.	VI. Diminution of bulk.	VII. Quantity of spirit per cent.	VIII. Decimal multipliers.
Sp. + W.								Sp. + W.							
100 + 0	,83582	100	—	100,00	—	100,00	1,0131	100 + 50	,90725	100	41,75	138,19	3,56	72,36	,7331
1	,83812	—	0,84	100,72	0,12	99,28	1,0000	51	,90814	—	42,58	138,97	3,61	71,95	,7291
2	,84036	—	1,67	101,45	0,22	98,57	,9988	52	,90901	—	43,42	139,76	3,66	71,55	,7249
3	,84256	—	2,51	102,18	0,33	97,87	,9916	53	,90987	—	44,25	140,55	3,70	71,15	,7208
4	,84470	—	3,34	102,91	0,43	97,18	,9845	54	,91072	—	45,09	141,33	3,76	70,75	,7169
100 + 5	,84677	—	4,18	103,64	0,54	96,49	,9775	100 + 55	,91155	—	45,93	142,12	3,81	70,36	,7129
6	,84880	—	5,01	104,38	0,63	95,80	,9706	56	,91237	—	46,76	142,91	3,85	69,97	,7090
7	,85077	—	5,85	105,12	0,73	95,13	,9638	57	,91320	—	47,59	143,70	3,89	69,59	,7051
8	,85269	—	6,68	105,87	0,81	94,46	,9570	58	,91399	—	48,43	144,48	3,95	69,21	,7012
9	,85458	—	7,51	106,61	0,90	93,80	,9503	59	,91477	—	49,26	145,27	3,99	68,83	,6974
100 + 10	,85639	—	8,35	107,36	0,99	93,15	,9437	100 + 60	,91556	—	50,10	146,06	4,04	68,46	,6936
11	,85818	—	9,19	108,11	1,08	92,50	,9372	61	,91632	—	50,93	146,85	4,08	68,09	,6899
12	,85993	—	10,02	108,86	1,16	91,86	,9307	62	,91706	—	51,77	147,65	4,12	67,73	,6862
13	,86165	—	10,86	109,61	1,25	91,23	,9243	63	,91781	—	52,61	148,43	4,18	67,37	,6825
14	,86333	—	11,69	110,37	1,32	90,61	,9180	64	,91853	—	53,44	149,23	4,21	67,01	,6789
100 + 15	,86496	—	12,53	111,12	1,41	89,99	,9117	100 + 65	,91925	—	54,28	150,02	4,26	66,66	,6753
16	,86656	—	13,37	111,88	1,49	89,38	,9055	66	,91997	—	55,11	150,81	4,30	66,31	,6718
17	,86814	—	14,19	112,64	1,55	88,78	,8994	67	,92068	—	55,95	151,61	4,34	65,96	,6683
18	,86968	—	15,03	113,40	1,63	88,18	,8934	68	,92137	—	56,78	152,40	4,38	65,61	,6648
19	,87120	—	15,86	114,16	1,70	87,59	,8874	69	,92205	—	57,61	153,19	4,42	65,27	,6614
100 + 20	,87268	—	16,70	114,93	1,77	87,01	,8815	100 + 70	,92274	—	58,45	153,99	4,46	64,94	,6579
21	,87413	—	17,53	115,69	1,84	86,43	,8757	71	,92340	—	59,28	154,78	4,50	64,61	,6546
22	,87556	—	18,37	116,45	1,92	85,86	,8699	72	,92406	—	60,11	155,58	4,53	64,28	,6512
23	,87697	—	19,21	117,22	1,99	85,30	,8642	73	,92470	—	60,95	156,37	4,58	63,95	,6479
24	,87835	—	20,04	117,99	2,05	84,75	,8586	74	,92536	—	61,79	157,17	4,62	63,63	,6446
100 + 25	,87971	—	20,88	118,76	2,12	84,20	,8530	100 + 75	,92599	—	62,63	157,96	4,67	63,31	,6414
26	,88103	—	21,71	119,53	2,18	83,66	,8476	76	,92661	—	63,46	158,75	4,71	62,99	,6382
27	,88235	—	22,55	120,30	2,25	83,13	,8422	77	,92723	—	64,29	159,55	4,74	62,67	,6349
28	,88363	—	23,38	121,07	2,31	82,60	,8368	78	,92784	—	65,13	160,34	4,79	62,36	,6318
29	,88489	—	24,22	121,84	2 38	82,07	,8315	79	,92844	—	65,97	161,14	4,83	62,06	,6288
100 + 30	,88613	—	25,05	122,62	2,43	81,55	,8262	100 + 80	,92906	—	66,80	161,93	4,87	61,75	,6256
31	,88735	—	25,88	123,39	2,49	81,05	,8210	81	,92965	—	67,63	162,73	4,90	61,45	,6226
32	,88856	—	26,72	124,17	2,55	80,54	,8159	82	,93023	—	68,47	163,53	4,94	61,15	,6196
33	,88975	—	27,56	124,95	2,61	80,04	,8109	83	,93080	—	69,31	164,33	4,98	60,85	,6166
34	,89091	—	28,39	125,72	2,67	79,55	,8059	84	,93138	—	70,14	165,13	5,01	60,56	,6136
100 + 35	,89205	—	29,23	126,49	2,74	79,06	,8010	100 + 85	,93194	—	70,98	165,92	5,06	60,27	,6106
36	,89318	—	30,07	127,27	2,80	78,58	,7961	86	,93249	—	71,81	166,72	5,09	59,98	,6077
37	,89429	—	30,90	128,04	2,86	78,10	,7913	87	,93303	—	72,65	167,52	5,13	59,70	,6048
38	,89538	—	31,73	128,82	2,91	77,62	,7865	88	,93357	—	73,48	168,32	5,16	59,41	,6019
39	,89645	—	32,56	129,60	2,96	77,16	,7817	89	,93410	—	74,32	169,12	5,20	59,13	,5991
100 + 40	,89750	—	33,40	130,38	3,02	76,70	,7771	100 + 90	,93461	—	75,15	169,92	5,23	58,85	,5962
41	,89855	—	34,23	131,15	3,08	76,25	,7725	91	,93513	—	75,98	170,72	5,26	58,57	,5934
42	,89958	—	35,07	131,93	3,14	75,80	,7679	92	,93563	—	76,82	171,52	5,30	58,30	,5906
43	,90059	—	35,91	132,71	3,20	75,35	,7634	93	,93613	—	77,65	172,33	5,32	58,03	,5879
44	,90160	—	36,74	133,49	3,25	74,91	,7590	94	,93663	—	78,49	173,13	5,36	57,76	,5852
100 + 45	,90259	—	37,57	134,27	3,30	74,48	,7545	100 + 95	,93712	—	79,33	173,93	5,40	57,49	,5825
46	,90353	—	38,41	135,05	3,36	74,04	,7502	96	,93760	—	80,15	174,72	5,43	57,24	,5799
47	,90450	—	39,25	135,84	3,41	73,62	,7460	97	,93807	—	80,98	175,52	5,46	56,97	,5772
48	,90543	—	40,08	136,62	3,46	73,20	,7417	98	,93854	—	81,82	176,32	5,50	56,71	,5746
49	,90635	—	40,92	137,40	3,52	72,78	,7373	99	,93900	—	82,66	177,12	5,54	56,46	,5719

TABLE II.   HEAT 37°.

I. Water and spirit by weight.	II. Specific gravity.	III. Spirit by measure.	IV. Water by measure.	V. Bulk of mixture.	VI. Diminution of bulk.	VII. Quantity of spirit per cent.	VIII. Decimal multipliers.
W.+Sp.							
100+100	,93947	100	83,51	177,93	5,58	56,19	,5694
99	,93991	—	84,36	178,75	5,61	55,94	,5667
98	,94037	—	85,22	179,59	5,63	55,68	,5641
97	,94082	—	86,10	180,44	5,66	55,42	,5615
96	,94127	—	86,99	181,30	5,69	55,16	,5588
100+95	,94172	—	87,90	182,18	5,72	54,88	,5561
94	,94219	—	88,83	183,08	5,75	54,62	,5534
93	,94266	—	89.79	184,00	5,79	54,35	,5506
92	,94313	—	90,77	184,95	5,82	54,06	,5478
91	,94361	—	91,76	185,91	5,85	53,79	,5450
100+90	,94408	—	92,79	186,90	5,89	53,51	,5421
89	,94459	—	93,82	187,91	5,91	53,22	,5392
88	,94508	—	94,88	188,94	5,94	52,93	,5363
87	,94558	—	95,98	190,00	5,98	52,64	,5333
86	,94608	—	97,10	191,08	6,02	52,34	,5303
100+85	,94658	—	98,24	192,18	6,06	52,04	,5272
84	,94710	—	99,40	193,31	6,09	51,73	,5241
83	,94761	—	100,61	194,47	6,14	51,42	,5210
82	,94812	—	101,84	195,66	6,18	51,10	,5178
81	,94863	—	103,09	196,88	6,21	50,79	,5146
100+80	,94913	—	104,38	198,14	6,24	50,47	,5113
79	,94964	—	105,70	199,42	6,28	50,14	,5080
78	,95016	—	107,06	200,74	6,32	49,81	,5047
77	,95068	—	108,45	202,09	6,36	49,48	,5013
76	,95120	—	109,87	203,49	6,38	49,14	,4979
100+75	,95171	—	111,33	204,91	6,42	48,80	,4944
74	,95224	—	112,83	206,38	6,45	48,45	,4909
73	,95275	—	114,38	207,89	6,49	48,10	,4873
72	,95328	—	115,97	209,45	6,52	47,74	,4837
71	,95380	—	117,60	211,06	6,54	47,39	,4801
100+70	,95433	—	119,29	212,69	6,60	47,01	,4703
69	,95486	—	121,01	214,39	6,62	46,65	,4720
68	,95541	—	122,79	216,13	6,66	46,26	,4688
67	,95595	—	124,63	217,93	6,70	45,88	,4649
66	,95649	—	126,52	219,78	6,74	45,50	,4610
100+65	,95704	—	128,47	221,69	6,78	45,11	,4570
64	,95759	—	130,47	223,66	6,81	44,71	,4530
63	,95814	—	132,53	225,69	6,84	44,30	,4489
62	,95870	—	134,68	227,79	6,89	43,90	,4447
61	,95925	—	136.88	229,97	6,91	43,48	,4405
100+60	,95980	—	139,16	232,22	6,94	43,00	,4303
59	,96035	—	141,52	234,54	6,98	42,63	,4319
58	,96090	—	143,96	236,94	7,02	42.20	,4275
57	,96144	—	146,49	239,45	7,04	41,76	,4231
56	,96199	—	149.11	242.04	7,07	41,31	,4186
100+55	,96252	—	151,81	244,71	7,10	40,86	,4140
54	,96307	—	154.60	247,49	7,11	40,40	,4093
53	,96361	—	157.51	250,39	7,12	39,93	,4046
52	,96415	—	160.55	253,40	7,15	39,46	,3998
51	,96467	—	163,72	256,52	7,20	38,98	,3949
W.+Sp.							
100+50	,96522	100	167,00	259,79	7,21	38,49	,3900
49	,96575	—	170,41	263,17	7,24	38,00	,3850
48	,96628	—	173,96	266,69	7,27	37,49	,3799
47	,96682	—	177,66	270,38	7,28	36,98	,3748
46	,96734	—	181,53	274,23	7,30	36,46	,3695
100+45	,96787	—	185,56	278,26	7,30	35,93	,3641
44	,96838	—	189,77	282,46	7,31	35,40	,3587
43	,96889	—	194,19	286,89	7,30	34,85	,3532
42	,96940	—	198,82	291,51	7,31	34,30	,3476
41	,96989	—	203,66	296,35	7,31	33,74	,3419
100+40	,97039	—	208,76	301,46	7,30	33,17	,3361
39	,97087	—	214,11	306,82	7,29	32,59	,3302
38	,97136	—	219,74	312,49	7,25	32,00	,3242
37	,97183	—	225,68	318,46	7,22	31,40	,3181
36	,97232	—	231,95	324,75	7,20	30,79	,3120
100+35	,97280	—	238,58	331,40	7,18	30,17	,3057
34	,97328	—	245,59	338,45	7,14	29,55	,2994
33	,97376	—	253,04	345,93	7,11	28,91	,2929
32	,97424	—	260,94	353,90	7,04	28,26	,2863
31	,97473	—	269,37	362,36	7,01	27,60	,2796
100+30	,97523	—	278,34	371,38	6,96	26,93	,2728
29	,97571	—	287,94	381,04	6,90	26,24	,2659
28	,97622	—	298,22	391,40	6,82	25,55	,2589
27	,97672	—	309,27	402,51	6,76	24,84	,2517
26	,97723	—	321,16	414,49	6,67	24,11	,2444
100+25	,97776	—	334,01	427,41	6,60	23,40	,2370
24	,97829	—	347,93	441,41	6,52	22,65	,2295
23	,97884	—	303,06	456,63	6,43	21,90	,2219
22	,97941	—	379,55	473,24	6,31	21,13	,2141
21	,97999	—	397,64	491,41	6,23	20,35	,2062
100+20	,98060	—	417,52	511,41	6,11	19,55	,1981
19	,98120	—	439,49	533,51	5,98	18,74	,1899
18	,98182	—	463,90	558,06	5,84	17,92	,1816
17	,98248	—	491,20	585,49	5,71	17,08	,1731
16	,98316	—	521,90	616,34	5,56	16,22	,1644
100+15	,98388	—	556,69	651,29	5,40	15,35	,1556
14	,98462	—	596,46	691,22	5,24	14,46	,1466
13	,98540	—	642,34	737,27	5,07	13,56	,1374
12	,98622	—	695,86	790,98	4,88	12,64	,1281
11	,98709	—	759,13	854,43	4,70	11,70	,1186
100+10	,98801	—	835,04	930,55	4,49	10,74	,1089
9	,98905	—	927,82	1023,51	4,27	9,77	,0990
8	,98998	—	1043,79	1139,77	4,02	8,78	,0889
7	,99107	—	1192,91	1289,11	3,80	7,76	,0786
6	,99221	—	1388,11	1488,11	3,57	6,72	,0681
100+5	,99344	—	1670,09	1706,78	3,31	5,66	,0573
4	,99475	—	2087,59	2184,58	3,01	4,58	,0464
3	,99615	—	2783,45	2880,72	2,73	3,47	,0352
2	,99765	—	4175,18	4272,71	2,47	2,34	,0237
1	,99924	—	8350,36	8448,20	2,16	1,18	,0120

TABLE I. HEAT 38°.

I. Spirit and water by weight.	II. Specific gravity.	III. Spirit by measure.	IV. Water by measure.	V. Bulk of mixture.	VI. Diminution of bulk.	VII. Quantity of spirit per cent.	VIII. Decimal multipliers
Sp. + W.							
100 + 0	,83536	100	—	100,00	—	100,00	1,0126
1	,83766	—	0,83	100,72	0,11	99,28	1,0054
2	,83990	—	1,67	101,45	0,22	98,57	,9982
3	,84209	—	2,51	102,18	0,33	97,87	,9911
4	,84424	—	3,34	102.01	0,43	97,18	,9840
100 + 5	,84631	—	4,18	103,64	0,54	96,49	,9770
6	,84834	—	5,01	104,38	0,63	95,80	,9701
7	,85032	—	5,85	105,12	0,73	95,13	,9633
8	,85224	—	6,68	105,86	0,82	94,40	,9505
9	,85413	—	7,51	106,61	0,90	93,80	,9498
100 + 10	,85594	—	8,35	107,36	0,99	93,15	,9432
11	,85774	—	9,18	108,10	1,08	92,50	,9367
12	,85949	—	10,01	108,86	1,15	91,86	,9302
13	,86121	—	10,85	109,60	1,25	91,23	,9238
14	,86288	—	11,68	110,36	1,32	90,61	,9175
100 + 15	,86451	—	12,52	111,12	1,40	89,99	,9112
16	,86611	—	13,36	111,88	1,48	89,38	,9051
17	,86769	—	14,19	112,64	1,55	88,78	,8990
18	,86924	—	15,02	113,40	1,62	88,18	,8930
19	,87075	—	15,85	114,16	1,69	87,59	,8870
100 + 20	,87224	—	16,69	114,93	1,76	87,01	,8310
21	,87369	—	17,52	115,69	1,83	86,43	,8752
22	,87512	—	18,36	116,45	1,91	85,86	,8694
23	,87653	—	19,20	117,22	1,98	85,31	,8638
24	,87791	—	20,03	117,99	2,04	84,75	,8582
100 + 25	,87926	—	20,86	118,70	2,10	84,20	,8526
26	,88059	—	21,69	119,53	2,16	83,66	,8472
27	,88191	—	22,53	120,30	2,23	83,13	,84:8
28	,88319	—	23,37	121,07	2,30	82,60	,8364
29	,88445	—	24,21	121,84	2,37	82,07	,8311
100 + 30	,88569	—	25,04	122,62	2,42	81,56	,8258
31	,88691	—	25,87	123,39	2,48	81,05	,8206
32	,88812	—	26,71	124,17	2,54	80,54	,8155
33	,88931	—	27,55	124,94	2,61	80.04.	,8105
34	,89047	—	28,38	125,71	2,67	79,55	,8055
100 + 35	,89161	—	29,21	126,48	2,73	79,06	,8006
36	,89274	—	30,05	127,26	2,79	78,58	,7957
37	,89385	—	30,88	128.03	2,85	78,10	,7909
38	,89494	—	31,72	128,81	2,91	77,63	,7861
39	,89600	—	32,54	129,59	2,95	77,16	,7813
100 + 40	,89705	—	33,39	130,37	3,02	76,70	,7767
41	,89811	—	34,22	131,15	3,07	76,25	,7721
42	,89914	—	35,05	131,93	3,12	75,80	,7675
43	,90015	—	35,89	132,71	3,18	75,35	,7630
44	,90116	—	36,72	133,48	3,24	74,91	,7586
100 + 45	,90215	—	37,55	134,26	3,29	74,48	,7542
46	,90310	—	38,39	135,05	3,34	74,05	,7499
47	,90407	—	39,22	135,83	3,39	73,62	,7456
48	,90500	—	40,06	136,61	3,45	73,20	,7413
49	,90592	—	40,90	137,39	3,51	72,78	,7370
100 + 50	,90682	100	41,73	138,18	3,55	72,37	,7328
51	,90771	—	42,56	138,96	3,60	71,96	,7287
52	,90858	—	43,39	139,75	3,64	71,56	,7246
53	,90944	—	44,23	140,54	3,69	71,16	,7205
54	,91029	—	45,07	141,32	3,75	70,76	,7165
100 + 55	,91112	—	45,91	142,11	3,80	70,37	,7125
56	,91194	—	46,73	142,90	3,83	69,98	,7086
57	,91277	—	47,57	143,69	3,88	69,60	,7047
58	,91357	—	48,40	144,47	3.93	69,22	,7009
59	,91435	—	49,24	145,26	3,98	68,84	,6971
100 + 60	,91513	—	50,07	146,05	4,02	68,47	,6933
61	,91590	—	50,90	146,84	4,06	68,10	,6896
62	,91664	—	51,74	147,63	4,11	67,74	,6859
63	,91739	—	52,58	148,42	4,16	67,38	,6822
64	,91811	—	53,41	149,21	4,20	67,02	,6786
100 + 65	,91883	—	54,25	150,00	4,25	66,66	,6750
66	,91955	—	55,08	150,79	4,29	66,31	,6715
67	,92027	—	55,92	151,59	4,33	65,97	,6680
68	,92096	—	56,75	152,38	4,37	65,62	,6645
69	,92164	—	57,58	153,17	4,41	65,28	,6611
100 + 70	,92233	—	58,42	153,97	4,45	64,95	,6576
71	,92300	—	59,25	154,77	4,48	64,62	,6543
72	,92365	—	60,08	155,56	4,52	64,29	,6509
73	,92430	—	60,92	156 35	4,57	63,99	,6470
74	,92495	—	61,75	157,15	4,60	63,64	,6443
100 + 75	,92558	—	62,59	157,94	4,65	63,32	,6411
76	,92620	—	63,43	158 73	4,70	63,00	,6379
77	,92682	—	64,26	159,53	4,73	62,68	,6340
78	,92743	—	65,09	160,32	4,77	62,37	,6315
79	,92802	—	65,93	161,12	4,81	62,07	,6265
100 + 80	,92805	—	66,76	161,91	4,85	61,70	,6254
81	,92924	—	67,59	162,71	4,88	61,46	,6223
82	,92982	—	68,43	163,51	4,92	61,16	,6193
83	,93039	—	69,27	104,31	4 96	60,86	,6103
84	,93098	—	70,10	165 11	4.99	60,57	,6133
100 + 85	,93154	—	70 94	165,90	5,04	60,28	,6103
86	,93209	—	71,77	166,70	5,07	59,99	,6075
87	,93263	—	72,61	167,50	5,11	59,70	,604c
88	,93317	—	73,44	168 30	5,14	59,42	,6017
89	,93370	—	74,28	169,10	5,18	59,14	,5986
100 + 90	,93421	—	75,11	109 90	5,21	58,66	,5900
91	,93473	—	75,94	170,70	5,24	58,58	,5932
92	,93523	—	76,78	171,50	5,28	58,31	,5904
93	,93573	—	77,61	172,31	5,30	58,04	,5877
94	,93522	—	78.45	173,11	5,34	57,77	5849
100 + 95	,93672	—	79,28	173,91	5,37	57,50	,5823
96	,93720	—	80,10	174,70	5,40	57,24	,5796
97	,93767	—	80,93	175,50	5,43	56,98	5770
98	,93814	—	81,77	176,30	5,47	56,72	,5744
99	,93860	—	82,61	177,10	5,51	56,46	,571;

TABLE II.   HEAT 38°.

I. Water and spirit by weight.	II. Specific gravity.	III. Spirit by measure.	IV. Water by measure.	V. Bulk of mixture.	VI. Diminution of bulk.	VII. Quantity of spirit per cent.	VIII. Decimal multipliers.
W. + Sp.							
100+100	,93907	100	83,46	177,91	5,55	56,20	,5691
99	,93952	—	84,31	178,73	5,58	55,95	,5665
98	,93998	—	85,17	179,56	5,61	55,69	,5639
97	,94043	—	86,05	180,41	5,64	55,42	,5613
96	,94088	—	86,94	181,27	5,67	55,16	,5586
100+95	,94134	—	87,85	182,15	5,70	54,89	,5559
94	,94181	—	88,78	183,05	5,73	54,62	,5532
93	,94228	—	89,74	183,97	5,77	54,36	,5504
92	,94275	—	90,71	184,92	5,79	54,08	,5476
91	,94323	—	91,71	185,89	5,82	53,80	,5447
100+90	,94370	—	92,73	186,87	5,86	53,51	,5418
89	,94421	—	93,77	187,88	5,89	53,22	,5390
88	,94470	—	94,83	188,91	5,92	52,93	,5361
87	,94520	—	95,93	189,97	5,96	52,64	,5331
86	,94571	—	97,04	191,05	5,99	52,35	,5301
100+85	,94621	—	98,19	192,15	6,04	52,04	,5270
84	,94673	—	99,35	193,28	6,07	51,74	,5239
83	,94724	—	100,55	194,44	6,11	51,43	,5208
82	,94775	—	101,78	195,63	6,15	51,11	,5176
81	,94826	—	103,03	196,85	6,18	50,80	,5144
100+80	,94876	—	104,32	198,11	6,21	50,48	,5111
79	,94928	—	105,64	199,39	6,25	50,15	,5078
78	,94957	—	107,00	200,71	6,29	49,82	,5045
77	,95031	—	108,39	202,06	6,33	49,49	,5011
76	,95083	—	109,81	203,46	6,35	49,15	,4977
100+75	,95134	—	111,27	204,88	6,39	48,81	,4942
74	,95187	—	112,77	206,35	6,42	48,46	,4907
73	,95239	—	114,32	207,86	6,46	48,11	,4871
72	,95292	—	115,91	209,41	6,50	47,75	,4835
71	,95345	—	117,54	211,02	6,52	47,40	,4799
100+70	,95398	—	119,22	212,65	6,57	47,02	,4761
69	,95451	—	120,94	214,35	6,59	46,65	,4724
68	,95506	—	122,72	216,09	6,63	46,27	,4686
67	,95560	—	124,56	217,89	6,67	45,89	,4647
66	,95615	—	126,45	219,74	6,71	45,51	,4608
100+65	,95670	—	128,40	221,65	6,75	45,12	,4568
64	,95725	—	130,40	223,62	6,78	44,72	,4528
63	,95780	—	132,46	225,65	6,81	44,31	,4487
62	,95836	—	134,60	227,74	6,86	43,91	,4446
61	,95891	—	136,81	229,92	6,89	43,49	,4404
100+60	,95947	—	139,08	232,17	6,91	43,07	,4361
59	,96002	—	141,44	234,49	6,95	42,64	,4318
58	,96057	—	143,88	236,89	6,99	42,21	,4274
57	,96112	—	146,41	239,40	7,01	41,77	,4230
56	,96167	—	149,03	241,99	7,04	41,32	,4185
100+55	,96221	—	151,73	244,66	7,07	40,87	,4139
54	,96276	—	154,52	247,44	7,08	40,41	,4092
53	,96331	—	157,42	250,33	7,09	39,94	,4045
52	,96385	—	160,46	253,34	7,12	39,47	,3997
51	,96438	—	163,63	256,46	7,17	38,99	,3948
100+50	,96493	100	166,91	259,73	7,18	38,50	,3899
49	,96546	—	170,31	263,10	7,21	38,01	,3849
48	,96600	—	173,86	266,62	7,24	37,51	,3798
47	,96654	—	177,56	270,31	7,25	37,00	,3747
46	,96707	—	181,43	274,16	7,27	36,47	,3694
100+45	,96760	—	185,46	278,19	7,27	35,94	,3640
44	,96812	—	189,67	282,39	7,28	35,41	,3586
43	,96863	—	194,08	286,81	7,27	34,86	,3531
42	,96915	—	198,71	291,43	7,28	34,31	,3475
41	,96965	—	203,55	296,27	7,28	33,75	,3418
100+40	,97015	—	208,64	301,37	7,27	33,18	,3360
39	,97064	—	213,99	306,73	7,26	32,60	,3301
38	,97114	—	219,62	312,39	7,23	32,01	,3242
37	,97162	—	225,55	318,35	7,20	31,41	,3181
36	,97211	—	231,82	324,64	7,18	30,80	,3119
100+35	,97260	—	238,45	331,29	7,16	30,18	,3056
34	,97309	—	245,45	338,33	7,12	29,56	,2993
33	,97357	—	252,90	345,81	7,09	28,92	,2928
32	,97406	—	260,80	353,76	7,04	28,27	,2862
31	,97456	—	269,22	362,23	6,99	27,61	,2795
100+30	,97506	—	278,19	371,24	6,95	26,94	,2727
29	,97555	—	287,78	380,89	6,89	26,25	,2658
28	,97607	—	298,06	391,25	6,81	25,56	,2588
27	,97658	—	309,10	402,35	6,75	24,85	,2517
26	,97709	—	320,98	414,32	6,66	24,13	,2444
100+25	,97763	—	333,82	427,23	6,59	23,41	,2370
24	,97817	—	347,74	441,22	6,52	22,66	,2295
23	,97873	—	362,86	456,43	6,43	21,91	,2219
22	,97931	—	379,34	473,03	6,31	21,14	,2141
21	,97990	—	397,42	491,19	6,23	20,36	,2061
100+20	,98052	—	417,29	511,18	6,11	19,56	,1981
19	,98112	—	439,25	533,26	5,99	18,75	,1898
18	,98175	—	463,65	557,80	5,85	17,93	,1815
17	,98242	—	490,93	585,21	5,72	17,09	,1730
16	,98311	—	521,61	616,04	5,57	16,23	,1643
100+15	,98383	—	556,38	650,97	5,41	15,36	,1555
14	,98458	—	596,13	690,87	5,26	14,47	,1466
13	,98536	—	641,99	736,89	5,10	13,57	,1374
12	,98619	—	695,48	790,57	4,91	12,65	,1281
11	,98707	—	758,71	853,99	4,72	11,71	,1186
100+10	,98799	—	834,58	930,05	4,53	10,75	,1089
9	,98885	—	927,31	1023,01	4,30	9,77	,0990
8	,98997	—	1043,22	1139,16	4,06	8,78	,0889
7	,99107	—	1192,25	1288,41	3,84	7,76	,0786
6	,99223	—	1390,97	1487,36	3,61	6,72	,0681
100+5	,99344	—	1609,17	1765,83	3,35	5,66	,0573
4	,99476	—	2086,44	2183,38	3,06	4,58	,0464
3	,99610	—	2781,92	2879,13	2,79	3,47	,0352
2	,99766	—	4172,88	4270,35	2,53	2,34	,0237
1	,99925	—	8345,76	8443,54	2,22	1,18	,0120

TABLE I.　　　　HEAT 39°.

I. Spirit and water by weight.	II. Specific gravity.	III. Spirit by measure.	IV. Water by measure.	V. Bulk of mixture.	VI. Diminution of bulk.	VII. Quantity of spirit per cent.	VIII. Decimal multipliers.	I. Spirit and water by weight.	II. Specific gravity.	III. Spirit by measure.	IV. Water by measure.	V. Bulk of mixture.	VI. Diminution of bulk.	VII. Quantity of spirit per cent.	VIII. Decimal multipliers.
Sp. + W.								Sp. + W.							
100 + 0	,83491	100	—	100,00	—	100,00	1,0120	100 + 50	,90639	100	41,70	138,17	3,53	72,37	,7324
1	,83720	—	0,83	100,72	0,11	99,28	1,0048	51	,90728	—	42,54	138,95	3,59	71,96	,7283
2	,83944	—	1,67	101,45	0,22	98,57	,9976	52	,90815	—	43,37	139,74	3,63	71,56	,7242
3	,84163	—	2,51	102,18	0,33	97,87	,9905	53	,90901	—	44,20	140,53	3,67	71,16	,7201
4	,84378	—	3,34	102,91	0,43	97,18	,9834	54	,90986	—	45,04	141,31	3,73	70,76	,7162
100 + 5	,84585	—	4,18	103,64	0,54	96,49	,9764	100 + 55	,91069	—	45,88	142,10	3,78	70,37	,7122
6	,84788	—	5,01	104,38	0,63	95,80	,9696	56	,91151	—	46,70	142,89	3,81	69,98	,7083
7	,84987	—	5,84	105,12	0,72	95,13	,9628	57	,91234	—	47,54	143,68	3,86	69,60	,7044
8	,85179	—	6,68	105,86	0,82	94,47	,9560	58	,91314	—	48,37	144,46	3,91	69,23	,7006
9	,85368	—	7,50	106,60	0,90	93,80	,9493	59	,91393	—	49,21	145,25	3,96	68,85	,6968
100 + 10	,85550	—	8,34	107,35	0,99	93,15	,9427	100 + 60	,91470	—	50,04	146,04	4,00	68,48	,6929
11	,85730	—	9,18	108,10	1,08	92,51	,9362	61	,91547	—	50,87	146,83	4,04	68,10	,6893
12	,85905	—	10,01	108,85	1,16	91,87	,9297	62	,91622	—	51,71	147,62	4,09	67,74	,6856
13	,86076	—	10,85	109,60	1,25	91,24	,9233	63	,91697	—	52,55	148,41	4,14	67,38	,6819
14	,86243	—	11,68	110,36	1,32	90,62	,9170	64	,91769	—	53,38	149,20	4,18	67,02	,6783
100 + 15	,86406	—	12,51	111,12	1,39	89,99	,9107	100 + 65	,91841	—	54,22	149,99	4,23	66,67	,6747
16	,86566	—	13,35	111,88	1,47	89,38	,9046	66	,91913	—	55,05	150,78	4,27	66,32	,6712
17	,86724	—	14,18	112,64	1,54	88,78	,8985	67	,91985	—	55,89	151,58	4,31	65,97	,6677
18	,86879	—	15,01	113,40	1,61	88,18	,8925	68	,92055	—	56,72	152,37	4,35	65,02	,6642
19	,87030	—	15,85	114,16	1,69	87,59	,8865	69	,92123	—	57,55	153,16	4,39	65,28	,6608
100 + 20	,87179	—	16,68	114,92	1,76	87,01	,8806	100 + 70	,92192	—	58,39	153,96	4,43	64,95	,6573
21	,87324	—	17,52	115,69	1,83	86,44	,8748	71	,92259	—	59,22	154,75	4,47	64,62	,6540
22	,87468	—	18,35	116,45	1,90	85,87	,8690	72	,92325	—	60,05	155,54	4,51	64,29	,6506
23	,87608	—	19,18	117,21	1,97	85,31	,8633	73	,92390	—	60,89	156,31	4,55	63,96	,6473
24	,87746	—	20,02	117,98	2,04	84,75	,8577	74	,92454	—	61,72	157,13	4,59	63,64	,6440
100 + 25	,87882	—	20,85	118,75	2,10	84,20	,8522	100 + 75	,92517	—	62,55	157,92	4,63	63,32	,6408
26	,88015	—	21,68	119,52	2,16	83,66	,8468	76	,92579	—	63,39	158,71	4,68	63,00	,6376
27	,88147	—	22,52	120,29	2,23	83,13	,8414	77	,92641	—	64,23	159,52	4,71	62,68	,6343
28	,88275	—	23,36	121,07	2,29	82,60	,8360	78	,92702	—	65,06	160,30	4,76	62,37	,6312
29	,88401	—	24,19	121,84	2,35	82,07	,8307	79	,92760	—	65,89	161,10	4,79	62,08	,6280
100 + 30	,88525	—	25,03	122,61	2,41	81,56	,8254	100 + 80	,92824	—	66,72	161,89	4,83	61,76	,6251
31	,88647	—	25,85	123,39	2,46	81,05	,8202	81	,92883	—	67,55	162,69	4,86	61,46	,6220
32	,88768	—	26,69	124,16	2,53	80,54	,8151	82	,92941	—	68,39	163,49	4,90	61,16	,6190
33	,88887	—	27,53	124,94	2,59	80,04	,8101	83	,92979	—	69,23	164,29	4,94	60,86	,6160
34	,89003	—	28,36	125,71	2,65	79,56	,8051	84	,93057	—	70,06	165,09	4,97	60,57	,6130
100 + 35	,89117	—	29,20	126,48	2,72	79,08	,8002	100 + 85	,93113	—	70,90	165,88	5,02	60,28	,6101
36	,89230	—	30,03	127,26	2,77	78,58	,7953	86	,93169	—	71,73	166,68	5,05	59,99	,6072
37	,89341	—	30,87	128,03	2,84	78,10	,7905	87	,93223	—	72,57	167,48	5,09	59,71	,6043
38	,89449	—	31,70	128,81	2,89	77,63	,7857	88	,93277	—	73,40	168,28	5,12	59,42	,6014
39	,89556	—	32,53	129,59	2,94	77,17	,7809	89	,93230	—	74,24	169,08	5,11	59,14	,5985
100 + 40	,89661	—	33,37	130,37	3,00	76,71	,7763	100 + 90	,93381	—	75,07	169,88	5,19	58,86	,5957
41	,89767	—	34,20	131,14	3,06	76,26	,7717	91	,93433	—	75,90	170,68	5,2.	58,59	,5929
42	,89870	—	35,03	131,92	3,11	75,81	,7671	92	,93483	—	76,74	171,48	5,20	58,31	,5901
43	,89971	—	35,87	132,70	3,17	75,36	,7626	93	,93533	—	77,57	172,29	5,28	58,4	,5874
44	,90072	...	36,70	133,48	3,22	74,92	,7582	94	,93583	—	78,41	173,09	5,3.	57,77	,5847
100 + 45	,90171	—	37,53	134,26	3,27	74,48	,7538	100 + 95	,93632	—	79,24	173,89	5,35	57,50	,5820
46	,90267	—	38,37	135,04	3,33	74,05	,7495	96	,93680	—	80,06	174,68	5,36	57,25	,5794
47	,90363	—	39,20	135,82	3,38	73,63	,7452	97	,93727	—	80,89	175,48	5,41	56,99	,5768
48	,90457	—	40,04	136,60	3,44	73,21	,7409	98	,93774	—	81,73	176,28	5,44	56,7.	,5742
49	,90549	—	40,87	137,39	3,48	72,79	,7366	99	,93771	—	82,57	177,08	5,49	56,4.	,5717

TABLE II.   HEAT 39°.

I. Water and spirit by weight.	II. Specific gravity.	III. Spirit by measure.	IV. Water by measure.	V. Bulk of mixture.	VI. Diminution of bulk.	VII. Quantity of spirit per cent.	VIII. Decimal multipliers.	I. Water and spirit by weight.	II. Specific gravity.	III. Spirit by measure.	IV. Water by measure.	V. Bulk of mixture.	VI. Diminution of bulk.	VII. Quantity of spirit per cent.	VIII. Decimal multipliers.
W. + Sp.								W. + Sp.							
100+100	,93867	100	83,42	177,89	5,53	56,21	,5689	100 + 50	,96465	100	166,82	259,67	7,15	38,51	,3897
99	,93913	—	84,27	178,71	5,56	55,95	,5662	49	,96517	—	170,22	263,03	7,19	38,02	,3848
98	,93959	—	85,13	179,54	5,59	55,69	,5636	48	,96572	—	173,77	266,55	7,22	37,51	,3797
97	,94005	—	86,00	180,39	5,61	55,43	,5610	47	,96626	—	177,46	270,24	7,22	37,01	,3746
96	,94050	—	86,89	181,25	5,64	55,17	,5583	46	,96680	—	181,33	274,09	7,24	36,48	,3693
100 + 95	,94096	—	87,80	182,12	5,68	54,90	,5557	100 + 45	,96733	—	185,36	278,12	7,24	35,95	,3639
94	,94143	—	88,73	183,02	5,71	54,63	,5530	44	,96786	—	189,57	282,32	7,25	35,42	,3585
93	,94190	—	89,69	183,95	5,74	54,36	,5502	43	,96838	—	193,98	286,73	7,25	34,87	,3530
92	,94237	—	90,66	184,90	5,76	54,08	,5474	42	,96890	—	198,60	291,34	7,26	34,32	,3474
91	,94285	—	91,66	185,86	5,80	53,80	,5445	41	,96941	—	203,44	296,19	7,25	33,76	,3417
100 + 90	,94332	—	92,68	186,84	5,84	53,52	,5416	100 + 40	,96991	—	208,53	301,28	7,25	33,19	,3359
89	,94383	—	93,72	187,85	5,87	53,23	,5388	39	,97041	—	213,87	306,64	7,23	32,61	,3300
88	,94433	—	94,78	188,88	5,90	52,94	,5359	38	,97092	—	219,50	312,29	7,21	32,02	,3241
87	,94483	—	95,87	189,94	5,93	52,65	,5329	37	,97141	—	225,43	318,25	7,18	31,42	,3180
86	,94534	—	96,99	191,02	5,97	52,36	,5299	36	,97191	—	231,69	324,53	7,16	30,81	,3119
100 + 85	,94584	—	98,13	192,12	6,01	52,05	,5268	100 + 35	,97240	—	238,32	331,18	7,14	30,19	,3056
84	,94636	—	99,29	193,25	6,04	51,75	,5237	34	,97290	—	245,32	338,21	7,11	29,57	,2993
83	,94687	—	100,50	194,41	6,09	51,44	,5206	33	,97339	—	252,76	345,69	7,07	28,93	,2928
82	,94738	—	101,72	195,60	6,12	51,12	,5174	32	,97388	—	260,66	353,63	7,03	28,28	,2862
81	,94789	—	102,97	196,82	6,15	50,80	,5142	31	,97439	—	269,07	362,09	6,98	27,62	,2795
100 + 80	,94839	—	104,27	198,08	6,19	50,49	,5109	100 + 30	,97489	—	278,04	371,10	6,94	26,95	,2727
79	,94891	—	105,58	199,36	6,22	50,16	,5076	29	,97539	—	287,62	380,75	6,87	26,26	,2658
78	,94942	—	106,94	200,68	6,26	49,83	,5043	28	,97592	—	297,90	391,09	6,81	25,57	,2588
77	,94994	—	108,33	202,03	6,30	49,50	,5009	27	,97644	—	308,93	402,19	6,74	24,86	,2517
76	,95046	—	109,75	203,43	6,32	49,15	,4975	26	,97696	—	320,81	414,15	6,66	24,14	,2443
100 + 75	,95097	—	111,21	204,85	6,36	48,81	,4940	100 + 25	,97750	—	333,64	427,05	6,59	23,41	,2370
74	,95150	—	112,71	206,32	6,39	48,47	,4905	24	,97805	—	347,55	441,04	6,51	22,67	,2295
73	,95203	—	114,26	207,82	6,44	48,12	,4869	23	,97862	—	362,66	456,24	6,42	21,92	,2218
72	,95256	—	115,85	209,38	6,47	47,76	,4833	22	,97921	—	379,14	472,82	6,32	21,15	,2140
71	,95310	—	117,48	210,98	6,50	47,40	,4797	21	,97981	—	397,20	490,97	6,23	20,37	,2061
100 + 70	,95363	—	119,15	212,61	6,54	47,03	,4760	100 + 20	,98043	—	417,06	510,95	6,11	19,57	,1981
69	,95417	—	120,87	214,31	6,56	46,66	,4722	19	,98104	—	439,01	533,01	6,00	18,76	,1898
68	,95472	—	122,65	216,05	6,60	46,28	,4684	18	,98168	—	463,40	557,54	5,86	17,94	,1815
67	,95526	—	124,49	217,85	6,64	45,90	,4646	17	,98236	—	490,66	584,93	5,73	17,10	,1730
66	,95581	—	126,38	219,70	6,68	45,52	,4607	16	,98306	—	521,32	615,74	5,58	16,24	,1643
100 + 65	,95636	—	128,33	221,61	6,72	45,13	,4567	100 + 15	,98378	—	556,08	650,65	5,43	15,37	,1555
64	,95691	—	130,33	223,58	6,75	44,73	,4527	14	,98454	—	595,80	690,53	5,27	14,48	,1466
63	,95746	—	132,39	225,60	6,79	44,32	,4486	13	,98533	—	641,64	736,51	5,13	13,58	,1374
62	,95802	—	134,52	227,70	6,82	43,91	,4445	12	,98611	—	695,10	790,17	4,93	12,65	,1281
61	,95857	—	136,73	229,88	6,85	43,50	,4403	11	,98705	—	758,29	853,55	4,74	11,71	,1186
100 + 60	,95913	—	139,00	232,12	6,88	43,08	,4360	100 + 10	,98797	—	834,12	929,56	4,56	10,76	,1089
59	,95969	—	141,36	234,44	6,92	42,65	,4317	9	,98894	—	926,80	1022,47	4,33	9,78	,0990
58	,96024	—	143,80	236,84	6,96	42,22	,4273	8	,98990	—	1042,65	1138,55	4,10	8,78	,0889
57	,96080	—	146,33	239,35	6,98	41,78	,4229	7	,99106	—	1191,60	1287,72	3,88	7,77	,0786
56	,96135	—	148,95	241,94	7,01	41,33	,4184	6	,99223	—	1390,20	1486,55	3,65	6,73	,0681
100 + 55	,96190	—	151,65	244,61	7,04	40,88	,4137	100 + 5	,99344	—	1668,25	1704,86	3,39	5,67	,0573
54	,96245	—	154,44	247,38	7,06	40,42	,4091	4	,99484	—	2085,30	2182,19	3,11	4,58	,0464
53	,96301	—	157,34	250,27	7,07	39,95	,4044	3	,99616	—	2780,39	2877,55	2,84	3,47	,0352
52	,96355	—	160,37	253,28	7,09	39,48	,3996	2	,99766	—	4170,59	4268,01	2,58	2,34	,0237
51	,96409	—	163,54	256,40	7,14	39,00	,3947	1	,99925	—	8341,18	8438,90	2,28	1,18	,0120

Mr. GILPIN's *Tables*

TABLE I.   HEAT 40°.

I. Spirit and water by weight.	II. Specific gravity	III. Spirit by measure.	IV. Water by measure.	V. Bulk of mixture.	VI. Diminution of bulk.	VII. Quantity of spirit per cent.	VIII. Decimal multipliers.	I. Spirit and water by weight.	II. Specific gravity.	III. Spirit by measure.	IV. Water by measure.	V. Bulk of mixture.	VI. Diminution of Bulk.	VII. Quantity of spirit per cent.	VIII. Decimal multipliers.
Sp. + W.								Sp. + W.							
100 + 0	,83445	100	—	100,00	—	100,00	1,0114	100 + 50	,90596	100	41,68	138,16	3,52	72,38	,7321
1	,83074	—	0,83	100,72	0,11	99,28	1,0042	51	,90685	—	42,52	138,95	3,57	71,97	,7280
2	,83898	—	1,67	101,45	0,22	98,57	,9970	52	,90772	—	43,35	139,73	3,62	71,57	,7239
3	,84117	—	2,50	102,18	0,32	97,87	,9899	53	,90858	—	44,18	140,52	3,66	71,17	,7198
4	,84331	—	3,33	102,91	0,42	97,18	,9829	54	,90943	—	45,02	141,30	3,72	70,77	,7158
100 + 5	,84539	—	4,17	103,64	0,53	96,49	,9759	100 + 55	,91026	—	45,85	142,09	3,76	70,38	,7118
6	,84742	—	5,00	104,38	0,62	95,81	,9690	56	,91109	—	46,68	142,88	3,80	69,99	,7079
7	,84941	—	5,84	105,12	0,72	95,13	,9622	57	,91191	—	47,52	143,67	3,85	69,61	,7040
8	,85134	—	6,67	105,86	0,81	94,47	,9555	58	,91271	—	48,35	144,45	3,90	69,23	,7002
9	,85323	—	7,50	106,60	0,90	93,81	,9488	59	,91350	—	49,19	145,24	3,95	68,85	,6964
100 + 10	,85507	—	8,34	107,35	0,99	93,16	,9422	100 + 60	,91428	—	50,02	146,03	3,99	68,48	,6926
11	,85686	—	9,17	108,10	1,07	92,51	,9357	61	,91504	—	50,85	146,82	4,03	68,11	,6889
12	,85860	—	10,00	108,85	1,15	91,87	,9292	62	,91579	—	51,69	147,61	4,08	67,75	,6852
13	,86031	—	10,84	109,60	1,24	91,24	,9228	63	,91654	—	52,52	148,40	4,12	67,39	,6816
14	,86198	—	11,67	110,36	1,31	90,62	,9165	64	,91727	—	53,35	149,19	4,16	67,03	,6780
100 + 15	,86361	—	12,50	111,12	1,38	90,00	,9103	100 + 65	,91799	—	54,19	149,98	4,21	66,67	,6744
16	,86521	—	13,34	111,87	1,47	89,38	,9041	66	,91871	—	55,02	150,77	4,25	66,32	,6708
17	,86679	—	14,17	112,63	1,54	88,78	,8980	67	,91943	—	55,86	151,57	4,29	65,98	,6673
18	,86834	—	15,01	113,39	1,62	88,19	,8920	68	,92013	—	56,69	152,36	4,33	65,63	,6639
19	,86985	—	15,84	114,15	1,69	87,60	,8860	69	,92082	—	57,52	153,15	4,37	65,29	,6605
100 + 20	,87134	—	16,67	114,92	1,75	87,02	,8801	100 + 70	,92151	—	58,36	153,94	4,42	64,96	,6570
21	,87280	—	17,51	115,68	1,83	86,44	,8743	71	,92218	—	59,19	154,73	4,46	64,63	,6537
22	,87423	—	18,34	116,45	1,89	85,87	,8686	72	,92284	—	60,02	155,53	4,49	64,30	,6503
23	,87564	—	19,17	117,21	1,96	85,31	,8629	73	,92349	—	60,86	156,32	4,54	63,97	,6470
24	,87702	—	20,01	117,98	2,03	84,76	,8573	74	,92413	—	61,69	157,11	4,58	63,64	,6437
100 + 25	,87838	—	20,84	118,75	2,09	84,21	,8518	100 + 75	,92476	—	62,52	157,91	4,61	63,33	,6405
26	,87971	—	21,67	119,52	2,15	83,67	,8463	76	,92538	—	63,36	158,70	4,66	63,01	,6373
27	,88102	—	22,51	120,29	2,22	83,14	,8409	77	,92600	—	64,19	159,50	4,69	62,69	,6341
28	,88231	—	23,34	121,06	2,28	82,61	,8355	78	,92661	—	65,03	160,29	4,74	62,38	,6310
29	,88357	—	24,18	121,83	2,35	82,08	,8302	79	,92722	—	65,86	161,09	4,77	62,08	,6279
100 + 30	,88481	—	25,01	122,61	2,40	81,57	,8250	100 + 80	,92783	—	66,69	161,88	4,81	61,77	,6248
31	,88603	—	25,84	123,38	2,46	81,06	,8198	81	,92842	—	67,52	162,68	4,84	61,47	,6217
32	,88723	—	26,68	124,16	2,52	80,55	,8147	82	,92900	—	68,36	163,48	4,86	61,17	,6187
33	,88842	—	27,51	124,93	2,58	80,05	,8097	83	,92958	—	69,19	164,27	4,92	60,87	,6157
34	,88959	—	28,35	125,70	2,65	79,56	,8047	84	,93016	—	70,03	165,07	4,96	60,58	,6127
100 + 35	,89073	—	29,18	126,47	2,71	79,07	,7998	100 + 85	,93072	—	70,86	165,86	5,00	60,29	,6098
36	,89185	—	30,01	127,25	2,76	78,59	,7949	86	,93128	—	71,69	166,66	5,03	60,00	,6069
37	,89296	—	30,85	128,02	2,83	78,11	,7901	87	,93183	—	72,53	167,46	5,0-	59,72	,6040
38	,89404	—	31,68	128,80	2,88	77,64	,7853	88	,93237	—	73,36	168,26	5,10	59,43	,6011
39	,89511	—	32,51	129,58	2,93	77,17	,7806	89	,93290	—	74,20	169,06	5,14	59,15	,5983
100 + 40	,89617	—	33,35	130,36	2,99	76,71	,7759	100 + 90	,93341	—	75,03	169,86	5,17	58,87	,5955
41	,89722	—	34,18	131,13	3,05	76,26	,7713	91	,93393	—	75,86	170,66	5,20	58,60	,5927
42	,89825	—	35,01	131,91	3,10	75,81	,7667	92	,93443	—	76,70	171,46	5,24	58,32	,5899
43	,89927	—	35,85	132,69	3,16	75,36	,7622	93	,93493	—	77,53	172,27	5,26	58,05	,5872
44	,90028	—	36,68	133,47	3,21	74,92	,7578	94	,93543	—	78,37	173,07	5,30	57,78	,5845
100 + 45	,90127	—	37,51	134,25	3,26	74,49	,7534	100 + 95	,93592	—	79,20	173,87	5,33	57,52	,5818
46	,90224	—	38,35	135,03	3,32	74,06	,7491	96	,93640	—	80,02	174,66	5,36	57,26	,5791
47	,90319	—	39,18	135,81	3,37	73,63	,7448	97	,93687	—	80,85	175,46	5,39	56,99	,5765
48	,90413	—	40,02	136,59	3,43	73,21	,7405	98	,93734	—	81,69	176,26	5,43	56,73	,5739
49	,90505	—	40,85	137,38	3,47	72,79	,7363	99	,93781	—	82,52	177,06	5,46	56,48	,5713

TABLE II.   HEAT 40°.

I. Water and spirit by weight.	II. Specific gravity.	III. Spirit by measure.	IV. Water by measure.	V. Bulk of mixture.	VI. Diminution of bulk.	VII. Quantity of spirit per cent.	VIII. Decimal multipliers.
W. + Sp.							
100+100	,93827	100	83,37	177,37	5,50	56,22	,5686
99	,93873	—	84,22	178,69	5,53	55,96	,5660
98	,93919	—	85,08	179,52	5,56	55,72	,5634
97	,93966	—	85,96	180,37	5,59	55,44	,5608
96	,94012	—	86,85	181,23	5,62	55,18	,5581
100+95	,94058	—	87,76	182,10	5,66	54,91	,5554
94	,94105	—	88,69	183,00	5,69	54,64	,5527
93	,94152	—	89,64	183,93	5,71	54,37	,5499
92	,94199	—	90,61	184,87	5,74	54,09	,5471
91	,94247	—	91,61	185,83	5,78	53,81	,5443
100+90	,94295	—	92,63	186,82	5,81	53,53	,5414
89	,94345	—	93.67	187,82	5,85	53,24	,5385
88	,94395	—	94,73	188,85	5,88	52,95	,5356
87	,94445	—	95,82	189,91	5,91	52,66	,5326
86	,94496	—	96,94	190,99	5,95	52,36	,5296
100+85	,94547	—	98,08	192,09	5,99	52,06	,5265
84	,94598	—	99,24	193,22	6,02	51,75	,5235
83	,94649	—	100,44	194,38	6,06	51,44	,5204
82	,94700	—	101,67	195,57	6,10	51,13	,5172
81	,94751	—	102,92	196,79	6,13	50,81	,5140
100+80	,94802	—	104,21	198,05	6,16	50,49	,5107
79	,94853	—	105,53	199,33	6,20	50,17	,5074
78	,94904	—	106,88	200,65	6,23	49,84	,5041
77	,94956	—	108,27	202,00	6,27	49,50	,5007
76	,95008	—	109,69	203,39	6,30	49,16	,4973
100+75	,95060	—	111,15	204,82	6,33	48,82	,4938
74	,95113	—	112,65	206,28	6,37	48,48	,4903
73	,95166	—	114,20	207,79	6,41	48,12	,4867
72	,95220	—	115,79	209,34	6,45	47,77	,4831
71	,95274	—	117,42	210,94	6,48	47,41	,4795
100+70	,95328	—	119,09	212,58	6,51	47,04	,4758
69	,95382	—	120,81	214,27	6,54	46,67	,4720
68	,95437	—	122,59	216,01	6,58	46,29	,4682
67	,95492	—	124,42	217,81	6,61	45,91	,4644
66	,95547	—	126,31	219,66	6,65	45,53	,4605
100+65	,95602	—	128,26	221,57	6,69	45,14	,4565
64	,95657	—	130,26	223,53	6,73	44,74	,4525
63	,95712	—	132,32	225,56	6,76	44,33	,4484
62	,95768	—	134,45	227,66	6,79	43,92	,4443
61	,95823	—	136,66	229,84	6,82	43,51	,4401
100+60	,95879	—	138,94	232,08	6,86	43,09	,4358
59	,95935	—	141,29	234,40	6,89	42,66	,4315
58	,95991	—	143,73	236,80	6,93	42,23	,4271
57	,96047	—	146,25	239,30	6,95	41,79	,4227
56	,96103	—	148,87	241,89	6,98	41,34	,4182
100+55	,96159	—	151,57	244,56	7,01	40,88	,4136
54	,96214	—	154,36	247,33	7,03	40,43	,4089
53	,96270	—	157,26	250,22	7,04	39,96	,4042
52	,96325	—	160,29	253,22	7,07	39,49	,3994
51	,96379	—	163,45	256,34	7,11	39,01	,3945
W. + Sp.							
100+50	,96432	100	166,73	259,59	7,14	38,52	,3896
49	,96489	—	170,13	262,97	7,16	38,03	,3846
48	,96544	—	173,68	266,49	7,19	37,52	,3795
47	,96598	—	177,37	270,17	7,20	37,02	,3744
46	,96652	—	181,23	274,02	7,21	36,49	,3691
100+45	,96706	—	185,26	278,04	7,22	35,96	,3638
44	,96759	—	189,47	282,24	7,23	35,43	,3584
43	,96812	—	193,88	286,65	7,23	34,88	,3529
42	,96864	—	198,49	291,26	7,23	34,33	,3473
41	,96919	—	203,33	296,10	7,23	33,77	,3416
100+40	,96967	—	208,42	301,19	7,23	33,20	,3358
39	,97018	—	213,76	306,55	7,21	32,62	,3299
38	,97069	—	219,38	312,19	7,19	32,03	,3240
37	,97119	—	225,31	318,14	7,17	31,43	,3179
36	,97170	—	231,57	324,42	7,15	30,82	,3118
100+35	,97220	—	238,19	331,06	7,13	30,21	,3055
34	,97270	—	245,19	338,09	7,10	29,58	,2992
33	,97320	—	252,62	345,57	7,05	28,94	,2927
32	,97370	—	260,52	353,50	7,02	28,29	,2861
31	,97421	—	268,92	361,95	6,97	27,63	,2794
100+30	,97472	—	277,89	370,97	6,92	26,96	,2726
29	,97523	—	287,47	380,60	6,87	26,27	,2657
28	,97576	—	297,74	390,93	6,81	25,58	,2587
27	,97629	—	308,76	402,03	6,73	24,87	,2516
26	,97682	—	320,64	413,98	6,66	24,15	,2443
100+25	,97737	—	333,46	426,88	6,58	23,42	,2369
24	,97793	—	347,36	440,86	6,50	22,68	,2294
23	,97851	—	362,46	456,05	6,41	21,93	,2218
22	,97910	—	378,94	472,62	6,32	21,16	,2140
21	,97971	—	396,98	490,76	6,22	20,37	,2061
100+20	,98033	—	416,83	510,72	6,11	19,58	,1980
19	,98096	—	438,77	532,77	6,00	18,77	,1898
18	,98161	—	463,15	557,28	5,87	17,94	,1815
17	,98229	—	490,39	584,65	5,74	17,10	,1730
16	,98300	—	521,04	615,44	5,60	16,25	,1643
100+15	,98373	—	555,78	650,33	5,45	15,38	,1555
14	,98449	—	595,47	690,19	5,28	14,49	,1466
13	,98529	—	641,28	736,14	5,14	13,58	,1374
12	,98613	—	694,72	789,77	4,95	12,66	,1281
11	,98702	—	757,87	853,11	4,76	11,72	,1186
100+10	,98795	—	833,66	929,00	4,59	10,76	,1089
9	,98892	—	926,29	1021,93	4,36	9,78	,0990
8	,98995	—	1042,08	1137,94	4,14	8,79	,0889
7	,99105	—	1190,95	1287,03	3,92	7,77	,0786
6	,99222	—	1389,44	1485,75	3,69	6,73	,0681
100+5	,99345	—	1667,33	1763,90	3,43	5,67	,0573
4	,99476	—	2084,16	2181,00	3,16	4,59	,0464
3	,99616	—	2778,87	2875,98	2,89	3,48	,0352
2	,99760	—	4168,31	4265,67	2,64	2,34	,0237
1	,99925	—	8336,63	8434,27	2,36	1,18	,0120

TABLE I.   HEAT 41°.

I. Spirit and water by weight.	II. Specific gravity.	III. Spirit by measure.	IV. Water by measure.	V. Bulk of mixture.	VI. Diminution of bulk.	VII. Quantity of Spirit per cent.	VIII. Decimal multipliers.	I. Spirit and Water by weight.	II. Specific gravity.	III. Spirit by measure.	IV. Water by measure.	V. Bulk of mixture.	VI. Diminution of bulk.	VII. Quantity of Spirit per cent.	VIII. Decimal multipliers.
Sp. + W.								**Sp. + W.**							
100 + 0	,83399	100	—	100,00	—	100,00	1,0109	100 + 50	,90553	100	41,66	138,15	3,51	72,39	,7317
1	,83628	—	0,83	100,72	0,11	99,28	1,0037	51	,90642	—	42,50	138,94	3,56	71,97	,7277
2	,83853	—	1,67	101,45	0,22	98,57	,9965	52	,90729	—	43,33	139,72	3,61	71,57	,7236
3	,84072	—	2,50	102,18	0,32	97,87	,9894	53	,90815	—	44,15	140,50	3,65	71,17	,7195
4	,84286	—	3,33	102,91	0,42	97,18	,9824	54	,90900	—	44,99	141,29	3,70	70,78	,7155
100 + 5	,84493	—	4,17	103,64	0,53	96,49	,9754	100 + 55	,90984	—	45,82	142,07	3,75	70,39	,7115
6	,84697	—	5,00	104,38	0,62	95,81	,9685	56	,91066	—	46,65	142,86	3,79	70,00	,7076
7	,84895	—	5,84	105,12	0,72	95,13	,9617	57	,91148	—	47,49	143,65	3,84	69,61	,7037
8	,85088	—	6,67	105,86	0,81	94,47	,9550	58	,91228	—	48,32	144,43	3,89	69,23	,6999
9	,85277	—	7,50	106,60	0,90	93,81	,9483	59	,91307	—	49,16	145,22	3,94	68,86	,6961
100 + 10	,85461	—	8,34	107,35	0,99	93,16	,9417	100 + 60	,91385	—	49,99	146,01	3,98	68,49	,6923
11	,85640	—	9,17	108,10	1,07	92,51	,9352	61	,91461	—	50,82	146,80	4,02	68,12	,6886
12	,85814	—	10,00	108,85	1,15	91,87	,9288	62	,91536	—	51,66	147,59	4,07	67,75	,6849
13	,85985	—	10,84	109,60	1,24	91,24	,9224	63	,91611	—	52,49	148,38	4,11	67,39	,6813
14	,86152	—	11,67	110,36	1,31	90,62	,9160	64	,91684	—	53,32	149,17	4,15	67,03	,6777
100 + 15	,86315	—	12,49	111,12	1,37	90,00	,9098	100 + 65	,91756	—	54,16	149,96	4,20	66,68	,6741
16	,86476	—	13,33	111,87	1,46	89,38	,9037	66	,91828	—	54,99	150,75	4,24	66,33	,6705
17	,86634	—	14,17	112,63	1,54	88,78	,8976	67	,91900	—	55,82	151,55	4,27	65,98	,6670
18	,86789	—	15,00	113,39	1,61	88,19	,8916	68	,91971	—	56,65	152,34	4,31	65,64	,6636
19	,86940	—	15,83	114,15	1,68	87,60	,8856	69	,92040	—	57,49	153,13	4,36	65,29	,6602
100 + 20	,87088	—	16,66	114,92	1,74	87,02	,8797	100 + 70	,92109	—	58,32	153,92	4,40	64,97	,6567
21	,87235	—	17,50	115,68	1,82	86,44	,8739	71	,92176	—	59,15	154,71	4,44	64,64	,6534
22	,87378	—	18,33	116,44	1,89	85,87	,8681	72	,92242	—	59,98	155,51	4,47	64,30	,6500
23	,87519	—	19,16	117,21	1,95	85,31	,8625	73	,92307	—	60,82	156,30	4,52	63,98	,6467
24	,87657	—	20,00	117,98	2,02	84,76	,8569	74	,92371	—	61,65	157,10	4,55	63,65	,6435
100 + 25	,87793	—	20,83	118,74	2,09	84,21	,8513	100 + 75	,92434	—	62,48	157,89	4,59	63,34	,6402
26	,87926	—	21,66	119,51	2,15	83,67	,8459	76	,92496	—	63,32	158,68	4,64	63,01	,6370
27	,88057	—	22,50	120,28	2,22	83,14	,8405	77	,92558	—	64,15	159,49	4,66	62,70	,6338
28	,88186	—	23,33	121,06	2,27	82,61	,8351	78	,92619	—	64,99	160,27	4,72	62,39	,6307
29	,88312	—	24,17	121,83	2,34	82,08	,8298	79	,92680	—	65,82	161,07	4,75	62,09	,6276
100 + 30	,88436	—	25,00	122,60	2,40	81,57	,8246	100 + 80	,92741	—	66,65	161,86	4,79	61,78	,6245
31	,88558	—	25,83	123,37	2,46	81,06	,8194	81	,92800	—	67,48	162,66	4,82	61,47	,6214
32	,88678	—	26,66	124,15	2,51	80,55	,8143	82	,92858	—	68,32	163,46	4,86	61,18	,6184
33	,88797	—	27,49	124,92	2,57	80,05	,8093	83	,92916	—	69,15	164,25	4,90	60,88	,6154
34	,88914	—	28,33	125,69	2,64	79,56	,8043	84	,92974	—	69,99	165,05	4,93	60,58	,6125
100 + 35	,89029	—	29,16	126,46	2,70	79,07	,7993	100 + 85	,93030	—	70,82	165,84	4,98	60,29	,6095
36	,89141	—	29,99	127,24	2,75	78,59	,7945	86	,93086	—	71,65	166,65	5,00	60,00	,6067
37	,89252	—	30,84	128,01	2,83	78,11	,7897	87	,93142	—	72,49	167,44	5,05	59,72	,6038
38	,89360	—	31,66	128,79	2,87	77,64	,7849	88	,93195	—	73,32	168,24	5,08	59,43	,6009
39	,89467	—	32,49	129,57	2,92	77,17	,7802	89	,93248	—	74,16	169,05	5,11	59,16	,5981
100 + 40	,89572	—	33,33	130,35	2,98	76,71	,7756	100 + 90	,93299	—	74,99	169,85	5,14	58,88	,5952
41	,89678	—	34,16	131,12	3,04	76,26	,7709	91	,93351	—	75,82	170,65	5,17	58,60	,5924
42	,89781	—	34,99	131,90	3,09	75,81	,7664	92	,93401	—	76,65	171,44	5,21	58,33	,5897
43	,89884	—	35,83	132,68	3,15	75,37	,7619	93	,93451	—	77,49	172,25	5,24	58,06	,5869
44	,89985	—	36,66	133,46	3,20	74,93	,7575	94	,93501	—	78,32	173,05	5,27	57,79	,5842
100 + 45	,90084	—	37,49	134,24	3,25	74,49	,7531	100 + 95	,93550	—	79,15	173,85	5,30	57,52	,5815
46	,90181	—	38,32	135,02	3,30	74,06	,7488	96	,93598	—	79,97	174,64	5,33	57,26	,5789
47	,90276	—	39,16	135,80	3,36	73,64	,7445	97	,93646	—	80,80	175,44	5,36	57,00	,5763
48	,90370	—	39,99	136,58	3,41	73,22	,7402	98	,93693	—	81,64	176,24	5,40	56,74	,5737
49	,90462	—	40,83	137,37	3,46	72,80	,7360	99	,93740	—	82,47	177,04	5,43	56,48	,5711

I. Water and spirit by weight. (W. + Sp.)	II. Specific gravity.	III. Spirit by measure.	IV. Water by measure.	V. Bulk of mixture.	VI. Diminution of bulk.	VII. Quantity of spirit per cent.	VIII. Decimal multipliers.
100+100	,93786	100	83,32	177,85	5,47	56,22	,5684
99	,93833	—	84,17	178,66	5,51	55,97	,5658
98	,93879	—	85,03	179,49	5,54	55,71	,5631
97	,93926	—	85,91	180,34	5,57	55,45	,5605
96	,93972	—	80,80	181,20	5,60	55,19	,5579
100+95	,94019	—	87,71	182,07	5,64	54,92	,5552
94	,94063	—	88,64	182,98	5,66	54,65	,5525
93	,94113	—	89,59	183,90	5,69	54,38	,5497
92	,94160	—	90,56	184,84	5,72	54,10	,5469
91	,94208	—	91,56	185,80	5,76	53,82	,5441
100+90	,94256	—	92,58	186,79	5,79	53,53	,5412
89	,94306	—	93,62	187,79	5,83	53,24	,5383
88	,94356	—	94,68	188,82	5,86	52,95	,5354
87	,94406	—	95,76	189,88	5,88	52,67	,5324
86	,94457	—	96,88	190,97	5,91	52,37	,5294
100+85	,94508	—	98,02	192,06	5,96	52,06	,5263
84	,94559	—	99,19	193,20	5,99	51,76	,5233
83	,94610	—	100,38	194,36	6,02	51,45	,5202
82	,94661	—	101,61	195,55	6,06	51,14	,5170
81	,94712	—	102,86	196,76	6,10	50,82	,5138
100+80	,94762	—	104,15	198,03	6,12	50,50	,5105
79	,94814	—	105,47	199,30	6,17	50,17	,5072
78	,94866	—	106,82	200,62	6,20	49,84	,5039
77	,94918	—	108,21	201,97	6,24	49,51	,5005
76	,94970	—	109,63	203,36	6,27	49,17	,4971
100+75	,9,023	—	111,09	204,79	6,30	48,83	,4930
74	,95076	—	112,59	206,25	6,34	48,48	,4901
73	,95129	—	114,14	207,75	6,39	48,13	,4865
72	,95183	—	115,72	209,30	6,42	47,78	,4830
71	,95237	—	117,35	210,90	6,45	47,42	,4794
100+70	,95291	—	119,02	212,55	6,47	47,05	,4750
69	,95346	—	120,74	214,24	6,50	46,68	,4719
68	,95401	—	122,52	215,97	6,55	46,30	,4681
67	,95456	—	124,35	217,77	6,58	45,92	,4643
66	,95511	—	126,24	219,62	6,62	45,53	,4604
100+65	,95567	—	128,18	221,52	6,66	45,14	,4564
64	,95622	—	130,28	223,48	6,69	44,74	,4524
63	,95677	—	132,24	225,52	6,72	44,34	,4483
62	,95733	—	134,37	227,61	6,76	43,93	,4442
61	,95788	—	136,58	229,80	6,78	43,52	,4400
100+60	,95845	—	138,86	232,03	6,83	43,10	,4357
59	,95901	—	141,21	234,36	6,85	42,67	,4314
58	,95957	—	143,65	236,75	6,90	42,24	,4270
57	,96014	—	146,17	239,25	6,92	41,80	,4226
56	,96070	—	148,79	241,83	6,96	41,35	,4181
100+55	,96125	—	151,48	244,50	6,98	40,90	,4135
54	,96182	—	154,27	247,27	7,00	40,43	,4088
53	,96238	—	157,17	250,17	7,00	39,97	,4041
52	,96293	—	160,20	253,17	7,03	39,50	,3993
51	,96348	—	163,36	256,28	7,08	39,02	,3944

I. Water and spirit by weight. (W. + Sp.)	II. Specific gravity.	III. Spirit by measure.	IV. Water by measure.	V. Bulk of mixture.	VI. Diminution of bulk.	VII. Quantity of spirit per cent.	VIII. Decimal multipliers.
100+50	,96404	100	166,64	259,53	7,11	38,53	,3895
49	,96459	—	170,04	262,90	7,14	38,04	,3845
48	,96514	—	173,58	266,42	7,16	37,53	,3794
47	,96569	—	177,27	270,10	7,17	37,02	,3743
46	,96623	—	181,13	273,95	7,18	36,50	,3690
100+45	,96678	—	185,16	277,96	7,20	35,97	,3637
44	,96731	—	189,36	282,16	7,20	35,43	,3583
43	,96785	—	193,77	286,58	7,19	34,89	,3528
42	,96838	—	198,38	291,19	7,19	34,34	,3472
41	,96890	—	203,22	296,01	7,21	33,78	,3415
100+40	,96941	—	208,30	301,10	7,20	33,21	,3357
39	,96994	—	213,64	306,45	7,19	32,63	,3299
38	,97045	—	219,26	312,09	7,17	32,04	,3239
37	,97096	—	225,19	318,04	7,15	31,44	,3179
36	,97148	—	231,44	324,31	7,13	30,83	,3117
100+35	,97198	—	238,06	330,95	7,11	30,22	,3054
34	,97249	—	245,05	337,97	7,08	29,59	,2991
33	,97300	—	252,48	345,45	7,03	28,95	,2927
32	,97351	—	260,38	353,38	7,00	28,30	,2861
31	,97403	—	268,77	361,82	6,95	27,64	,2794
100+30	,97455	—	277,74	370,83	6,91	26,97	,2726
29	,97507	—	287,31	380,46	6,85	26,28	,2657
28	,97560	—	297,57	390,78	6,79	25,59	,2587
27	,97614	—	308,59	401,87	6,72	24,88	,2516
26	,97668	—	320,46	413,82	6,64	24,16	,2443
100+25	,97722	—	333,28	426,70	6,58	23,43	,2369
24	,97780	—	347,17	440,67	6,50	22,69	,2294
23	,97839	—	362,26	455,85	6,41	21,93	,2218
22	,97898	—	378,73	472,41	6,32	21,16	,2140
21	,97960	—	396,76	490,54	6,22	20,38	,2061
100+20	,98023	—	416,60	510,49	6,11	19,59	,1980
19	,98087	—	438,53	532,52	6,01	18,78	,1898
18	,98152	—	462,40	557,02	5,88	17,95	,1815
17	,98221	—	490,12	584,37	5,75	17,11	,1730
16	,98293	—	520,76	615,14	5,62	16,26	,1643
100+15	,98366	—	555,48	650,01	5,47	15,39	,1555
14	,98443	—	595,15	689,85	5,30	14,50	,1465
13	,98524	—	640,92	735,78	5,14	13,59	,1374
12	,98608	—	694,34	789,37	4,97	12,67	,1281
11	,98698	—	757,46	852,67	4,79	11,73	,1186
100+10	,98790	—	833,21	928,59	4,62	10,77	,1089
9	,98889	—	925,79	1021,40	4,39	9,79	,0990
8	,98993	—	1041,51	1137,34	4,17	8,79	,0889
7	,99103	—	1190,30	1286,34	3,96	7,77	,0786
6	,99221	—	1388,69	1484,95	3,74	6,73	,0681
100+5	,99344	—	1626,43	1762,94	3,49	5,67	,0573
4	,99475	—	2083,03	2179,81	3,22	4,59	,0464
3	,99615	—	2777,36	2874,60	2,98	3,68	,0352
2	,99765	—	4166,03	4263,33	2,70	2,35	,0237
1	,99925	—	8332,10	8429,66	2,44	1,19	,0120

Mr. GILPIN's *Tables*

TABLE I.   HEAT 42°.

I. Spirit and water by weight.	II. Specific gravity.	III. Spirit by measure.	IV. Water by measure.	V. Bulk of mixture.	VI. Diminution of bulk.	VII. Quantity of spirit per cent.	VIII. Decimal multipliers.	I. Spirit and water by weight.	II. Specific gravity.	III. Spirit by measure.	IV. Water by measure.	V. Bulk of mixture.	VI. Diminution of bulk.	VII. Quantity of spirit per cent.	VIII. Decimal multipliers.
Sp. + W.								Sp. + W.							
100 + 0	,83353	100	—	100,00	—	100,00	1,0104	100 + 50	,90510	100	41,64	138,14	3,50	72,39	,7314
1	,83582	—	0,83	100,72	0,11	99,28	1,0032	51	,90599	—	42,47	138,93	3,54	71,98	,7273
2	,83807	—	1,67	101,45	0,23	98,57	,9960	52	,90686	—	43,30	139,71	3,59	71,58	,7232
3	,84026	—	2,50	102,18	0,32	97,87	,9889	53	,90772	—	44,13	140,49	3,64	71,18	,7191
4	,84240	—	3,33	102,91	0,42	97,18	,9819	54	,90857	—	44,97	141,28	3,69	70,78	,7151
100 + 5	,84447	—	4,17	103,64	0,53	96,49	,9749	100 + 55	,90941	—	45,80	142,06	3,74	70,39	,7112
6	,84651	—	5,00	104,38	0,62	95,81	,9680	56	,91023	—	46,63	142,85	3,78	70,00	,7073
7	,84849	—	5,83	105,12	0,71	95,13	,9612	57	,91105	—	47,46	143,64	3,82	69,62	,7034
8	,85042	—	6,66	105,86	0,80	94,47	,9545	58	,91185	—	48,30	144,42	3,88	69,24	,6996
9	,85231	—	7,49	106,60	0,89	93,81	,9478	59	,91263	—	49,14	145,21	3,93	68,86	,6958
100 + 10	,85415	—	8,33	107,35	0,98	93,16	,94.2	100 + 60	,91341	—	49,96	146,00	3,96	68,49	,6920
11	,85594	—	9,16	108,10	1,06	92,51	,9347	61	,91418	—	50,80	146,79	4,01	68,12	,6883
12	,85768	—	10,00	108,85	1,15	91,87	,9283	62	,91493	—	51,63	147,58	4,05	67,76	,6846
13	,85939	—	10,83	109,60	1,23	91,24	,9219	63	,91568	—	52,46	148,37	4,09	67,40	,6810
14	,86106	—	11,66	110,36	1,30	90,62	,9156	64	,91641	—	53,29	149,16	4,13	67,04	,6773
100 + 15	,86269	—	12,48	111,11	1,37	90,00	,9093	100 + 65	,91713	—	54,13	149,95	4,18	66,68	,6737
16	,86430	—	13,32	111,87	1,45	89,38	,9032	66	,91785	—	54,96	150,74	4,22	66,33	,6702
17	,86588	—	14,16	112,63	1,53	88,78	,8971	67	,91857	—	55,79	151,54	4,25	65,99	,6667
18	,86743	—	14,99	113,39	1,60	88,19	,8911	68	,91928	—	56,62	152,33	4,29	65,64	,6633
19	,86895	—	15,82	114,15	1,67	87,60	,8851	69	,91998	—	57,46	153,12	4,34	65,30	,6598
100 + 20	,87043	—	16,65	114,91	1,74	87,02	,8792	100 + 70	,92066	—	58,29	153,91	4,38	64,97	,6564
21	,87190	—	17,49	115,68	1,81	86,45	,8734	71	,92133	—	59,12	154,70	4,42	64,64	,6531
22	,87333	—	18,32	116,44	1,88	85,88	,8677	72	,92200	—	59,95	155,50	4,45	64,31	,6497
23	,87474	—	19,15	117,20	1,95	85,32	,8620	73	,92265	—	60,79	156,29	4,50	63,98	,6464
24	,87612	—	19,98	117,97	2,01	84,76	,8564	74	,92329	—	61,62	157,08	4,54	63,66	,6432
100 + 25	,87748	—	20,82	118,74	2,08	84,22	,8509	100 + 75	,92391	—	62,45	157,88	4,57	63,34	,6399
26	,87881	—	21,65	119,51	2,14	83,68	,8454	76	,92454	—	63,29	158,67	4,62	63,02	,6367
27	,88012	—	22,49	120,28	2,21	83,14	,8400	77	,92516	—	64,12	159,48	4,64	62,71	,6335
28	,88141	—	23,32	121,05	2,27	82,61	,8347	78	,92577	—	64,96	160,26	4,70	62,40	,6304
29	,88267	—	24,15	121,82	2,33	82.09	,8294	79	,92638	—	65,79	161,05	4,74	62,09	,6273
100 + 30	,88390	—	24,98	122,69	2,39	81,57	,8241	100 + 80	,92699	—	66,62	161,85	4,77	61,78	,6242
31	,88513	—	25,81	123,37	2,44	81,06	,8190	81	,92758	—	67,45	162,65	4,80	61,48	,6211
32	,88633	—	26,65	124,14	2,51	80,56	,8139	82	,92816	—	68,28	163,45	4,83	61,18	,6181
33	,88752	—	27,48	124,92	2,56	80,06	,8089	83	,92874	—	69,12	164,24	4,88	60,88	,6151
34	,88869	—	28,32	125,68	2,64	79,57	,8039	84	,92932	—	69,95	165,04	4,91	60,59	,6122
100 + 35	,88984	—	29,15	126,46	2,69	79,08	,7989	100 + 85	,92988	—	70,78	165,83	4,95	60,30	,6093
36	,89096	—	29,98	127,23	2,75	78,60	,7941	86	,93044	—	71,61	166,63	4,98	60,01	,6064
37	,89207	—	30,82	128,01	2,81	78,12	,7893	87	,93101	—	72,45	167,43	5,02	59,73	,6035
38	,89316	—	31,65	128,78	2,87	77,65	,7845	88	,93153	—	73,28	168,23	5,05	59,44	,6006
39	,89423	—	32,47	129,56	2,91	77,18	,7798	89	,93206	—	74,11	169,03	5,08	59,16	,5978
100 + 40	,89528	—	33,31	130,34	2,97	76,72	,7752	100 + 90	,93257	—	74,95	169,83	5,12	58,88	,5950
41	,89634	—	34,14	131,11	3,03	76,27	,7706	91	,93309	—	75,78	170,63	5,15	58,61	,5922
42	,89738	—	34,97	131,89	3,08	75,82	,7661	92	,93359	—	76,61	171,43	5,18	58,33	,5894
43	,89840	—	35,81	132,67	3,14	75,37	,7616	93	,93409	—	77,45	172,23	5,22	58,06	,5866
44	,89941	—	36,64	133,45	3,19	74,93	,7571	94	,93459	—	78,28	173,03	5,25	57,79	,5840
100 + 45	,90041	—	37,47	134,23	3,24	74,50	,7527	100 + 95	,93508	—	79,11	173,83	5,28	57,53	,5813
46	,90138	—	38,30	135,01	3,29	74,07	,7484	96	,93556	—	79,93	174,62	5,31	57,27	,5786
47	,90232	—	39,14	135,79	3,35	73,64	,7441	97	,93604	—	80,76	175,42	5,34	57,01	,5760
48	,90327	—	39,97	136,57	3,40	73,22	,7399	98	,93651	—	81,59	176,21	5,38	56,75	,5734
49	,90419	—	40,80	137,35	3,45	72,80	,7356	99	,93699	—	82,43	177,02	5,41	56,49	,5708

TABLE II.   HEAT 42°.

TABLE II.   HEAT 42°.

I. Water and spirit by weight.	II. Specific gravity	III. Spirit by measure.	IV. Water by measure.	V. Bulk of mixture.	VI. Diminution of bulk.	VII. Quantity of spirit per cent.	VIII. Decimal multipliers.
W. + Sp.							
100+100	,93744	100	83,28	177,83	5,45	56,23	,5681
99	,93792	—	84,13	178.64	5 49	55,97	,5655
98	,93838	—	84,99	179.47	5,52	55,72	,5629
97	,93885	—	85,86	180,32	5,54	55 45	,5603
96	,93932	—	86,75	181,18	5,57	55,19	,5576
100+95	,93980	—	87,66	182,05	5,61	54,92	,5549
94	,94026	—	88,59	182.95	5,64	54,66	,5522
93	,94074	—	89,54	183,88	5,66	54,39	,5495
92	,94120	—	90,51	184,82	5,69	54,11	,5467
91	,94168	—	91,51	185,78	5,73	53,82	,5439
100+90	,94216	—	92,53	186,77	5,76	53,54	,5410
89	,94260	—	93.57	187,76	5,81	53.25	,5381
88	,94316	—	94.63	188,80	5,83	52.96	,5352
87	,94366	—	95,71	189,86	5,85	52,67	,5322
86	,94417	—	96,83	190.94	5,89	52,37	,5292
100+85	,94468	—	97.97	192,04	5,93	52,07	,5261
84	,94519	—	99,13	193,17	5,96	51,77	,5231
83	,94570	—	100 33	194,33	6,00	51,46	,5199
82	,94621	—	101,56	195,52	6,04	51,14	,5168
81	,94672	—	102,81	196.74	6,07	50,83	,5133
100+80	,94722	—	104,09	198,00	6,09	50,51	,5103
79	,94775	—	105,41	199,28	6,13	50,18	,5070
78	,94827	—	106,76	200,59	6,17	49,85	,5037
77	,94879	—	108,15	201,94	6,21	49,52	,5003
76	,94932	—	109.57	203,33	6,24	49,18	,4969
100+75	,94985	—	111,03	204.76	6,27	48,84	,4934
74	,95038	—	112,53	206,22	6,31	48,49	,4899
73	,95091	—	114,08	207,72	6,36	48,14	,4863
72	,95145	—	115,66	209.27	6,39	47,78	,4828
71	,95200	—	117,28	210,87	6,41	47,42	,4792
100+70	,95254	—	118,96	212,51	6,45	47,05	,4754
69	,95309	—	120,67	214,20	6,47	46,68	,4717
68	,95364	—	122,45	215,94	6,51	46,30	,4679
67	,95419	—	124,28	217,73	6,55	45,92	,4641
66	,95475	—	126,17	219,58	6,59	45,54	,4602
100+65	,95531	—	128,11	221,48	6,63	45,15	,4562
64	,95586	—	130,21	223,45	6,66	44,75	,4522
63	,95641	—	132,17	225,48	6,69	44,35	,4481
62	,95698	—	134,30	227,57	6,73	43,94	,4440
61	,95754	—	136,51	229.75	6,76	43,52	,4398
100+60	,95810	—	138,78	231,99	6,79	43,11	,4355
59	,95866	—	141,14	234,31	6,83	42,68	,4312
58	,95923	—	143.57	236,70	6,87	42,24	,4268
57	,95981	—	146,09	239,20	6,89	41,80	,4224
56	,96039	—	148,71	241,78	6,93	41,36	,4179
100+55	,96092	—	151,40	244,45	6,95	40,90	,4133
54	,96149	—	154,19	247,24	6,97	40,44	,4087
53	,96261	—	157,08	250,11	6,97	39,98	,4040
52	,96261	—	160,11	253,11	7,00	39,51	,3992
51	96317	—	163 27	256,22	7,05	39,03	,3943

I. Water and spirit by weight.	II. Specific gravity	III. Spirit by measure.	IV. Water by measure.	V. Bulk of mixture.	VI. Diminution of bulk.	VII. Quantity of spirit per cent.	VIII. Decimal multipliers.
W. + Sp.							
100+50	,96373	100	166,55	259,47	7,08	38,54	,3894
49	,96428	—	169,95	262,84	7,11	38,05	,3844
48	,96484	—	173,48	266,36	7,12	37,54	,3793
47	,96539	—	177,17	270,03	7,14	37,03	,3742
46	,96594	—	181,03	273 88	7,15	36,51	,3689
100+45	,96650	—	185,06	277,89	7,17	35,98	,3636
44	,96703	—	189,26	282,08	7,18	35,44	,3582
43	,96757	—	193,66	286,50	7,16	34,90	,3527
42	,96811	—	198,27	291,11	7,16	34,35	,3471
41	,96864	—	203,11	295,93	7,18	33,79	,3414
100+40	,96916	—	208,18	301,01	7,17	33,22	,3356
39	,96969	—	213,52	306,36	7,16	32,64	,3298
38	,97021	—	219,14	311,99	7,15	32,05	,3238
37	,97072	—	225,07	317,94	7,13	31,45	,3178
36	,97125	—	231,31	324.21	7,10	30,84	,3116
100+35	,97176	—	237,93	330,84	7,09	30,23	,3054
34	,97228	—	244,92	337,86	7,06	29,60	,2991
33	,97280	—	252,34	345,34	7,00	28,96	,2926
32	,97332	—	260,24	353,20	6,98	28,31	,2860
31	,97384	—	268,62	361,69	6,93	27,65	,2793
100+30	,97438	—	277,59	370,69	6,90	26,98	,2725
29	,97490	—	287,15	380,32	6,83	26,29	,2656
28	,97543	—	297,41	390,64	6,77	25,60	,2586
27	,97598	—	308,42	401,72	6,70	24,89	,2515
26	,97653	—	320,29	413,66	6,63	24,17	,2442
100+25	,97708	—	333,10	426,53	6,57	23,44	,2369
24	,97766	—	346,98	440,48	6,50	22,70	,2294
23	,97826	—	362,06	455,66	6,40	21,94	,2218
22	,97886	—	378,52	472,20	6,32	21,17	,2140
21	,97946	—	396,54	490,32	6,22	20,39	,2061
100+20	,98013	—	416,37	510,26	6,11	19,60	,1980
19	,98077	—	438,29	532,28	6,01	18,79	,1898
18	,98143	—	462,65	556,76	5,89	17,96	,1815
17	,98212	—	489,85	584,09	5,76	17,12	,1730
16	,98286	—	520,48	614 84	5,64	16,26	,1643
100+15	,98359	—	555,18	649,69	5,49	15,39	,1555
14	,98436	—	594,83	689,51	5,32	14,50	,1465
13	,98516	—	640,56	735,42	5,14	13,59	,1374
12	,98603	—	693,96	788,97	4,99	12,67	,1281
11	,98693	—	757,05	852,24	4,81	11,73	,1186
100+10	,98786	—	832,50	928,11	4,65	10 77	,1089
9	,98885	—	925,29	1020,87	4,42	9,79	,0990
8	,98990	—	1040,95	1136,74	4,21	8,80	,0889
7	,99101	—	1189,66	1285,66	4,00	7,78	,0786
6	,99219	—	1387,94	1484,15	3 79	6,74	,0681
100+5	,99342	—	1605,53	1701,99	3 54	5,58	,0573
4	,99474	—	2081 90	2178,62	3,21	4 59	,0464
3	,99614	—	2775,86	2872,83	3 03	3 48	,0352
2	,99766	—	4163,75	4261,01	2,74	2,35	,0237
1	,99923	—	8320 59	8425 08	2,51	1,19	,0120

TABLE I. HEAT 43°.

I. Spirit and water by weight.	II. Specific gravity.	III. Spirit by measure.	IV. Water by measure.	V. Bulk of mixture.	VI. Diminution of bulk.	VII. Quantity of spirit per cent.	VIII. Decimal multipliers.
Sp. + W.							
100 + 0	,83307	100	—	100,00	—	100,00	1,0098
1	,83536	—	0,83	100,72	0,11	99,28	1,0026
2	,83761	—	1,67	101,45	0,22	98,57	,9955
3	,83980	—	2,50	102,17	0,33	97,87	,9884
4	,84194	—	3,33	102,90	0,43	97,18	,9814
100 + 5	,84401	—	4,17	103,64	0,53	96,49	,9744
6	,84605	—	5,00	104,38	0,62	95,81	,9675
7	,84803	—	5,83	105,12	0,71	95,13	,9607
8	,84996	—	6,66	105,86	0,80	94,47	,9540
9	,85185	—	7,49	106,60	0,89	93,81	,9473
100 + 10	,85369	—	8,32	107,35	0,97	93,16	,9407
11	,85548	—	9,16	108,10	1,06	92,51	,9342
12	,85722	—	9,99	108,85	1,14	91,87	,9278
13	,85893	—	10,82	109,60	1,22	91,24	,9214
14	,86060	—	11,65	110,35	1,30	90,62	,9151
100 + 15	,86223	—	12,48	111,11	1,37	90,00	,9088
16	,86384	—	13,32	111,87	1,45	89,39	,9027
17	,86542	—	14,15	112,63	1,52	88,79	,8966
18	,86697	—	14,98	113,39	1,59	88,19	,8906
19	,86850	—	15,81	114,15	1,66	87,60	,8847
100 + 20	,86997	—	16,65	114,91	1,74	87,02	,8787
21	,87145	—	17,48	115,67	1,81	86,45	,8729
22	,87288	—	18,31	116,43	1,88	85,88	,8672
23	,87429	—	19,14	117,20	1,94	85,32	,8616
24	,87567	—	19,97	117,97	2,00	84,77	,8559
100 + 25	,87703	—	20,80	118,73	2,07	84,22	,8504
26	,87836	—	21,64	119,50	2,14	83,68	,8450
27	,87967	—	22,47	120,27	2,20	83,15	,8396
28	,88096	—	23,30	121,05	2,25	82,62	,8343
29	,88222	—	24,14	121,81	2,33	82,09	,8290
100 + 30	,88345	—	24,97	122,59	2,38	81,58	,8237
31	,88468	—	25,80	123,36	2,44	81,07	,8186
32	,88588	—	26,63	124,14	2,49	80,56	,8135
33	,88707	—	27,46	124,91	2,55	80,06	,8085
34	,88824	—	28,30	125,68	2,62	79,57	,8035
100 + 35	,88939	—	29,13	126,45	2,68	79,08	,7985
36	,89052	—	29,96	127,23	2,73	78,60	,7937
37	,89163	—	30,80	128,00	2,80	78,12	,7889
38	,89271	—	31,63	128,78	2,85	77,65	,7841
39	,89379	—	32,46	129,56	2,90	77,18	,7794
100 + 40	,89484	—	33,30	130,34	2,96	76,72	,7748
41	,89590	—	34,13	131,11	3,02	76,27	,7702
42	,89694	—	34,95	131,89	3,06	75,82	,7657
43	,89796	—	35,79	132,66	3,13	75,38	,7612
44	,89897	—	36,62	133,44	3,18	74,94	,7567
100 + 45	,89997	—	37,45	134,22	3,23	74,50	,7523
46	,90095	—	38,28	135,00	3,28	74,07	,7480
47	,90189	—	39,12	135,78	3,34	73,65	,7437
48	,90284	—	39,95	136,56	3,39	73,23	,7395
49	,90376	—	40,78	137,34	3,44	72,81	,7353
100 + 50	,90467	100	41,61	138,13	3,48	72,40	,7310
51	,90556	—	42,45	138,91	3,54	71,99	,7270
52	,90643	—	43,28	139,70	3,58	71,58	,7229
53	,90729	—	44,10	140,48	3,62	71,18	,7188
54	,90814	—	44,94	141,27	3,67	70,79	,7148
100 + 55	,90898	—	45,78	142,05	3,73	70,40	,7108
56	,90980	—	46,60	142,84	3,76	70,01	,7069
57	,91062	—	47,44	143,63	3,81	69,62	,7030
58	,91141	—	48,27	144,41	3,86	69,24	,6993
59	,91220	—	49,11	145,20	3,91	68,87	,6955
100 + 60	,91297	—	49,93	145,99	3,94	68,50	,6917
61	,91375	—	50,77	146,78	3,99	68,13	,6880
62	,91450	—	51,60	147,57	4,03	67,76	,6843
63	,91525	—	52,43	148,36	4,07	67,40	,6807
64	,91598	—	53,26	149,15	4,11	67,04	,6770
100 + 65	,91670	—	54,10	149,94	4,16	66,69	,6734
66	,91742	—	54,93	150,73	4,20	66,34	,6699
67	,91814	—	55,76	151,52	4,24	65,99	,6664
68	,91885	—	56,59	152,32	4,27	65,65	,6630
69	,91955	—	57,43	153,10	4,33	65,31	,6595
100 + 70	,92023	—	58,26	153,89	4,37	64,98	,6561
71	,92090	—	59,09	154,68	4,41	64,65	,6527
72	,92157	—	59,92	155,48	4,44	64,32	,6494
73	,92222	—	60,75	156,27	4,48	63,99	,6461
74	,92286	—	61,59	157,07	4,52	63,67	,6429
100 + 75	,92348	—	62,42	157,86	4,56	63,35	,6397
76	,92412	—	63,25	158,65	4,60	63,03	,6364
77	,92474	—	64,08	159,46	4,62	62,71	,6332
78	,92535	—	64,92	160,24	4,68	62,40	,6301
79	,92596	—	65,75	161,04	4,71	62,10	,6270
100 + 80	,92656	—	66,58	161,83	4,75	61,79	,6239
81	,92715	—	67,41	162,63	4,78	61,49	,6208
82	,92774	—	68,24	163,43	4,81	61,19	,6178
83	,92831	—	69,08	164,22	4,86	60,89	,6148
84	,92889	—	69,91	165,02	4,89	60,60	,6119
100 + 85	,92945	—	70,74	165,81	4,93	60,31	,6090
86	,93001	—	71,57	166,62	4,95	60,02	,6061
87	,93059	—	72,41	167,41	5,00	59,74	,6032
88	,93110	—	73,24	168,21	5,03	59,45	,6003
89	,93164	—	74,07	169,01	5,06	59,17	,5975
100 + 90	,93215	—	74,91	169,81	5,10	58,89	,5947
91	,93267	—	75,74	170,61	5,13	58,62	,5919
92	,93317	—	76,57	171,41	5,16	58,34	,5891
93	,93367	—	77,41	172,21	5,20	58,07	,5864
94	,93417	—	78,23	173,01	5,22	57,80	,5837
100 + 95	,93466	—	79,06	173,81	5,25	57,54	,5810
96	,93514	—	79,88	174,60	5,28	57,28	,5784
97	,93562	—	80,71	175,40	5,31	57,01	,5758
98	,93610	—	81,55	176,19	5,36	56,75	,5732
99	,93657	—	82,38	177,00	5,38	56,50	,5705

I. Water and spirit by weight.	II. Specific gravity.	III. Spirit by measure.	IV. Water by measure.	V. Bulk of mixture.	VI. Diminution of bulk.	VII. Quantity of spirit *per cent.*	VIII. Decimal multipliers.	I. Water and spirit by weight.	II. Specific gravity.	III. Spirit by measure.	IV. Water by measure.	V. Bulk of mixture.	VI. Diminution of bulk.	VII. Quantity of spirit *per cent.*	VIII. Decimal multipliers.
W. + Sp.								W. + Sp.							
100 + 100	,93703	100	83,23	177,81	5,42	56,24	,5679	100 + 50	,96342	100	166,46	259,41	7,05	38,55	,3893
99	,93751	—	84,09	178,62	5,47	55,98	,5653	49	,96398	—	169,86	262,78	7,08	38,06	,3843
98	,93798	—	84,94	179,45	5,49	55,72	,5626	48	,96454	—	173,39	266,30	7,09	37,55	,3792
97	,93845	—	85,81	180,29	5,52	55,46	,5600	47	,96509	—	177,08	269,97	7,11	37,04	,3741
96	,93892	—	86,70	181,15	5,55	55,20	,5574	46	,96565	—	180,93	273,81	7,12	36,52	,3688
100 + 95	,93940	—	87,61	182,02	5,59	54,93	,5547	100 + 45	,96621	—	184,90	277,82	7,14	35,99	,3635
94	,93987	—	88,54	182,93	5,61	54,66	,5520	44	,96675	—	189,16	282,01	7,15	35,45	,3581
93	,94034	—	89,49	183,85	5,64	54,39	,5493	43	,96729	—	193,56	286,42	7,14	34,91	,3526
92	,94080	—	90,46	184,79	5,67	54,11	,5465	42	,96784	—	198,16	291,03	7,13	34,36	,3470
91	,94128	—	91,46	185,75	5,71	53,83	,5437	41	,96838	—	203,00	295,85	7,15	33,80	,3413
100 + 90	,94176	—	92,48	186,74	5,74	53,55	,5407	100 + 40	,96890	—	208,07	300,92	7,15	33,23	,3356
89	,94226	—	93,52	187,73	5,78	53,26	,5379	39	,96944	—	213,40	306,27	7,13	32,65	,3297
88	,94276	—	94,58	188,77	5,81	52,97	,5350	38	,96997	—	219,02	311,90	7,12	32,06	,3238
87	,94326	—	95,66	189,83	5,83	52,68	,5320	37	,97049	—	224,95	317,84	7,11	31,46	,3177
86	,94377	—	96,77	190,92	5,85	52,38	,5290	36	,97102	—	231,19	324,11	7,08	30,85	,3116
100 + 85	,94428	—	97,91	192,01	5,90	52,08	,5259	100 + 35	,97154	—	237,80	330,73	7,07	30,24	,3053
84	,94479	—	99,08	193,15	5,93	51,78	,5228	34	,97207	—	244,79	337,75	7,04	29,61	,2990
83	,94530	—	100,27	194,31	5,96	51,47	,5197	33	,97260	—	252,20	345,22	6,98	28,97	,2925
82	,94581	—	101,50	195,50	6,00	51,15	,5166	32	,97313	—	260,10	353,14	6,96	28,32	,2860
81	,94632	—	102,75	196,71	6,04	50,83	,5134	31	,97366	—	268,48	361,56	6,92	27,66	,2793
100 + 80	,94683	—	104,03	197,97	6,06	50,51	,5101	100 + 30	,97420	—	277,44	370,55	6,89	26,98	,2725
79	,94736	—	105,35	199,25	6,10	50,19	,5068	29	,97473	—	286,99	380,18	6,81	26,30	,2656
78	,94788	—	106,70	200,56	6,14	49,86	,5035	28	,97527	—	297,25	390,49	6,76	25,61	,2586
77	,94841	—	108,09	201,91	6,18	49,52	,5001	27	,97582	—	308,26	401,57	6,69	24,90	,2515
76	,94894	—	109,51	203,30	6,21	49,18	,4967	26	,97638	—	320,12	413,50	6,62	24,18	,2442
100 + 75	,94947	—	110,97	204,72	6,25	48,84	,4932	100 + 25	,97694	—	332,92	426,35	6,57	23,45	,2368
74	,95000	—	112,47	206,19	6,28	48,50	,4897	24	,97753	—	346,79	440,30	6,49	22,71	,2293
73	,95054	—	114,02	207,68	6,34	48,15	,4861	23	,97813	—	361,86	455,47	6,39	21,95	,2217
72	,95108	—	115,59	209,24	6,35	47,79	,4826	22	,97874	—	378,32	472,00	6,32	21,18	,2139
71	,95163	—	117,22	210,83	6,39	47,43	,4790	21	,97937	—	396,33	490,11	6,22	20,40	,2060
100 + 70	,95217	—	118,89	212,48	6,41	47,06	,4752	100 + 20	,98002	—	416,15	510,03	6,12	19,61	,1980
69	,95272	—	120,61	214,17	6,44	46,69	,4715	19	,98067	—	438,05	532,04	6,01	18,80	,1898
68	,95327	—	122,38	215,90	6,48	46,31	,4677	18	,98134	—	462,40	556,50	5,90	17,97	,1814
67	,95383	—	124,21	217,69	6,52	45,93	,4639	17	,98204	—	489,59	583,82	5,77	17,13	,1729
66	,95439	—	126,10	219,54	6,56	45,55	,4600	16	,98277	—	520,20	614,55	5,65	16,27	,1643
100 + 65	,95495	—	128,04	221,44	6,60	45,16	,4560	100 + 15	,98352	—	554,88	649,38	5,50	15,40	,1555
64	,95550	—	130,14	223,41	6,63	44,76	,4520	14	,98430	—	594,51	689,17	5,34	14,51	,1465
63	,95606	—	132,10	225,44	6,66	44,36	,4479	13	,98512	—	640,20	735,06	5,14	13,60	,1374
62	,95663	—	134,23	227,53	6,70	43,95	,4438	12	,98598	—	693,59	788,57	5,02	12,68	,1280
61	,95719	—	136,44	229,71	6,73	43,53	,4396	11	,98688	—	756,64	851,81	4,83	11,74	,1185
100 + 60	,95775	—	138,71	231,95	6,76	43,11	,4353	100 + 10	,98782	—	832,31	927,63	4,68	10,78	,1089
59	,95831	—	141,06	234,27	6,79	42,68	,4310	9	,98882	—	924,79	1020,34	4,45	9,80	,0990
58	,95889	—	143,49	236,66	6,83	42,25	,4266	8	,98987	—	1040,39	1136,14	4,25	8,80	,0889
57	,95948	—	146,01	239,15	6,86	41,81	,4222	7	,99099	—	1189,02	1284,98	4,04	7,78	,0786
56	,96002	—	148,63	241,73	6,90	41,36	,4177	6	,99217	—	1387,19	1483,36	3,83	6,74	,0681
100 + 55	,96059	—	151,32	244,40	6,92	40,91	,4132	100 + 5	,99342	—	1664,03	1761,04	3,59	5,08	,0573
54	,96116	—	154,11	247,17	6,94	40,45	,4086	4	,99473	—	2080,77	2177,44	3,33	4,59	,0464
53	,96173	—	157,00	250,05	6,95	39,99	,4039	3	,99613	—	2773,36	2871,27	3,09	3,48	,0352
52	,96229	—	160,02	253,05	6,97	39,52	,3991	2	,99763	—	4161,48	4258,69	2,79	2,35	,0237
51	,96286	—	163,18	256,16	7,02	39,04	,3942	1	,99922	—	8323,10	8420,51	2,59	1,19	,0120

TABLE I.   HEAT 44°.

I. Spirit and water by weight.	II. Specific gravity.	III. Spirit by measure.	IV. Water by measure.	V. Bulk of mixture.	VI. Diminution of bulk.	VII. Quantity of spirit per cent.	VIII. Decimal multipliers.
Sp. + W.							
100 + 0	,83261	100	—	100,00	—	100,00	1,0092
1	,83490	—	0,83	100,72	0,11	99,28	1,0020
2	,83715	—	1,66	101,44	0,22	98,57	,9949
3	,83934	—	2,50	102,17	0,33	97,87	,9878
4	,84148	—	3,33	102,90	0,43	97,18	,9808
100 + 5	,84355	—	4,17	103,64	0,53	96,49	,9738
6	,84559	—	4,99	104,37	0,62	95,81	,9670
7	,84757	—	5,82	105,11	0,71	95,14	,9602
8	,84950	—	6,66	105,85	0,81	94,47	,9535
9	,85139	—	7,49	106,60	0,89	93,81	,9468
100 + 10	,85323	—	8,32	107,34	0,98	93,16	,9402
11	,85502	—	9,15	108,09	1,06	92,52	,9337
12	,85676	—	9,99	108,84	1,15	91,88	,9273
13	,85847	—	10,82	109,59	1,23	91,25	,9209
14	,86014	—	11,65	110,35	1,30	90,62	,9146
100 + 15	,86177	—	12,47	111,11	1,36	90,00	,9083
16	,86338	—	13,31	111,86	1,45	89,39	,9022
17	,86496	—	14,14	112,62	1,52	88,79	,8961
18	,86651	—	14,97	113,38	1,59	88,20	,8901
19	,86804	—	15,80	114,14	1,66	87,61	,8842
100 + 20	,86952	—	16,64	114,91	1,73	87,03	,8783
21	,87100	—	17,47	115,67	1,80	86,45	,8725
22	,87243	—	18,30	116,43	1,87	85,88	,8668
23	,87384	—	19,13	117,19	1,94	85,32	,8612
24	,87522	—	19,96	117,96	2,00	84,77	,8556
100 + 25	,87658	—	20,79	118,73	2,06	84,23	,8500
26	,87793	—	21,63	119,50	2,13	83,69	,8445
27	,87922	—	22,46	120,27	2,19	83,15	,8392
28	,88051	—	23,29	121,04	2,25	82,62	,8339
29	,88177	—	24,13	121,81	2,32	82,09	,8286
100 + 30	,88300	—	24,96	122,58	2,38	81,58	,8233
31	,88423	—	25,78	123,36	2,42	81,07	,8182
32	,88543	—	26,61	124,13	2,48	80,56	,8131
33	,88662	—	27,44	124,90	2,54	80,07	,8081
34	,88779	—	28,29	125,67	2,62	79,58	,8031
100 + 35	,88894	—	29,12	126,45	2,67	79,09	,7981
36	,89007	—	29,95	127,22	2,73	78,61	,7933
37	,89118	—	30,78	127,99	2,79	78,13	,7885
38	,89227	—	31,61	128,77	2,84	77,66	,7837
39	,89335	—	32,44	129,55	2,89	77,19	,7790
100 + 40	,89440	—	33,28	130,33	2,95	76,73	,7744
41	,89546	—	34,11	131,10	3,01	76,28	,7698
42	,89650	—	34,94	131,88	3,06	75,83	,7653
43	,89752	—	35,77	132,65	3,12	75,38	,7608
44	,89853	—	36,60	133,43	3,17	74,94	,7563
100 + 45	,89953	—	37,43	134,21	3,22	74,51	,7519
46	,90051	—	38,26	134,99	3,27	74,08	,7476
47	,90145	—	39,10	135,77	3,33	73,65	,7433
48	,90240	—	39,93	136,55	3,38	73,23	,7391
49	,90333	—	40,76	137,33	3,43	72,81	,7349
100 + 50	,90424	100	41,59	138,12	3,47	72,40	,7307
51	,90513	—	42,42	138,90	3,52	71,99	,7266
52	,90600	—	43,25	139,69	3,56	71,59	,7225
53	,90686	—	44,08	140,47	3,61	71,19	,7184
54	,90771	—	44,92	141,26	3,66	70,79	,7145
100 + 55	,90855	—	45,75	142,04	3,71	70,40	,7105
56	,90937	—	46,58	142,83	3,75	70,01	,7066
57	,91019	—	47,41	143,62	3,79	69,63	,7027
58	,91098	—	48,24	144,40	3,84	69,25	,6990
59	,91176	—	49,08	145,19	3,89	68,87	,6952
100 + 60	,91254	—	49,90	145,98	3,92	68,50	,6913
61	,91331	—	50,74	146,77	3,97	68,13	,6877
62	,91407	—	51,57	147,56	4,01	67,77	,6840
63	,91482	—	52,40	148,35	4,05	67,41	,6803
64	,91555	—	53,23	149,14	4,09	67,05	,6766
100 + 65	,91627	—	54,07	149,93	4,14	66,70	,6731
66	,91699	—	54,90	150,72	4,18	66,35	,6696
67	,91771	—	55,73	151,51	4,22	66,00	,6661
68	,91842	—	56,56	152,30	4,26	65,66	,6627
69	,91912	—	57,40	153,09	4,31	65,32	,6592
100 + 70	,91980	—	58,23	153,88	4,35	64,98	,6558
71	,92047	—	59,06	154,67	4,39	64,65	,6524
72	,92114	—	59,89	155,47	4,42	64,32	,6491
73	,92179	—	60,72	156,26	4,46	63,99	,6458
74	,92243	—	61,55	157,05	4,50	63,67	,6426
100 + 75	,92305	—	62,39	157,85	4,54	63,35	,6394
76	,92370	—	63,22	158,64	4,58	63,03	,6361
77	,92432	—	64,05	159,44	4,61	62,72	,6330
78	,92493	—	64,88	160,23	4,65	62,41	,6299
79	,92553	—	65,71	161,02	4,69	62,10	,6268
100 + 80	,92613	—	66,55	161,82	4,73	61,80	,6237
81	,92673	—	67,37	162,61	4,76	61,49	,6206
82	,92731	—	68,20	163,41	4,79	61,19	,6176
83	,92788	—	69,05	164,21	4,84	60,89	,6146
84	,92846	—	69,87	165,01	4,86	60,60	,6116
100 + 85	,92902	—	70,70	165,79	4,91	60,31	,6087
86	,92958	—	71,53	166,60	4,93	60,02	,6058
87	,93017	—	72,37	167,40	4,97	59,74	,6029
88	,93067	—	73,20	168,19	5,01	59,46	,6001
89	,93121	—	74,03	168,99	5,04	59,18	,5973
100 + 90	,93173	—	74,87	169,79	5,08	58,90	,5944
91	,93225	—	75,70	170,59	5,11	58,62	,5916
92	,93175	—	76,53	171,39	5,14	58,35	,5888
93	,93325	—	77,37	172,19	5,18	58,08	,5861
94	,93375	—	78,19	172,99	5,20	57,81	,5834
100 + 95	,93424	—	79,02	173,79	5,23	57,54	,5807
96	,93472	—	79,84	174,58	5,26	57,28	,5781
97	,93520	—	80,67	175,38	5,29	57,02	,5755
98	,93568	—	81,50	176,17	5,33	56,76	,5729
99	,93615	—	82,34	176,98	5,36	56,50	,5703

TABLE II.　　　　HEAT 44°.

I. Water and Spirit by weight.	II. Specific gravity.	III. Spirit by measure.	IV. Water by measure.	V. Bulk of mixture.	VI. Diminution of bulk.	VII. Quantity of spirit per cent.	VIII. Decimal multipliers.	I. Water and spirit by weight.	II. Specific gravity.	III. Spirit by measure.	IV. Water by measure.	V. Bulk of mixture.	VI. Diminution of bulk.	VII. Quantity of spirit per cent.	VIII. Decimal multipliers.
W. + Sp.								W. + Sp.							
100+100	,93662	100	83,18	177,79	5,39	56,24	,5676	100 + 50	,96311	100	166,37	259,35	7,02	38,56	,3891
99	,93710	—	84,04	178,60	5,44	55,99	,5650	49	,96368	—	169,77	262,72	7,05	38,06	,3842
98	,93757	—	84,89	179,43	5,46	55,73	,5624	48	,96424	—	173,30	266,24	7,06	37,56	,3791
97	,93804	—	85,76	180,27	5,49	55,47	,5598	47	,96480	—	176,99	269,91	7,08	37,05	,3740
96	,93852	—	86,65	181,13	5,52	55,21	,5571	46	,96536	—	180,83	273,74	7,09	36,53	,3687
100+ 95	,93900	—	87,56	182,00	5,56	54,94	,5545	100+ 45	,96592	—	184,86	277,75	7,11	35,00	,3634
94	,93947	—	88,49	182,90	5,59	54,67	,5518	44	,96647	—	189,06	281,94	7,12	35,46	,3580
93	,93994	—	89,44	183,82	5,62	54,40	,5491	43	,96702	—	193,46	286,34	7,12	34,92	,3525
92	,94041	—	90,41	184,77	5,64	54,12	,5462	42	,96757	—	198,06	290,95	7,11	34,37	,3469
91	,94088	—	91,41	185,73	5,68	53,84	,5434	41	,96812	—	202,89	295,77	7,12	33,81	,3412
100+ 90	,94136	—	92,43	186,72	5,71	53,55	,5405	100+ 40	,96865	—	207,96	300,83	7,13	33,24	,3355
89	,94186	—	93,47	187,71	5,76	53,26	,5377	39	,96919	—	213,29	306,18	7,11	32,66	,3296
88	,94236	—	94,53	188,75	5,78	52,97	,5348	38	,96973	—	218,90	311,81	7,09	32,07	,3237
87	,94286	—	95,61	189,80	5,81	52,68	,5318	37	,97026	—	224,83	317,74	7,09	31,47	,3177
86	,94337	—	96,72	190,89	5,83	52,39	,5288	36	,97079	—	231,07	324,01	7,06	30,86	,3115
100+ 85	,94388	—	97,86	191,98	5,88	52,09	,5257	100+ 35	,97132	—	237,67	330,62	7,05	30,25	,3052
84	,94439	—	99,03	193,12	5,91	51,79	,5226	34	,97186	—	244,66	337,64	7,02	29,62	,2989
83	,94490	—	100,22	194,28	5,94	51,48	,5195	33	,97240	—	252,07	345,10	6,97	28,98	,2925
82	,94540	—	101,45	195,47	5,98	51,16	,5164	32	,97294	—	259,96	353,02	6,94	28,33	,2859
81	,94593	—	102,69	196,68	6,01	50,84	,5132	31	,97348	—	268,34	361,43	6,91	27,67	,2792
100+ 80	,94644	—	103,97	197,94	6,03	50,52	,5099	100+ 30	,97402	—	277,29	370,41	6,88	27,00	,2724
79	,94697	—	105,29	199,22	6,07	50,19	,5066	29	,97456	—	286,84	380,04	6,80	26,31	,2655
78	,94750	—	106,64	200,53	6,11	49,86	,5033	28	,97511	—	297,09	390,34	6,75	25,62	,2586
77	,94803	—	108,03	201,88	6,15	49,53	,4999	27	,97567	—	308,09	401,42	6,67	24,91	,2515
76	,94855	—	109,45	203,27	6,18	49.19	,4965	26	,97623	—	319,95	413,34	6,61	24,19	,2442
100+ 75	,94909	—	110,91	204,69	6,22	48,85	,4930	100+ 25	,97680	—	332,74	426,18	6,56	23,46	,2368
74	,94963	—	112,41	206,16	6,25	48,50	,4895	24	,97740	—	346,60	440,12	6,48	22,72	,2293
73	,95017	—	113,96	207,65	6,31	48,15	,4859	23	,97800	—	361,67	455,28	6,39	21,96	,2217
72	,95071	—	115,53	209,20	6,33	47,80	,4824	22	,97862	—	378,12	471,80	6,32	21,19	,2139
71	,95126	—	117,16	210,80	6,36	47,44	,4788	21	,97926	—	396,12	489,90	6,22	20,41	,2060
100+ 70	,95180	—	118,83	212,44	6,39	47,07	,4751	100+ 20	,97991	—	415,93	509,80	6,13	19,62	,1980
69	,95236	—	120,55	214,13	6,42	46,70	,4713	19	,98057	—	437,82	531,80	6,02	18,81	,1897
68	,95292	—	122,32	215,86	6,46	46,32	,4675	18	,98125	—	462,15	556,24	5,91	17,98	,1814
67	,95347	—	124,14	217,65	6,49	45,94	,4637	17	,98196	—	489,33	583,55	5,78	17,14	,1729
66	,95403	—	126,03	219,50	6,53	45,56	,4598	16	,98269	—	519,92	614,26	5,66	16,28	,1643
100+ 65	,95459	—	127,97	221,40	6,57	45,17	,4558	100+ 15	,98345	—	554,58	649,07	5,51	15,41	,1555
64	,95515	—	130,07	223,37	6,60	44,77	,4519	14	,98424	—	594,19	688,83	5,36	14,52	,1465
63	,95571	—	132,03	225,40	6,63	44,37	,4478	13	,98506	—	639,85	734,70	5,15	13,61	,1374
62	,95628	—	134,16	227,49	6,67	43,96	,4436	12	,98593	—	693,22	788,18	5,04	12,69	,1280
61	,95684	—	136,37	229,67	6,70	43,54	,4394	11	,98683	—	756,23	851,38	4,85	11,74	,1185
100+ 60	,95740	—	138,64	231,90	6,74	43,12	,4352	100+ 10	,98778	—	831,86	927,16	4,70	10,78	,1088
59	,95797	—	140,99	234,22	6,77	42,69	,4309	9	,98879	—	924,29	1019,81	4,48	9,80	,0990
58	,95855	—	143,41	236,61	6,80	42,26	,4265	8	,98984	—	1039,83	1135,54	4,29	8,81	,0889
57	,95913	—	145,93	239,10	6,83	41,82	,4220	7	,99097	—	1188,38	1284,30	4,08	7,79	,0786
56	,95969	—	148,55	241,68	6,87	41,37	,4176	6	,99215	—	1386,44	1482,57	3,87	6,74	,0681
100+ 55	,96026	—	151,24	244,35	6,89	40,92	,4130	100+ 5	,99340	—	1663,73	1760,09	3,64	5,68	,0573
54	,96084	—	154,03	247,11	6,92	40,46	,4085	4	,99472	—	2079,65	2176,26	3,39	4,60	,0464
53	,96141	—	156,91	249,99	6,92	40,00	,4038	3	,99612	—	2772,87	2869,71	3,16	3,48	,0352
52	,96197	—	159,94	252,99	6,95	39,53	,3990	2	,99762	—	4159,23	4256,39	2,84	2,35	,0237
51	,96255	—	163,09	256,10	6,99	39,05	,3941	1	,99921	—	8318,63	8415,96	2,67	1,19	,0120

TABLE I.   HEAT 45°.

I. Spirit and water by weight.	II. Specific gravity.	III. Spirit by measure.	IV. Water by measure.	V. Bulk of mixture.	VI. Diminution of bulk.	VII. Quantity of spirit per cent.	VIII. Decimal multipliers.	I. Spirit and water by weight.	II. Specific gravity.	III. Spirit by measure.	IV. Water by measure.	V. Bulk of mixture.	VI. Diminution of bulk.	VII. Quantity of spirit per cent.	VIII. Decimal multipliers.
Sp. + W.								Sp. + W.							
100 + 0	,83214	100	—	100,00	—	100,00	1,0087	100 + 50	,90380	100	41,57	138,11	3,46	72,41	,7303
1	,83444	—	0,83	100,72	0,11	99,28	1,0014	51	,90469	--	42,40	138,89	3,51	72,00	,7262
2	,83669	—	1,66	101,44	0,22	98,58	,9943	52	,90557	—	43,23	139,68	3,55	71,59	,7221
3	,83888	—	2,49	102,17	0,32	97,88	,9872	53	,90643	—	44,06	140,46	3,60	71,19	,7181
4	,84102	—	3,32	102,90	0,42	97,18	,9802	54	,90726	—	44,89	141,25	3,64	70,80	,7141
100 + 5	,84310	—	4,16	103,63	0,53	96,49	,9733	100 + 55	,90812	—	45,73	142,03	3,70	70,41	,7102
6	,84513	—	4,99	104,37	0,62	95,81	,9664	56	,90894	—	46,56	142,82	3,74	70,02	,7063
7	,84711	—	5,82	105,11	0,71	95,14	,9596	57	,90975	—	47,39	143,61	3,78	69,63	,7024
8	,84904	—	6,65	105,85	0,80	94,47	,9529	58	,91054	—	48,22	144,39	3,83	69,25	,6986
9	,85093	—	7,48	106,59	0,89	93,81	,9463	59	,91132	—	49,05	145,18	3,87	68,88	,6948
100 + 10	,85277	—	8,31	107,34	0,97	93,16	,9397	100 + 60	,91211	—	49,88	145,97	3,91	68,51	,6910
11	,85456	—	9,14	108,09	1,05	92,52	,9332	61	,91287	—	50,72	146,76	3,96	68,14	,6873
12	,85630	—	9,98	108,84	1,14	91,88	,9268	62	,91363	—	51,55	147,55	4,00	67,77	,6836
13	,85801	—	10,81	109,59	1,22	91,25	,9204	63	,91438	—	52,38	148,34	4,04	67,41	,6799
14	,85968	—	11,64	110,35	1,29	90,63	,9141	64	,91511	—	53,21	149,13	4,08	67,05	,6763
100 + 15	,86131	—	12,47	111,10	1,37	90,01	,9078	100 + 65	,91584	—	54,04	149,92	4,12	66,70	,6728
16	,86292	—	13,30	111,86	1,44	89,39	,9017	66	,91656	—	54,87	150,71	4,16	66,35	,6693
17	,86450	—	14,13	112,62	1,51	88,79	,8956	67	,91728	—	55,70	151,50	4,20	66,00	,6658
18	,86605	—	14,96	113,38	1,58	88,20	,8896	68	,91799	—	56,53	152,29	4,24	65,66	,6623
19	,86758	—	15,79	114,14	1,65	87,61	,8837	69	,91869	--	57,37	153,08	4,29	65,33	,6589
100 + 20	,86905	—	16,63	114,90	1,73	87,03	,8778	100 + 70	,91937	—	58,20	153,87	4,33	64,99	,6555
21	,87054	—	17,46	115,66	1,80	86,46	,8721	71	,92004	—	59,03	154,66	4,37	64,66	,6521
22	,87198	—	18,29	116,42	1,87	85,89	,8664	72	,92071	—	59,86	155,46	4,40	64,33	,6488
23	,87339	—	19,12	117,19	1,93	85,33	,8607	73	,92136	—	60,69	156,25	4,44	64,00	,6455
24	,87477	—	19,95	117,96	1,99	84,78	,8551	74	,92200	—	61,52	157,04	4,48	63,68	,6423
100 + 25	,87613	—	20,78	118,72	2,06	84,23	,8496	100 + 75	,92264	—	62,36	157,83	4,53	63,36	,6391
26	,87746	—	21,62	119,49	2,13	83,68	,8441	76	,92327	—	63,19	158,63	4,56	63,04	,6359
27	,87877	—	22,45	120,26	2,19	83,15	,8387	77	,92389	—	64,02	159,42	4,60	62,73	,6327
28	,88005	—	23,28	121,03	2,25	82,62	,8334	78	,92450	—	64,85	160,22	4,63	62,42	,6296
29	,88131	—	24,11	121,80	2,31	82,10	,8281	79	,92510	—	65,68	161,01	4,67	62,11	,6265
100 + 30	,88255	—	24,94	122,57	2,37	81,58	,8229	100 + 80	,92570	--	66,51	161,81	4,70	61,80	,6234
31	,88377	—	25,77	123,35	2,42	81,07	,8177	81	,92629	—	67,34	162,60	4,74	61,50	,6203
32	,88498	—	26,60	124,12	2,48	80,57	,8126	82	,92688	—	68,17	163,40	4,77	61,20	,6173
33	,88617	—	27,43	124,89	2,54	80,07	,8076	83	,92745	—	69,01	164,19	4,82	60,90	,6143
34	,88734	—	28,27	125,66	2,61	79,58	,8026	84	,92803	—	69,84	164,99	4,85	60,60	,6113
100 + 35	,88849	—	29,10	126,44	2,66	79,09	,7977	100 + 85	,92859	—	70,67	165,78	4,89	60,32	,6084
36	,88962	—	29,93	127,21	2,72	78,61	,7929	86	,92915	—	71,50	166,58	4,92	60,03	,6055
37	,89073	—	30,76	127,99	2,77	78,13	,7881	87	,92970	—	72,33	167,38	4,95	59,75	,6026
38	,89182	—	31,59	128,76	2,83	77,66	,7833	88	,93024	—	73,16	168,17	4,99	59,46	,5998
39	,89290	—	32,42	129,54	2,88	77,19	,7786	89	,93078	—	73,99	168,97	5,02	59,18	,5970
100 + 40	,89396	—	33,26	130,32	2,94	76,73	,7740	100 + 90	,93131	—	74,83	169,77	5,06	58,90	,5941
41	,89502	—	34,09	131,09	3,00	76,28	,7694	91	,93182	—	75,66	170,57	5,09	58,63	,5913
42	,89606	—	34,92	131,87	3,05	75,83	,7649	92	,93233	—	76,49	171,37	5,12	58,36	,5886
43	,89708	—	35,75	132,65	3,10	75,39	,7604	93	,93283	—	77,32	172,17	5,15	58,08	,5859
44	,89809	—	36,58	133,42	3,16	74,95	,7560	94	,93333	—	78,15	172,97	5,18	57,81	,5832
100 + 45	,89909	—	37,41	134,20	3,21	74,51	,7516	100 + 95	,93382	--	78,98	173,77	5,21	57,55	,5805
46	,90007	—	38,24	134,98	3,26	74,08	,7472	96	,93430	—	79,80	174,56	5,24	57,29	,5778
47	,90102	—	39,08	135,76	3,32	73,66	,7429	97	,93478	—	80,63	175,35	5,28	57,03	,5752
48	,90196	—	39,91	136,54	3,37	73,24	,7387	98	,93526	—	81,46	176,15	5,31	56,77	,5726
49	,90289	—	40,74	137,32	3,42	72,82	,7345	99	,93574	--	82,30	176,96	5,34	56,51	,5700

TABLE II.   HEAT 45°.

I. Water and spirit by weight.	II. Specific gravity.	III. Spirit by measure.	IV. Water by measure.	V. Bulk of mixture.	VI. Diminution of bulk.	VII. Quantity of spirit per cent.	VIII. Decimal multipliers.
W. + Sp.							
100+100	,93621	100	83,14	177,76	5,38	56,25	,5674
99	,93669	—	83,99	178,58	5,41	56,00	,5648
98	,93717	—	84,85	179,41	5,44	55,74	,5622
97	,93764	—	85,72	180,25	5,47	55,48	,5596
96	,93812	—	86,61	181,11	5,50	55,22	,5569
100 + 95	,93860	—	87,52	181,98	5,54	54,95	,5543
94	,93907	—	88,45	182,88	5,57	54,68	,5515
93	,93954	—	89,40	183,80	5,60	54,41	,5488
92	,94001	—	90,37	184,75	5,62	54,13	,5460
91	,94048	—	91,36	185,71	5,65	53,85	,5432
100 + 90	,94096	—	92,38	186,69	5,69	53,56	,5403
89	,94146	—	93,42	187,69	5,73	53,27	,5374
88	,94196	—	94,48	188,72	5,76	52,98	,5345
87	,94246	—	95,56	189,78	5,78	52,69	,5315
86	,94297	—	96,67	190,86	5,81	52,39	,5285
100 + 85	,94348	—	97,81	191,96	5,85	52,09	,5254
84	,94399	—	98,98	193,09	5,89	51,79	,5223
83	,94450	—	100,17	194,25	5,92	51,48	,5192
82	,94501	—	101,39	195,44	5,95	51,17	,5161
81	,94553	—	102,64	196,66	5,98	50,85	,5129
100 + 80	,94605	—	103,92	197,91	6,01	50,53	,5097
79	,94658	—	105,24	199,19	6,05	50,20	,5064
78	,94711	—	106,59	200,50	6,09	49,87	,5031
77	,94764	—	107,98	201,85	6,13	49,54	,4997
76	,94817	—	109,40	203,24	6,16	49,20	,4963
100 + 75	,94871	—	110,86	204,66	6,20	48,86	,4928
74	,94925	—	112,35	206,12	6,23	48,51	,4893
73	,94979	—	113,89	207,62	6,27	48,16	,4858
72	,95033	—	115,47	209,17	6,30	47,81	,4822
71	,95088	—	117,10	210,77	6,33	47,45	,4786
100 + 70	,95143	—	118,77	212,41	6,36	47,08	,4749
69	,95199	—	120,49	214,09	6,40	46,71	,4711
68	,95254	—	122,26	215,82	6,44	46,33	,4673
67	,95310	—	124,08	217,61	6,47	45,95	,4635
66	,95367	—	125,96	219,46	6,50	45,56	,4596
100 + 65	,95423	—	127,90	221,36	6,54	45,17	,4557
64	,95479	—	129,90	223,32	6,58	44,78	,4517
63	,95535	—	131,96	225,35	6,61	44,38	,4476
62	,95592	—	134,09	227,45	6,64	43,97	,4434
61	,95649	—	136,30	229,02	6,68	43,55	,4392
100 + 60	,95705	—	138,57	231,85	6,72	43,13	,4350
59	,95762	—	140,91	234,17	6,74	42,70	,4307
58	,95820	—	143,34	236,57	6,77	42,27	,4263
57	,95878	—	145,86	239,05	6,81	41,83	,4219
56	,95935	—	148,47	241,63	6,84	41,38	,4174
100 + 55	,95993	—	151,16	244,30	6,86	40,93	,4129
54	,96051	—	153,93	247,06	6,87	40,47	,4083
53	,96108	—	156,83	249,93	6,90	40,01	,4036
52	,96165	—	159,86	252,93	6,93	39,54	,3988
51	,96223	—	163,01	256,05	6,96	39,06	,3939

I. Water and spirit by weight.	II. Specific gravity.	III. Spirit by measure.	IV. Water by measure.	V. Bulk of mixture.	VI. Diminution of bulk.	VII. Quantity of spirit per cent.	VIII. Decimal multipliers.
W. + Sp.							
100 + 50	,96280	100	166,28	259,29	6,99	38,57	,3890
49	,96337	—	169,68	262,66	7,02	38,07	,3840
48	,96394	—	173,21	266,18	7,03	37,57	,3789
47	,96450	—	176,90	269,85	7,05	37,06	,3738
46	,96507	—	180,74	273,68	7,06	36,54	,3686
100 + 45	,96563	—	184,76	277,68	7,08	36,01	,3633
44	,96619	—	188,96	281,87	7,09	35,47	,3579
43	,96674	—	193,36	286,26	7,10	34,93	,3524
42	,96730	—	197,96	290,87	7,09	34,38	,3468
41	,96785	—	202,78	295,69	7,09	33,82	,3411
100 + 40	,96840	—	207,85	300,75	7,10	33,25	,3354
39	,96894	—	213,18	306,09	7,09	32,67	,3295
38	,96948	—	218,79	311,71	7,08	32,08	,3236
37	,97002	—	224,71	317,64	7,07	31,48	,3176
36	,97056	—	230,95	323,91	7,04	30,87	,3114
100 + 35	,97110	—	237,55	330,52	7,03	30,26	,3052
34	,97164	—	244,53	337,53	7,00	29,63	,2988
33	,97219	—	251,94	344,98	6,96	28,99	,2924
32	,97274	—	259,82	352,90	6,92	28,34	,2859
31	,97329	—	268,20	361,31	6,89	27,68	,2792
100 + 30	,97384	—	277,14	370,28	6,86	27,01	,2724
29	,97439	—	286,69	379,90	6,79	26,32	,2655
28	,97494	—	296,93	390,20	6,73	25,63	,2585
27	,97551	—	307,93	401,27	6,66	24,92	,2514
26	,97608	—	319,78	413,19	6,59	24,20	,2441
100 + 25	,97666	—	332,57	426,01	6,52	23,47	,2368
24	,97726	—	346,42	439,94	6,48	22,73	,2293
23	,97787	—	361,48	455,09	6,39	21,97	,2217
22	,97850	—	377,92	471,60	6,32	21,20	,2139
21	,97914	—	395,91	489,69	6,22	20,42	,2060
100 + 20	,97980	—	415,71	509,58	6,13	19,62	,1979
19	,98047	—	437,59	531,56	6,03	18,81	,1897
18	,98116	—	461,90	555,99	5,91	17,98	,1814
17	,98187	—	489,07	583,28	5,79	17,14	,1729
16	,98261	—	519,63	613,97	5,66	16,29	,1643
100 + 15	,98338	—	554,28	648,76	5,52	15,41	,1555
14	,98417	—	593,87	688,50	5,37	14,52	,1465
13	,98500	—	639,55	734,34	5,21	13,61	,1374
12	,98587	—	692,85	787,79	5,06	12,69	,1280
11	,98678	—	755,83	850,95	4,88	11,75	,1185
100 + 10	,98774	—	831,42	926,69	4,73	10,79	,1088
9	,98875	—	923,80	1019,29	4,51	9,81	,0990
8	,98981	—	1039,27	1134,95	4,32	8,81	,0889
7	,99094	—	1187,74	1283,62	4,12	7,79	,0786
6	,99213	—	1385,70	1481,78	3,92	6,75	,0681
100 + 5	,99338	—	1662,83	1759,14	3,69	5,69	,0573
4	,99470	—	2078,54	2175,09	3,45	4,60	,0464
3	,99611	—	2771,39	2868,16	3,23	3,49	,0352
2	,99761	—	4156,99	4254,09	2,90	2,35	,0237
1	,99919	—	8314,17	8411,41	2,76	1,19	,0120

I. Spirit and water by weight.	II. Specific gravity.	III. Spirit by measure.	IV. Water by measure.	V. Bulk of mixture.	VI. Diminution of bulk.	VII. Quantity of spirit per cent.	VIII. Decimal multipliers.
Sp. + W.							
100 + 0	,83167	100	—	100,00	—	100,00	1,0081
1	,83397	—	0,83	100,72	0,11	99,28	1,0009
2	,83622	—	1,66	101,44	0,22	98,58	,9938
3	,83842	—	2,49	102,17	0,32	97,88	,9867
4	,84056	—	3,32	102,90	0,42	97,18	,9797
100 + 5	,84263	—	4,16	103,63	0,53	96,49	,9727
6	,84467	—	4,99	104,37	0,62	95,81	,9659
7	,84665	—	5,82	105,11	0,71	95,14	,9591
8	,84858	—	6,65	105,85	0,80	94,47	,9524
9	,85046	—	7,48	106 59	0,89	93,81	,9458
100 + 10	,85230	—	8,31	107,34	0,97	93,16	,9392
11	,85410	—	9,14	108,09	1,05	92,52	,9327
12	,85584	—	9,97	108,84	1,13	91,88	,9263
13	,85755	—	10,81	109,59	1,22	91,25	,9199
14	,85922	—	11,63	110,35	1,28	90,63	,9136
100 + 15	,86085	—	12,46	111,10	1,36	90,01	,9074
16	,86247	—	13,30	111,86	1.44	89,40	,9013
17	,86404	—	14,12	112,62	1,50	88,79	,8952
18	,86559	—	14,95	113.38	1,57	88,20	8.92
19	,86712	—	15,78	114.14	1,64	87,61	,8833
100 + 20	,86859	—	16,62	114,90	1,72	87,03	,8774
21	,87008	—	17,45	115,66	1,79	86,46	,8716
22	,87152	—	18,28	116,42	1,86	85,89	,8660
23	,87293	—	19,11	117,18	1,93	85,33	,8603
24	,87431	—	19,94	117,95	1,99	84,78	,8547
100 + 25	,87568	—	20,77	118,72	2,05	84,23	,8491
26	,87700	—	21,61	119 48	2,13	83,69	,8437
27	,87832	—	22,44	120.26	2,18	83,16	,8383
28	,87960	—	23,27	121,03	2,24	82,63	,8330
29	,88086	—	24,10	121,80	2,30	82,10	,8277
100 + 30	,88210	—	24,93	122,57	2,36	81,59	,8225
31	,88332	—	25,76	123,34	2,42	81,08	,8173
32	,88454	—	26,59	124,11	2,48	80,57	,8122
33	,88573	—	27,42	124,88	2,54	80,07	,8072
34	,88690	—	28,25	125,65	2,60	79,58	,8022
100 + 35	,88804	—	29,09	126,43	2,66	79 10	,7973
36	,88917	—	29,91	127,20	2,71	78,62	,7925
37	,89029	—	30,74	127,98	2,76	78.14	,7877
38	,89138	—	31,57	128,75	2,82	77,67	,7829
39	,89246	—	32,40	129.53	2,87	77,20	,7782
100 + 40	,89351	—	33,24	130,31	2,93	76,74	,7736
41	,89458	—	34,07	131.08	2,99	76,29	,7691
42	,89562	—	34,90	131,86	3,04	75,84	,7646
43	,89664	—	35,73	132,64	3,09	75,39	,7601
44	,89765	—	36,56	133,41	3,15	74,95	,7557
100 + 45	,89864	—	37,39	134,19	3,20	74,52	,7512
46	,89962	—	38,22	134 97	3,25	74,09	,7469
47	,90058	—	39 05	135,75	3,30	73,66	,7426
48	,90152	—	39,88	136,53	3,35	73,24	,7384
49	,90245	—	40,72	137,31	3,41	72,82	,7342
Sp. + W.							
100 + 50	,90336	100	41,55	138,09	3,46	72,41	,7300
51	,90426	—	42,38	138,87	3,51	72 01	,7259
52	,90514	—	43,20	139,66	3,54	71,60	,7218
53	,90600	—	44,03	140.44	3,59	71,20	,7178
54	,90685	—	44.87	141,23	3,64	70.81	,7138
100 + 55	,90768	—	45.70	142,01	3,69	70.42	,7098
56	,90851	—	46 53	142,80	3,73	70,23	,7059
57	,90932	—	47.36	143,59	3,77	69,64	,7021
58	,91011	—	48,19	144,37	3 82	69,26	,6983
59	,91089	—	49,02	145,16	3,86	68,89	,6945
100 + 60	,91169	—	49 85	145,95	3,90	68,51	,6907
61	,91244	—	50,69	146,74	3 95	68.14	,6870
62	,91320	—	51.52	147,53	3 99	67.78	,6833
63	,91395	—	52.35	148.32	4,03	67.42	,6796
64	,91468	—	53,18	149,11	4,07	67,06	,6760
100 + 65	,91542	—	54.01	149,90	4,11	66,71	,6725
66	,91614	—	54,84	150,69	4,15	66,36	,6690
67	,91686	—	55,67	151.48	4,19	66,01	,6655
68	,91757	—	56.50	152,27	4 23	65,67	,6620
69	,91826	—	57,33	153,06	4,27	65,33	,6586
100 + 70	,91895	—	58 16	153,85	4,31	65,00	,6552
71	,91962	—	59,00	154,64	4,36	64 67	,6519
72	,92028	—	59.82	155,44	4,38	64.34	,6486
73	,92094	—	60.65	156.23	4,42	64.01	,6453
74	,92156	—	61.48	157.02	4,46	63,69	,6420
100 + 75	,92221	—	62,32	157.81	4,51	63,37	,6388
76	,92285	—	63.15	158 61	4,54	63,05	,6356
77	,92347	—	63,98	159.40	4 58	62,73	,6324
78	,92408	—	64.81	160,20	4.61	62,42	,6293
79	,92468	—	65,64	160,99	4,65	62,11	,6262
100 + 80	,92527	—	66,47	161.79	4.68	61,81	,6231
81	,92587	—	67,30	162,58	4,72	61,50	,6200
82	,92640	—	68.13	163,38	4,75	61,20	,6170
83	,92703	—	68,97	164,18	4.80	60,91	,6140
84	,92761	—	69,80	164 98	4,82	60,61	,6111
100 + 85	,92815	—	70.63	165,76	4,87	60,32	,6081
86	,92873	—	71,46	166,57	4,89	60,03	,6053
87	,92928	—	72,29	167,36	4,93	59 75	,6024
88	,92982	—	73,12	168.16	4,98	59,47	,5996
89	,93036	—	73,95	168,95	5,00	59,19	,5967
100 + 90	,93089	—	74,79	169,75	5,04	58,91	,5939
91	,93140	—	75,62	170.55	5,07	58,64	,5910
92	,93192	—	76,45	171,34	5,11	58,36	,5883
93	,93242	—	77,28	172,15	5,13	58,09	,5856
94	,93292	—	78.10	172 95	5,15	57.82	,5830
100 + 95	,93341	—	78,93	173,74	5,19	57,56	,5802
96	,93389	—	79.75	174,53	5.22	57,30	,5776
97	,93437	—	80,59	175,33	5,26	57,04	,5750
98	,93485	—	81,41	176,13	5,28	56,78	,5724
99	,93533	—	82,25	176,93	5,32	56,52	,5698

I. Water and spirit by weight.	II. Specific gravity.	III. Spirit by measure.	IV. Water by measure.	V. Bulk of mixture.	VI. Diminution of bulk.	VII. Quantity of spirit per cent.	VIII. Decimal multipliers.
W. + Sp.							
100+100	,93580	100	83,09	177,74	5,35	56,26	,5672
99	,93629	—	83,94	178,55	5,39	56,00	,5646
98	,93676	—	84,80	179,38	5,42	55,74	,5619
97	,93724	—	85,67	180,22	5,45	55,48	,5593
96	,93772	—	86,56	181,08	5,48	55,22	,5567
100+95	,93820	—	87,47	181,95	5,52	54,95	,5540
94	,93867	—	88,40	182,85	5,55	54,69	,5513
93	,93914	—	89,35	183,77	5,58	54,41	,5486
92	,93961	—	90,32	184,72	5,60	54,14	,5458
91	,94008	—	91,31	185,68	5,63	53,86	,5430
100+90	,94057	—	92,33	186,67	5,66	53,57	,5400
89	,94100	—	93,37	187,66	5,71	53,28	,5372
88	,94156	—	94,43	188,70	5,73	52,99	,5343
87	,94207	—	95,51	189,75	5,76	52,70	,5313
86	,94258	—	96,62	190,84	5,78	52,40	,5283
100+85	,94309	—	97,75	191,93	5,82	52,10	,5252
84	,94360	—	98,92	193,07	5,85	51,80	,5221
83	,94411	—	100,12	194,22	5,90	51,49	,5190
82	,94462	—	101,33	195,41	5,92	51,17	,5159
81	,94514	—	102,58	196,63	5,95	50,86	,5127
100+80	,94560	—	103,86	197,88	5,98	50,54	,5094
79	,94620	—	105,18	199,16	6,02	50,21	,5062
78	,94673	—	106,53	200,47	6,06	49,88	,5029
77	,94727	—	107,92	201,81	6,11	49,54	,4995
76	,94780	—	109,34	203,21	6,13	49,21	,4961
100+75	,94834	—	110,80	204,62	6,18	48,87	,4926
74	,94888	—	112,29	206,08	6,21	48,52	,4892
73	,94942	—	113,82	207,58	6,24	48,17	,4857
72	,94996	—	115,40	209,13	6,27	47,81	,4821
71	,95051	—	117,03	210,73	6,30	47,46	,4784
100+70	,95106	—	118,70	212,37	6,33	47,09	,4747
69	,95162	—	120,42	214,05	6,37	46,72	,4710
68	,95218	—	122,19	215,78	6,41	46,34	,4672
67	,95274	—	124,01	217,57	6,44	45,96	,4634
66	,95331	—	125,89	219,42	6,47	45,57	,4595
100+65	,95387	—	127,83	221,31	6,53	45,18	,4555
64	,95444	—	129,83	223,28	6,55	44,79	,4515
63	,95500	—	131,89	225,31	6,58	44,38	,4475
62	,95557	—	134,02	227,41	6,61	43,98	,4433
61	,95615	—	136,22	229,58	6,64	43,56	,4391
100+60	,95671	—	138,47	231,81	6,68	43,14	,4349
59	,95728	—	140,83	234,12	6,71	42,71	,4307
58	,95787	—	143,26	236,52	6,74	42,28	,4262
57	,95845	—	145,78	239,00	6,78	41,84	,4218
56	,95902	—	148,39	241,58	6,81	41,39	,4173
100+55	,95959	—	151,08	244,25	6,83	40,94	,4127
54	,96016	—	153,85	247,01	6,84	40,48	,4082
53	,96076	—	156,74	249,87	6,87	40,02	,4035
52	,96134	—	159,77	252,87	6,90	39,55	,3987
51	,96192	—	162,92	255,99	6,93	39,06	,3938
100+50	,96250	100	166,19	259,22	6,97	38,57	,3889
49	,96307	—	169,59	262,60	6,99	38,08	,3839
48	,96364	—	173,11	266,11	7,00	37,58	,3788
47	,96421	—	176,80	269,78	7,02	37,07	,3737
46	,96478	—	180,64	273,61	7,03	36,55	,3685
100+45	,96535	—	184,66	277,60	7,06	36,02	,3632
44	,96591	—	188,86	281,79	7,07	35,48	,3578
43	,96647	—	193,25	286,18	7,07	34,94	,3523
42	,96703	—	197,85	290,78	7,07	34,39	,3467
41	,96759	—	202,67	295,60	7,07	33,83	,3410
100+40	,96814	—	207,74	300,66	7,08	33,26	,3353
39	,96869	—	213,06	305,99	7,07	32,68	,3295
38	,96923	—	218,67	311,61	7,06	32,09	,3235
37	,96978	—	224,59	317,53	7,06	31,49	,3175
36	,97033	—	230,82	323,80	7,02	30,88	,3113
100+35	,97087	—	237,42	330,41	7,01	30,27	,3051
34	,97142	—	244,40	337,42	6,98	29,64	,2988
33	,97198	—	251,80	344,86	6,94	29,00	,2924
32	,97253	—	259,68	352,77	6,91	28,35	,2858
31	,97309	—	268,05	361,18	6,87	27,69	,2792
100+30	,97364	—	276,99	370,14	6,85	27,02	,2724
29	,97420	—	286,53	379,76	6,77	26,33	,2655
28	,97476	—	296,77	390,05	6,72	25,64	,2585
27	,97534	—	307,76	401,11	6,65	24,93	,2514
26	,97592	—	319,60	413,02	6,58	24,21	,2441
100+25	,97650	—	332,39	425,83	6,56	23,48	,2367
24	,97712	—	346,23	439,75	6,48	22,74	,2293
23	,97773	—	361,28	454,89	6,39	21,98	,2217
22	,97837	—	377,71	471,39	6,32	21,21	,2139
21	,97902	—	395,70	489,47	6,23	20,43	,2060
100+20	,97968	—	415,48	509,35	6,13	19,63	,1979
19	,98036	—	437,35	531,32	6,03	18,82	,1897
18	,98106	—	461,65	555,73	5,92	17,99	,1814
17	,98177	—	488,81	583,00	5,81	17,15	,1729
16	,98252	—	519,35	613,68	5,67	16,29	,1643
100+15	,98329	—	553,98	648,44	5,54	15,42	,1555
14	,98409	—	593,55	688,16	5,39	14,53	,1465
13	,98493	—	639,21	733,97	5,24	13,62	,1374
12	,98580	—	692,48	787,39	5,06	12,70	,1280
11	,98672	—	755,42	850,52	4,92	11,75	,1185
100+10	,98768	—	830,97	926,21	4,76	10,79	,1088
9	,98870	—	923,30	1018,76	4,54	9,81	,0990
8	,98976	—	1038,71	1134,36	4,35	8,81	,0890
7	,99090	—	1187,10	1282,95	4,15	7,79	,0789
6	,99210	—	1384,96	1481,00	3,96	6,75	,0681
100+5	,99334	—	1661,94	1758,21	3.73	5,69	,0573
4	,99467	—	2173,42	2173,93	3,49	4,60	,0464
3	,99608	—	2769,91	2866,63	3,28	3,49	,0352
2	,99758	—	4154,78	4251,80	2,98	2,35	,0237
1	,99916	—	8309,71	8406,88	2,83	1,19	,0120

I. Spirit and water by weight	II. Specific gravity	III. Spirit by measure	IV. Water by measure	V. Bulk of mixture	VI. Diminution of bulk	VII. Quantity of spirit per cent	VIII. Decimal multipliers
Sp. + W.							
100 + 0	,83120	100	—	100,00	—	100,00	1,0075
1	,83350	—	0,83	100,72	0,11	99,28	1,0004
2	,83575	—	1,66	101,44	0,22	98,58	,9933
3	,83795	—	2,49	102,17	0,32	97,88	,9862
4	,84009	—	3,32	102,90	0,42	97,18	,9792
100 + 5	,84216	—	4,16	103,63	0,53	96,49	,9722
6	,84420	—	4,98	104,37	0,61	95,81	,9654
7	,84618	—	5,81	105,11	0,70	95,14	,9586
8	,84811	—	6,64	105,85	0,80	94,47	,9519
9	,84999	—	7,47	106,59	0,88	93,81	,9453
100 + 10	,85183	—	8,30	107,34	0,96	93,16	,9387
11	,85363	—	9,13	108,09	1,04	92,52	,9322
12	,85538	—	9,97	108,84	1,13	91,88	,9258
13	,85709	—	10,80	109,59	1,21	91,25	,9194
14	,85876	—	11,63	110,34	1,27	90,63	,9131
100 + 15	,86039	—	12,46	111,09	1,37	90,01	,9069
16	,86201	—	13,29	111,85	1,44	89,40	,9008
17	,86358	—	14,11	112,61	1,50	88,80	,8947
18	,86513	—	14,94	113,37	1,57	88,21	,8887
19	,86666	—	15,77	114,13	1,64	87,62	,8828
100 + 20	,86813	—	16,61	114,89	1,72	87,04	,8769
21	,86962	—	17,44	115,65	1,79	86,46	,8712
22	,87106	—	18,27	116,41	1,86	85,89	,8655
23	,87247	—	19,10	117,18	1,92	85,33	,8598
24	,87385	—	19,93	117,95	1,98	84,79	,8542
100 + 25	,87522	—	20,76	118,71	2,05	84,24	,8487
26	,87655	—	21,60	119,48	2,12	83,70	,8432
27	,87787	—	22,43	120,25	2,18	83,16	,8378
28	,87915	—	23,20	121,02	2,24	82,63	,8325
29	,88041	—	24,09	121,79	2,30	82,11	,8273
100 + 30	,88165	—	24,92	122,56	2,36	81,59	,8221
31	,88288	—	25,74	123,33	2,41	81,08	,8169
32	,88409	—	26,57	124,10	2,47	80,58	,8118
33	,88528	—	27,40	124,87	2,53	80,08	,8068
34	,88645	—	28,24	125,64	2,60	79,59	,8018
100 + 35	,88760	—	29,07	126,42	2,65	79,10	,7969
36	,88873	—	29,90	127,19	2,71	78,62	,7921
37	,88984	—	30,73	127,97	2,76	78,14	,7873
38	,89093	—	31,56	128,74	2,82	77,67	,7825
39	,89202	—	32,39	129,52	2,87	77,20	,7778
100 + 40	,89307	—	33,22	130,30	2,92	76,74	,7732
41	,89413	—	34,05	131,07	2,98	76,29	,7687
42	,89517	—	34,88	131,85	3,03	75,84	,7642
43	,89619	—	35,71	132,63	3,08	75,40	,7597
44	,89720	—	36,54	133,40	3,14	74,96	,7553
100 + 45	,89819	—	37,37	134,18	3,19	74,52	,7509
46	,89917	—	38,20	134,96	3,24	74,09	,7466
47	,90014	—	39,03	135,74	3,29	73,67	,7422
48	,90108	—	39,86	136,52	3,34	73,25	,7380
49	,90201	—	40,70	137,30	3,40	72,83	,7338

I. Spirit and water by weight	II. Specific gravity	III. Spirit by measure	IV. Water by measure	V. Bulk of mixture	VI. Diminution of bulk	VII. Quantity of spirit per cent	VIII. Decimal multipliers
Sp. + W.							
100 + 50	,90292	100	41,52	138,08	3,44	72,42	,7296
51	,90382	—	42,35	138,86	3,49	72,01	,7255
52	,90470	—	43,18	139,65	3,53	71,61	,7214
53	,90557	—	44,01	140,43	3,58	71,21	,7174
54	,90642	—	44,84	141,22	3,62	70,81	,7135
100 + 55	,90725	—	45,68	142,00	3,68	70,42	,7095
56	,90808	—	46,51	142,79	3,72	70,03	,7056
57	,90889	—	47,33	143,58	3,75	69,64	,7018
58	,90969	—	48,17	144,36	3,81	69,27	,6980
59	,91047	—	49,00	145,15	3,85	68,89	,6942
100 + 60	,91126	—	49,82	145,94	3,88	68,52	,6903
61	,91202	—	50,66	146,73	3,93	68,15	,6867
62	,91277	—	51,49	147,51	3,98	67,79	,6830
63	,91353	—	52,32	148,31	4,01	67,43	,6793
64	,91426	—	53,15	149,10	4,05	67,07	,6757
100 + 65	,91499	—	53,98	149,89	4,09	66,72	,6721
66	,91571	—	54,81	150,68	4,13	66,37	,6687
67	,91643	—	55,64	151,47	4,17	66,02	,6652
68	,91714	—	56,47	152,25	4,22	65,67	,6617
69	,91783	—	57,30	153,05	4,25	65,34	,6583
100 + 70	,91852	—	58,13	153,84	4,29	65,00	,6549
71	,91919	—	58,96	154,63	4,33	64,67	,6516
72	,91985	—	59,79	155,43	4,36	64,34	,6483
73	,92051	—	60,62	156,22	4,40	64,01	,6450
74	,92115	—	61,45	157,01	4,44	63,69	,6417
100 + 75	,92178	—	62,29	157,80	4,49	63,37	,6385
76	,92242	—	63,12	158,59	4,53	63,05	,6353
77	,92304	—	63,95	159,39	4,56	62,74	,6321
78	,92365	—	64,78	160,18	4,60	62,43	,6290
79	,92425	—	65,61	160,97	4,64	62,12	,6259
100 + 80	,92484	—	66,44	161,77	4,67	61,82	,6228
81	,92544	—	67,27	162,57	4,70	61,51	,6198
82	,92604	—	68,09	163,36	4,73	61,21	,6168
83	,92661	—	68,93	164,15	4,78	60,91	,6138
84	,92718	—	69,76	164,96	4,80	60,62	,6108
100 + 85	,92773	—	70,59	165,75	4,84	60,33	,6079
86	,92831	—	71,42	166,55	4,87	60,04	,6050
87	,92886	—	72,25	167,34	4,91	59,76	,6021
88	,92940	—	73,08	168,14	4,94	59,48	,5993
89	,92994	—	73,91	168,93	4,98	59,20	,5965
100 + 90	,93047	—	74,75	169,73	5,02	58,92	,5936
91	,93098	—	75,58	170,53	5,05	58,64	,5908
92	,93150	—	76,41	171,32	5,09	58,37	,5881
93	,93201	—	77,23	172,12	5,11	58,10	,5854
94	,93251	—	78,06	172,92	5,14	57,83	,5827
100 + 95	,93301	—	78,89	173,72	5,17	57,56	,5800
96	,93349	—	79,71	174,51	5,20	57,30	,5773
97	,93397	—	80,54	175,30	5,24	57,04	,5747
98	,93445	—	81,37	176,10	5,27	56,78	,5721
99	,93492	—	82,21	176,91	5,30	56,53	,5696

TABLE II.   HEAT 47°.

I. Water and spirit by weight.	II. Specific gravity.	III. Spirit by measure.	IV. Water by measure.	V. Bulk of mixture.	VI. Diminution of bulk.	VII. Quantity of spirit per cent.	VIII. Decimal multipliers.
W. + Sp.							
100+100	,93539	100	83,05	177,71	5,34	56,27	,5669
99	,93588	—	83,90	178,53	5,37	56,01	,5643
98	,93636	—	84,76	179,36	5,40	55,75	,5617
97	,93683	—	85,63	180,20	5,43	55,49	,5592
96	,93731	—	86,52	181,06	5,46	55,23	,5565
100 + 95	,93780	—	87,42	181,93	5,49	54,96	,5538
94	,93827	—	88,35	182,83	5,52	54,70	,5511
93	,93874	—	89,30	183,75	5,55	54,42	,5483
92	,93921	—	90,27	184,70	5,57	54,15	,5455
91	,93968	—	91,26	185,66	5,60	53,87	,5427
100 + 90	,94017	—	92,28	186,64	5,64	53,58	,5398
89	,94066	—	93,32	187,64	5,68	53,29	,5369
88	,94116	—	94,38	188,67	5,71	53,00	,5340
87	,94167	—	95,46	189,73	5,73	52,71	,5310
86	,94218	—	96,57	190,81	5,76	52,41	,5280
100 + 85	,94269	—	97,70	191,90	5,80	52,11	,5250
84	,94320	—	98,87	193,04	5,83	51,81	,5219
83	,94372	—	100,06	194,19	5,87	51,50	,5188
82	,94424	—	101,28	195,38	5,90	51,18	,5157
81	,94475	—	102,53	196,60	5,93	50,87	,5125
100 + 80	,94528	—	103,81	197,84	5,97	50,54	,5092
79	,94582	—	105,13	199,12	6,01	50,22	,5060
78	,94635	—	106,47	200,43	6,04	49,89	,5027
77	,94689	—	107,86	201,78	6,08	49,55	,4993
76	,94742	—	109,28	203,17	6,11	49,22	,4959
100 + 75	,94797	—	110,74	204,59	6,15	48,88	,4925
74	,94850	—	112,23	206,05	6,18	48,53	,4890
73	,94904	—	113,76	207,55	6,21	48,18	,4855
72	,94959	—	115,34	209,09	6,25	47,82	,4819
71	,95014	—	116,97	210,70	6,27	47,47	,4782
100 + 70	,95069	—	118,64	212,33	6,31	47,10	,4745
69	,95125	—	120,35	214,01	6,34	46,73	,4708
68	,95181	—	122,12	215,74	6,38	46,35	,4670
67	,95237	—	123,94	217,53	6,41	45,97	,4632
66	,95294	—	125,83	219,37	6,45	45,58	,4593
100 + 65	,95351	—	127,76	221,27	6,49	45,19	,4553
64	,95408	—	129,76	223,24	6,52	44,79	,4513
63	,95464	—	131,82	225,26	6,56	44,39	,4473
62	,95522	—	133,95	227,36	6,59	43,99	,4431
61	,95580	—	136,14	229,53	6,61	43,57	,4389
100 + 60	,95637	—	138,42	231,70	6,66	43,15	,4347
59	,95695	—	140,75	234,07	6,68	42,72	,4304
58	,95753	—	143,19	236,47	6,72	42,29	,4260
57	,95812	—	145,70	238,94	6,76	41,85	,4216
56	,95870	—	148,31	241,52	6,79	41,40	,4172
100 + 55	,95927	—	151,00	244,19	6,81	40,95	,4126
54	,95987	—	153,77	246,95	6,82	40,49	,4080
53	,96045	—	156,66	249,81	6,85	40,02	,4033
52	,96102	—	159,68	252,81	6,87	39,55	,3985
51	,96161	—	162,83	255,92	6,91	39,07	,3937
100 + 50	,96219	100	166,10	259,16	6,94	38,58	,3888
49	,96276	—	169,50	262,53	6,97	38,09	,3838
48	,96334	—	173,02	266,04	6,98	37,59	,3787
47	,96391	—	176,70	269,70	7,00	37,08	,3736
46	,96449	—	180,55	273,53	7,02	36,56	,3684
100 + 45	,96507	—	184,56	277,53	7,03	36,03	,3630
44	,96562	—	188,76	281,71	7,05	35,49	,3577
43	,96619	—	193,15	286,10	7,05	34,95	,3522
42	,96675	—	197,74	290,70	7,04	34,40	,3466
41	,96732	—	202,56	295,52	7,04	33,84	,3409
100 + 40	,96788	—	207,63	300,57	7,06	33,27	,3352
39	,96843	—	212,95	305,89	7,06	32,69	,3294
38	,96898	—	218,55	311,51	7,04	32,10	,3234
37	,96953	—	224,47	317,43	7,04	31,50	,3174
36	,97000	—	230,70	323,68	7,02	30,89	,3112
100 + 35	,97064	—	237,29	330,29	7,00	30,28	,3050
34	,97119	—	244,27	337,30	6,97	29,65	,2987
33	,97176	—	251,67	344,73	6,94	29,01	,2923
32	,97231	—	259,54	352,64	6,90	28,36	,2858
31	,97288	—	267,91	361,04	6,87	27,70	,2791
100 + 30	,97344	—	276,84	370,00	6,84	27,03	,2723
29	,97401	—	286,38	379,61	6,77	26,34	,2654
28	,97458	—	296,61	389,90	6,71	25,65	,2584
27	,97516	—	307,59	400,94	6,65	24,94	,2513
26	,97675	—	319,43	412,85	6,58	24,22	,2440
100 + 25	,97634	—	332,21	425,66	6,55	23,49	,2367
24	,97697	—	346,05	439,56	6,49	22,75	,2292
23	,97759	—	361,09	454,69	6,40	22,09	,2216
22	,97823	—	377,51	471,19	6,32	21,22	,2138
21	,97889	—	395,49	489,25	6,24	20,44	,2059
100 + 20	,97956	—	415,26	509,12	6,14	19,64	,1979
19	,98024	—	437,11	531,08	6,03	18,83	,1897
18	,98095	—	461,40	555,47	5,93	18,00	,1814
17	,98167	—	488,55	582,73	5,82	17,16	,1729
16	,98242	—	519,07	613,39	5,68	16,30	,1643
100 + 15	,98320	—	553,68	648,13	5,55	15,43	,1555
14	,98401	—	593,23	687,82	5,41	14,54	,1465
13	,98485	—	638,87	733,61	5,26	13,63	,1373
12	,98573	—	692,11	787,00	5,11	12,70	,1280
11	,98666	—	755,02	850,09	4,93	11,76	,1185
100 + 10	,98762	—	830,53	925,74	4,79	10,80	,1088
9	,98864	—	922,81	1018,23	4,58	9,82	,0990
8	,98971	—	1018,15	1133,77	4,38	8,82	,0889
7	,99085	—	1186,47	1282,28	4,19	7,80	,0786
6	,99206	—	1384,22	1480,22	4,00	6,76	,0681
100 + 5	,99330	—	1661,05	1757,26	3,77	5,69	,0573
4	,99403	—	2076,31	2172,78	3,53	4,60	,0464
3	,99504	—	2768,43	2865,10	3,33	3,49	,0352
2	,99604	—	4152,58	4249,52	3,06	2,35	,0237
1	,99913	—	8305,28	8402,37	2,91	1,19	,0120

Mr. GILPIN's *Tables*

TABLE I.   HEAT 48°.

I. Spirit and water by weight.	II. Specific gravity.	III. Spirit by measure.	IV. Water by measure.	V. Bulk of mixture.	VI. Diminution of bulk.	VII. Quantity of Spirit per cent.	VIII. Decimal multipliers.
Sp. + W.							
100 + 0	,83073	100	—	100,00	—	100,00	1,0069
1	,83303	—	0,83	100,72	0,11	99,28	,9998
2	,83528	—	1,66	101,44	0,22	98,58	,9927
3	,83748	—	2,49	102,17	0,32	97,88	,9857
4	,83962	—	3,32	102,90	0,42	97,19	,9787
100 + 5	,84169	—	4,16	103,63	0,53	96,50	,9717
6	,84373	—	4,98	104,37	0,61	95,81	,9649
7	,84571	—	5,81	105,11	0,70	95,14	,9581
8	,84764	—	6,64	105,85	0,79	94,47	,9514
9	,84952	—	7,47	106,59	0,88	93,81	,9448
100 + 10	,85136	—	8,30	107,34	0,96	93,16	,9381
11	,85316	—	9,13	108,09	1,04	92,52	,9316
12	,85491	—	9,96	108,84	1,12	91,88	,9253
13	,85663	—	10,80	109,59	1,21	91,25	,9189
14	,85830	—	11,62	110,34	1,28	90,63	,9126
100 + 15	,85994	—	12,45	111,09	1,36	90,01	,9064
16	,86155	—	13,28	111,85	1,43	89,40	,9003
17	,86312	—	14,11	112,61	1,50	88,80	,8943
18	,86467	—	14,93	113,37	1,56	88,21	,8883
19	,86620	—	15,76	114,13	1,63	87,62	,8823
100 + 20	,86767	—	16,60	114,89	1,71	87,04	,8764
21	,86916	—	17,43	115,65	1,78	86,47	,8707
22	,87060	—	18,26	116,41	1,85	85,90	,8650
23	,87201	—	19,09	117,17	1,92	85,34	,8593
24	,87339	—	19,92	117,94	1,98	84,79	,8537
100 + 25	,87476	—	20,75	118,71	2,04	84,24	,8482
26	,87609	—	21,58	119,47	2,11	83,70	,8428
27	,87741	—	22,41	120,24	2,17	83,17	,8374
28	,87869	—	23,25	121,02	2,23	82,64	,8321
29	,87996	—	24,07	121,78	2,29	82,11	,8269
100 + 30	,88120	—	24,90	122,55	2,35	81,60	,8210
31	,88243	—	25,73	123,33	2,40	81,09	,8165
32	,88364	—	26,56	124,10	2,46	80,58	,8114
33	,88483	—	27,39	124,87	2,52	80,08	,8064
34	,88600	—	28,22	125,64	2,58	79,59	,8014
100 + 35	,88716	—	29,05	126,41	2,64	79,11	,7965
36	,88822	—	29,88	127,19	2,69	78,63	,7917
37	,88940	—	30,71	127,96	2,75	78,15	,7869
38	,89049	—	31,54	128,74	2,80	77,68	,7821
39	,89157	—	32,37	129,51	2,86	77,21	,7774
100 + 40	,89262	—	33,21	130,29	2,92	76,75	,7728
41	,89368	—	34,04	131,07	2,97	76,29	,7683
42	,89472	—	34,87	131,84	3,03	75,85	,7638
43	,89574	—	35,69	132,62	3,07	75,40	,7593
44	,89675	—	36,52	133,39	3,13	74,96	,7549
100 + 45	,89774	—	37,35	134,17	3,18	74,53	,7505
46	,89872	—	38,18	134,95	3,23	74,10	,7462
47	,89969	—	39,01	135,73	3,28	73,67	,7418
48	,90064	—	39,84	136,51	3,33	73,26	,7376
49	,90157	—	40,67	137,29	3,38	72,84	,7335

I. Spirit and Water by weight.	II. Specific gravity.	III. Spirit by measure.	IV. Water by measure.	V. Bulk of mixture.	VI. Diminution of bulk.	VII. Quantity of Spirit per cent.	VIII. Decimal multipliers.
Sp. + W.							
100 + 50	,90248	100	41,50	138.07	3,43	72,43	,7293
51	,90338	—	42,33	138,85	3,48	72,02	,7252
52	,90427	—	43,16	139,63	3,53	71,62	,7211
53	,90513	—	43,99	140,42	3,57	71,21	,7171
54	,90599	—	44,82	141,21	3,61	70,82	,7131
100 + 55	,90682	—	45,65	141,98	3,67	70,43	,7091
56	,90765	—	46,48	142,77	3,71	70,04	,7053
57	,90846	—	47,31	143,56	3,75	69,65	,7014
58	,90926	—	48,14	144,35	3,79	69,27	,6976
59	,91004	—	48,97	145,13	3,84	68,90	,6938
100 + 60	,91083	—	49,80	145,92	3,88	68,53	,6900
61	,91159	—	50,64	146,71	3,93	68,16	,6864
62	,91235	—	51,46	147,50	3,96	67,79	,6827
63	,91310	—	52,29	148,29	4,00	67,43	,6790
64	,91383	—	53,12	149,08	4,04	67,07	,6754
100 + 65	,91456	—	53,95	149,87	4,08	66,72	,6719
66	,91528	—	54,78	150,66	4,12	66,37	,6684
67	,91600	—	55,61	151,46	4,15	66,02	,6649
68	,91671	—	56,44	152,24	4,20	65,68	,6614
69	,91740	—	57,27	153,03	4,24	65,35	,6580
100 + 70	,91809	—	58,10	153,82	4,28	65,01	,6540
71	,91876	—	58,93	154,61	4,32	64,68	,6513
72	,91942	—	59,76	155,41	4,35	64,35	,6480
73	,92008	—	60,59	156,20	4,39	64,02	,6447
74	,92072	—	61,42	156,99	4,43	63,70	,6414
100 + 75	,92136	—	62,25	157,78	4,47	63,38	,6382
76	,92199	—	63,08	158,57	4,51	63,06	,6350
77	,92261	—	63,91	159,37	4,54	62,75	,6318
78	,92323	—	64,74	160,17	4,57	62,44	,6287
79	,92382	—	65,57	160.95	4,62	62,13	,6256
100 + 80	,92442	—	66,40	161,75	4,65	61,82	,6225
81	,92502	—	67,23	162,55	4,68	61,52	,6195
82	,92561	—	68,05	163,35	4,70	61,22	,6165
83	,92619	—	68,89	164,14	4,75	60,92	,6135
84	,92676	—	69,72	164,94	4,78	60,63	,6105
100 + 85	,92731	—	70,55	165,73	4,82	60,34	,0070
86	,92789	—	71,38	166,53	4,85	60,05	,6047
87	,92844	—	72,21	167,32	4,89	59,77	,6018
88	,92898	—	73,04	168,12	4,92	59,48	,5990
89	,92952	—	73,87	168,91	4,96	59,20	,5962
100 + 90	,93005	—	74,71	169,71	5,00	58,92	,5933
91	,93056	—	75,54	170,51	5,03	58,65	,5905
92	,93108	—	76,37	171,30	5,07	58,37	,5878
93	,93159	—	77,19	172,10	5,09	58,10	,5851
94	,93210	—	78,02	172,90	5,12	57,84	,5824
100 + 95	,93260	—	78,85	173,70	5,15	57,57	,5797
96	,93308	—	79,67	174,48	5,19	57,31	,5771
97	,93357	—	80,50	175,28	5,22	57,05	,5745
98	,93404	—	81,33	176,08	5,25	56,79	,5719
99	,93451	—	82,16	176,88	5,28	56,53	,5693

I. Water and spirit by weight.	II. Specific gravity.	III. Spirit by measure.	IV. Water by measure.	V. Bulk of mixture.	VI. Diminution of bulk.	VII. Quantity of spirit per cent.	VIII. Decimal multipliers.
W. + Sp.							
100+100	,93499	100	83.01	177,69	5.32	56,27	,5667
99	,93548	—	83,85	178,50	5,35	56,02	,5641
98	,93595	—	84,71	179.33	5,38	55,76	,5614
97	,93643	—	85,58	180,17	5,41	55,50	,5589
96	,93691	—	86,47	181,03	5,44	55,24	,5563
100+95	,93740	—	87,37	181,90	5,47	54,97	,5536
94	,93786	—	88,30	182,80	5,50	54,70	,5509
93	,93834	—	89,25	183,72	5,53	54,43	,5481
92	,93881	—	90,22	184,67	5,55	54,15	,5453
91	,93928	—	91,21	185,63	5,58	53.87	,5425
100+90	,93977	—	92,23	186,62	5,61	53,59	,5396
89	,94026	—	93,27	187,61	5,66	53,30	,5367
88	,94076	—	94,33	188,65	5,68	53,01	,5338
87	,94127	—	95,41	189,70	5,71	52,71	,5308
86	,94178	—	96,52	190,78	5,74	52,42	,5278
100+85	,94229	—	97,65	191,07	5,78	52,12	,5248
84	,94280	—	98,81	193,01	5,80	51,82	,5217
83	,94333	—	100,01	194,16	5,85	51,51	,5186
82	,94385	—	101,22	195,34	5,88	51,19	,5155
81	,94437	—	102,47	196,56	5,91	50,87	,5123
100+80	,94490	—	103,75	197,81	5,94	50,55	,5090
79	,94544	—	105,07	199,09	5,98	50,23	,5058
78	,94597	—	106,42	200,40	6,02	49,90	,5025
77	,94651	—	107,80	201,74	6,05	49,56	,4991
76	,94705	—	109,22	203,14	6,08	49,22	,4957
100+75	,94759	—	110,68	204,55	6,11	48,88	,4923
74	,94812	—	112,17	206,02	6,15	48,53	,4888
73	,94867	—	113,70	207,51	6,19	48,18	,4853
72	,94922	—	115,28	209,06	6,22	47,83	,4817
71	,94977	—	116,91	210.66	6,25	47,47	,4780
100+70	,95032	—	118,58	212,29	6,29	47,11	,4743
69	,95088	—	120,29	213,97	6,32	46,73	,4706
68	,95144	—	122,06	215,70	6,36	46,36	,4668
67	,95201	—	123,88	217,49	6,39	45,98	,4630
66	,95258	—	125,76	219,33	6,43	45,59	,4591
100+65	,95315	—	127,70	221,23	6,47	45,20	,4551
64	,95372	—	129,69	223,19	6,50	44,80	,4511
63	,95429	—	131,75	225,21	6,53	44,40	,4471
62	,95487	—	133,88	227,32	6,56	43,99	,4429
61	,95545	—	136,07	229,48	6,59	43,58	,4388
100+60	,95602	—	138,35	231,71	6,64	43,15	,4346
59	,95661	—	140,68	234,02	6,66	42,73	,4303
58	,95720	—	143,11	236,41	6,70	42,30	,4259
57	,95779	—	145,62	238,89	6,73	41,86	,4215
56	,95837	—	148,23	241,47	6,76	41,41	,4170
100+55	,95894	—	150,92	244,13	6,79	40,96	,4124
54	,95952	—	153,69	246,89	6,80	40,50	,4079
53	,96013	—	156,58	249,75	6,83	40,03	,4032
52	,96071	—	159,59	252,74	6,85	39,56	,3984
51	,96130	—	162,74	255,86	6,88	39,08	,3936

I. Water and spirit by weight.	II. Specific gravity.	III. Spirit by measure.	IV. Water by measure.	V. Bulk of mixture.	VI. Diminution of bulk.	VII. Quantity of spirit per cent.	VIII. Decimal multipliers.
W. + Sp.							
100+50	,96188	100	166,01	259,10	6,91	38,59	,3886
49	,96246	—	169.41	262,46	6,95	38,10	,3836
48	,96304	—	172,93	265,97	6,96	37,60	,3786
47	,96362	—	176,60	269,63	6,97	37,09	,3735
46	,96420	—	180,45	273,46	6,99	36,57	,3683
100+45	,96479	—	184,46	277,46	7,00	36,04	,3629
44	,96534	—	188,66	281,63	7,03	35,50	,3575
43	,96571	—	193,04	286,02	7,02	34,96	,3521
42	,96648	—	197,63	290,61	7,02	34,41	,3465
41	,96705	—	202,45	295.43	7,02	33,85	,3408
100+40	,96762	—	207,52	300,48	7,04	33,28	,3351
39	,96817	—	212,84	305,80	7,04	32,70	,3293
38	,96873	—	218,43	311,41	7,02	32,11	,3233
37	,96929	—	224,35	317,32	7,03	31,51	,3173
36	,96985	—	230.57	323,57	7,00	30,90	,3112
100+35	,97041	—	237,16	330,18	6,98	30,29	,3050
34	,97097	—	244,14	337,18	6,96	29,66	,2987
33	,97154	—	251,53	344,61	6,92	29,02	,2922
32	,97210	—	259,40	352,51	6,89	28,37	,2857
31	,97267	—	267,76	360,91	6,85	27,71	,2790
100+30	,97324	—	276,09	369,87	6,82	27,04	,2722
29	,97382	—	286,23	379,47	6,76	26,35	,2653
28	,97440	—	296,45	389,74	6,71	25,66	,2583
27	,97499	—	307,43	400,78	6,65	24,95	,2512
26	,97659	—	319,26	412,68	6,58	24,23	,2440
100+25	,97619	—	332,03	425,48	6,55	23,50	,2367
24	,97682	—	345,80	439,38	6,48	22,76	,2292
23	,97745	—	360,90	454,49	6,41	22,00	,2216
22	,97810	—	377,31	470,98	6,33	21,23	,2138
21	,97876	—	395.28	489,03	6,25	20,45	,2059
100+20	,97944	—	415,04	508,89	6,15	19,65	,1979
19	,98013	—	436,88	530,84	6.04	18,84	,1897
18	,98084	—	461,15	555,21	5,94	18,01	,1813
17	,98157	—	488,29	582,46	5,83	17,17	,1728
16	,98232	—	518,79	613,10	5,69	16,31	,1642
100+15	,98311	—	553,38	647,82	5,55	15,44	,1554
14	,98393	—	592,91	687,49	5,42	14,55	,1464
13	,98477	—	638,53	733,25	5,28	13,64	,1373
12	,98566	—	691,74	786,61	5,13	12,71	,1280
11	,98660	—	754,62	849.66	4,96	11,77	,1185
100+10	,98858	—	830,08	925,27	4,81	10,80	,1088
9	,98858	—	922,32	1017,70	4,62	9.82	,0990
8	,98960	—	1037,60	1133,18	4,42	8,82	,0889
7	,99060	—	1185,84	1281,61	4,23	7,80	,0786
6	,99201	—	1383,48	1479,14	4,04	6,76	,0681
100+5	,99320	—	1660 17	1750,30	3,81	5,69	,0573
4	,99459	—	2075,71	2171,63	3,58	4,61	,0464
3	,99600	—	2766,95	2863,58	3,37	3 49	,0352
2	,99750	—	4150.38	4247,26	3,12	2,36	,0237
1	,99909	—	8300 85	8397,88	2,97	1,19	,0120

I.	II.	III.	IV.	V.	VI.	VII.	VIII.	I.	II.	III.	IV.	V.	VI.	VII.	VIII.
Spirit and water by weight.	Specific gravity.	Spirit by measure.	Water by measure.	Bulk of mixture.	Diminution of bulk.	Quantity of spirit *per cent.*	Decimal multipliers.	Spirit and water by weight.	Specific gravity.	Spirit by measure.	Water by measure.	Bulk of mixture.	Diminution of bulk.	Quantity of spirit *per cent.*	Decimal multipliers.
Sp. + W.								Sp. + W.							
100 + 0	,83025	100	—	100,00	—	100,00	1,0064	100 + 50	,90204	100	41,48	138,06	3,42	72,43	,7289
1	,83256	—	0,83	100,72	0,11	99,28	,9992	51	,90294	—	42,31	138,84	3,47	72,02	,7248
2	,83481	—	1,66	101,44	0,22	98,58	,9921	52	,90383	—	43,14	139,62	3,52	71,62	,7207
3	,83701	—	2,49	102,17	0,32	97,88	,9851	53	,90470	—	43,97	140,41	3,56	71,22	,7167
4	,83915	—	3,32	102,90	0,42	97,19	,9781	54	,90556	—	44,80	141,19	3,61	70,82	,7127
100 + 5	,84122	—	4,16	103,63	0,53	96,50	,9711	100 + 55	,90639	—	45,63	141,97	3,66	70,43	,7088
6	,84326	—	4,98	104,37	0,61	95,81	,9643	56	,90722	—	46,46	142,76	3,70	70,04	,7050
7	,84524	—	5,81	105,11	0,70	95,14	,9576	57	,90803	—	47,28	143,55	3,73	69,66	,7011
8	,84717	—	6,64	105,84	0,80	94,48	,9509	58	,90803	—	48,12	144,34	3,78	69,28	,6973
9	,84905	—	7,47	106,59	0,88	93,82	,9443	59	,90962	—	48,94	145,12	3,82	68,90	,6935
100 + 10	,85089	—	8,30	107,33	0,97	93,17	,9370	100 + 60	,91040	—	49,77	145,91	3,86	68,53	,6897
11	,85269	—	9,13	108,08	1,05	92,53	,9311	61	,91116	—	50,61	146,70	3,91	68,17	,6861
12	,85444	—	9,96	108,83	1,13	91,89	,9247	62	,91192	—	51,43	147,49	3,94	67,80	,6824
13	,85616	—	10,79	109,58	1,21	91,26	,9184	63	,91267	—	52,26	148,28	3,98	67,44	,6787
14	,85784	—	11,62	110,34	1,28	90,63	,9121	64	,91340	—	53,09	149,07	4,02	67,08	,6751
100 + 15	,85948	—	12,45	111,09	1,36	90,02	,9059	100 + 65	,91413	—	53,92	149,86	4,06	66,73	,6715
16	,86109	—	13,28	111,85	1,43	89,41	,8998	66	,91485	—	54,75	150,65	4,10	66,38	,6681
17	,86266	—	14,10	112,60	1,50	88,80	,8938	67	,91557	—	55,58	151,44	4,14	66,03	,6646
18	,86421	—	14,93	113,36	1,57	88,21	,8878	68	,91628	—	56,41	152,23	4,18	65,69	,6611
19	,86574	—	15,75	114,12	1,63	87,62	,8819	69	,91697	—	57,24	153,01	4,23	65,35	,6577
100 + 20	,86721	—	16,59	114,89	1,70	87,04	,8760	100 + 70	,91766	—	58,07	153,81	4,26	65,02	,6543
21	,86870	—	17,42	115,64	1,78	86,47	,8702	71	,91833	—	58,90	154,60	4,30	64,69	,6510
22	,87014	—	18,25	116,41	1,84	85,90	,8645	72	,91899	—	59,73	155,39	4,34	64,36	,6477
23	,87155	—	19,08	117,17	1,91	85,34	,8589	73	,91965	—	60,56	156,19	4,37	64,03	,6444
24	,87293	—	19,91	117,94	1,97	84,79	,8533	74	,92029	—	61,39	156,97	4,42	63,70	,6411
100 + 25	,87430	—	20,74	118,70	2,04	84,25	,8478	100 + 75	,92094	—	62,22	157,77	4,45	63,38	,6379
26	,87564	—	21,57	119,47	2,10	83,71	,8423	76	,92156	—	63,05	158,55	4,50	63,06	,6347
27	,87695	—	22,40	120,24	2,16	83,17	,8369	77	,92218	—	63,88	159,35	4,53	62,75	,6315
28	,87824	—	23,24	121,01	2,23	82,64	,8317	78	,92280	—	64,71	160,15	4,56	62,44	,6284
29	,87951	—	24,06	121,78	2,28	82,12	,8265	79	,92340	—	65,54	160,93	4,61	62,14	,6253
100 + 30	,88075	—	24,89	122,55	2,34	81,60	,8212	100 + 80	,92400	—	66,37	161,73	4,64	61,83	,6222
31	,88198	—	25,71	123,32	2,39	81,09	,8161	81	,92459	—	67,19	162,53	4,66	61,53	,6192
32	,88319	—	26,54	124,09	2,45	80,59	,8110	82	,92518	—	68,02	163,33	4,69	61,23	,6162
33	,88438	—	27,37	124,86	2,51	80,09	,8060	83	,92577	—	68,85	164,12	4,73	60,93	,6132
34	,88555	—	28,21	125,63	2,58	79,60	,8010	84	,92633	—	69,68	164,92	4,76	60,63	,6102
100 + 35	,88671	—	29,03	126,40	2,63	79,11	,7961	100 + 85	,92689	—	70,51	165,71	4,80	60,34	,6073
36	,88784	—	29,87	127,18	2,69	78,63	,7913	86	,92746	—	71,34	166,51	4,83	60,05	,6044
37	,88895	—	30,70	127,95	2,75	78,15	,7865	87	,92801	—	72,17	167,30	4,87	59,77	,6015
38	,89004	—	31,53	128,73	2,80	77,68	,7817	88	,92855	—	73,00	168,10	4,90	59,49	,5987
39	,89113	—	32,36	129,50	2,86	77,21	,7770	89	,92909	—	73,83	168,89	4,94	59,21	,5959
100 + 40	,89218	—	33,19	130,28	2,91	76,75	,7724	100 + 90	,92962	—	74,67	169,69	4,98	58,93	,5930
41	,89324	—	34,02	131,06	2,96	76,30	,7679	91	,93014	—	75,50	170,49	5,01	58,65	,5902
42	,89427	—	34,85	131,83	3,02	75,85	,7634	92	,93066	—	76,33	171,28	5,05	58,38	,5875
43	,89530	—	35,67	132,61	3,06	75,41	,7589	93	,93118	—	77,15	172,08	5,07	58,11	,5848
44	,89630	—	36,50	133,39	3,11	74,97	,7545	94	,93169	—	77,98	172,87	5,11	57,84	,5821
100 + 45	,89729	—	37,33	134,17	3,16	74,53	,7501	100 + 95	,93219	—	78,81	173,68	5,13	57,58	,5794
46	,89827	—	38,16	134,94	3,22	74,10	,7458	96	,93267	—	79,63	174,46	5,17	57,32	,5768
47	,89924	—	38,99	135,72	3,27	73,68	,7415	97	,93316	—	80,46	175,26	5,20	57,06	,5742
48	,90019	—	39,82	136,50	3,32	73,26	,7373	98	,93364	—	81,29	176,05	5,24	56,80	,5716
49	,90113	—	40,65	137,28	3,37	72,84	,7331	99	,93411	—	82,12	176,86	5,26	56,54	,5690

TABLE II. HEAT 49°.

I. Water and spirit by weight.	II. Specific gravity.	III. Spirit by measure.	IV. Water by measure.	V. Bulk of mixture.	VI. Diminution of bulk.	VII. Quantity of spirit per cent.	VIII. Decimal multipliers.
W. + Sp.							
100+100	,93459	100	82.96	177,67	5,29	56,28	,5664
99	,93507	—	83,81	178,48	5,33	56,02	,5638
98	,93555	—	84,67	179,31	5,36	55,76	,5612
97	,93602	—	85,54	180,15	5,39	55,50	,5586
96	,93651	—	86,44	181,01	5,43	55,24	,5560
100+95	,93699	—	87,32	181,88	5,44	54,98	,5533
94	,93746	—	88,25	182,78	5,47	54,71	,5506
93	,93794	—	89,20	183,70	5,50	54,44	,5478
92	,93841	—	90,17	184,65	5,52	54,16	,5450
91	,93888	—	91,16	185,61	5,55	53,88	,5422
100+90	,93937	—	92,18	186,59	5,59	53,59	,5393
89	,93986	—	93,22	187,59	5,63	53,30	,5365
88	,94036	—	94,28	188,62	5,66	53,01	,5336
87	,94087	—	95,36	189,67	5,69	52,72	,5306
86	,94138	—	96,47	190,75	5.72	52,42	,5276
100+85	,94189	—	97,60	191,85	5,75	52,12	,5245
84	,94241	—	98,76	192,98	5,78	51,82	,5215
83	,94294	—	99,95	194,13	5,82	51,51	,5184
82	,94346	—	101,17	195,31	5,86	51,20	,5153
81	,94399	—	102,42	196,53	5,89	50,88	,5121
100+80	,94452	—	103,70	197,78	5,92	50,56	,5088
79	,94506	—	105,02	199,05	5,97	50,24	,5056
78	,94559	—	106,36	200,36	6,00	49,91	,5023
77	,94613	—	107,74	201,71	6,03	49,57	,4989
76	,94667	—	109,16	203,10	6,06	49,23	,4955
100+75	,94721	—	110,62	204,52	6,10	48,89	,4921
74	,94775	—	112,11	205,98	6,13	48,54	,4886
73	,94830	—	113,64	207,48	6,16	48,19	,4851
72	,94885	—	115,22	209,02	6,20	47,84	,4815
71	,94940	—	116,85	210,62	6,23	47,48	,4778
100+70	,94995	—	118,52	212,25	6,27	47,11	,4741
69	,95051	—	120,23	213,93	6,30	46,74	,4704
68	,95108	—	122,00	215,66	6,34	46,37	,4666
67	,95165	—	123,82	217,45	6,37	45,99	,4628
66	,95222	—	125,70	219,29	6,41	45,60	,4589
100+65	,95279	—	127,63	221,19	6,44	45,21	,4550
64	,95336	—	129,62	223,15	6,47	44,81	,4510
63	,95394	—	131,68	225,17	6,51	44,41	,4469
62	,95452	—	133,81	227,26	6,55	44,00	,4428
61	,95510	—	136,00	229,43	6,57	43,59	,4386
100+60	,95568	—	138,27	231,66	6,61	43,17	,4344
59	,95627	—	140,60	233,97	6,63	42,74	,4301
58	,95687	—	143,04	236,36	6,68	42,31	,4257
57	,95746	—	145,54	238,83	6,71	41,87	,4214
56	,95804	—	148,15	241,41	6,74	41,42	,4169
100+55	,95862	—	150,84	244,07	6,77	40,97	,4123
54	,95923	—	153,61	246,83	6,78	40,51	,4078
53	,95981	—	156,50	249,69	6,81	40,04	,4031
52	,96040	—	159,51	252,68	6,83	39,57	,3983
51	,96099	—	162,65	255,80	6,85	39,09	,3935

I. Water and spirit by weight.	II. Specific gravity.	III. Spirit by measure.	IV. Water by measure.	V. Bulk of mixture.	VI. Diminution of bulk.	VII. Quantity of spirit per cent.	VIII. Decimal multipliers.
W. + Sp.							
100+50	,96157	100	165,92	259,03	6,89	38,60	,3885
49	,96210	—	169,32	262,39	6,93	38,11	,3835
48	,96274	—	172,84	265,90	6,94	37,61	,3785
47	,96333	—	176,51	269,56	6,95	37,10	,3734
46	,96391	—	180,35	273,38	6,97	36,58	,3681
100+45	,96451	—	184,36	277,38	6,98	36,05	,3628
44	,96506	—	188,56	281,55	7,01	35,51	,3574
43	,96564	—	192,94	285,94	7,00	34,97	,3520
42	,96621	—	197,53	290,53	7,00	34,42	,3464
41	,96678	—	202,34	295,34	7,00	33,86	,3407
100+40	,96736	—	207,41	300,39	7,02	33,29	,3350
39	,96791	—	212,73	305,71	7,02	32,71	,3292
38	,96848	—	218,32	311,32	7,00	32,12	,3233
37	,96905	—	224,23	317,22	7,01	31,52	,3172
36	,96962	—	230,45	323,46	6,99	30,91	,3111
100+35	,97018	—	237,03	330,06	6,97	30,30	,3049
34	,97075	—	244,01	337,06	6,95	29,67	,2986
33	,97132	—	251,40	344,48	6,92	29,03	,2922
32	,97189	—	259,26	352,38	6,88	28,38	,2857
31	,97246	—	267,62	360,77	6,85	27,72	,2790
100+30	,97304	—	276,54	369,73	6,81	27,05	,2722
29	,97363	—	286,08	379,32	6,76	26,36	,2653
28	,97422	—	296,29	389,59	6,70	25,67	,2583
27	,97482	—	307,27	400,62	6,65	24,96	,2512
26	,97643	—	319,09	412,51	6,58	24,24	,2439
100+25	,97604	—	331,85	425,31	6,54	23,51	,2366
24	,97667	—	345,68	439,20	6,48	22,77	,2292
23	,97731	—	360,71	454,30	6,41	22,01	,2215
22	,97797	—	377,11	470,78	6,33	21,24	,2137
21	,97864	—	395,07	488,82	6,25	20,46	,2058
100+20	,97932	—	414,82	508,66	6,16	19,66	,1978
19	,98002	—	436,65	530,60	6,05	18,85	,1896
18	,98073	—	460,91	554,96	5,95	18,02	,1813
17	,98147	—	488,03	582,19	5,64	17,18	,1728
16	,98223	—	518,52	612,81	5,71	16,32	,1642
100+15	,98279	—	553,00	647,51	5,58	15,44	,1554
14	,98385	—	592,60	687,16	5,44	14,55	,1464
13	,98470	—	638,19	732,89	5,30	13,64	,1373
12	,98559	—	691,37	786,22	5,15	12,72	,1280
11	,98654	—	754,22	849,23	4,99	11,78	,1185
100+10	,98750	—	829,64	924,80	4,84	10,81	,1088
9	,98853	—	921,83	1017,18	4,65	9,83	,0989
8	,98961	—	1037,05	1132,60	4,45	8,83	,0888
7	,99075	—	1185,21	1280,94	4,27	7,81	,0786
6	,99196	—	1382,74	1478,67	4,07	6,76	,0681
100+5	,99321	—	1659,29	1755,44	3,85	5,70	,0573
4	,99454	—	2074,11	2170,49	3,62	4,61	,0464
3	,99581	—	2765,48	2862,07	3,41	3,49	,0352
2	,99746	—	4148,20	4245,02	3,18	2,36	,0237
1	,99905	—	8296,45	8393,42	3,03	1,19	,0120

TABLE I. HEAT 50°.

I. Spirit and water by weight.	II. Specific gravity.	III. Spirit by measure.	IV. Water by measure.	V. Bulk of mixture.	VI. Diminution of bulk.	VII. Quantity of spirit per cent.	VIII. Decimal multipliers.	I. Spirit and water by weight.	II. Specific gravity.	III. Spirit by measure.	IV. Water by measure.	V. Bulk of mixture.	VI. Diminution of bulk.	VII. Quantity of spirit per cent.	VIII. Decimal multipliers.
Sp. + W.								Sp. + W.							
100 + 0	,82977	100	—	100,00	—	100,00	1,0058	100 + 50	,90160	100	41,46	138,05	3,41	72,44	,7286
1	,83208	—	0,83	100,72	0,11	99,29	,9985	51	,90250	—	42,29	138,83	3,46	72,03	,7245
2	,83434	—	1,66	101,44	0,22	98,58	,9915	52	,90339	—	43,12	139,61	3,51	71,63	,7204
3	,83654	—	2,49	102,16	0,33	97,83	,9845	53	,90426	—	43,95	140,40	3,55	71,23	,7164
4	,83868	—	3,32	102,89	0,43	97,19	,9776	54	,90512	—	44,78	141,18	3,60	70,83	,7124
100 + 5	,84076	—	4,15	103,63	0,52	96,50	,9706	100 + 55	,90596	—	45,61	141,96	3,65	70,44	,7085
6	,84279	—	4,97	104,36	0,61	95,82	,9637	56	,90679	—	46,43	142,75	3,68	70,05	,7046
7	,84477	—	5,80	105,10	0,70	95,15	,9570	57	,90763	—	47,26	143,54	3,72	69,67	,7007
8	,84670	—	6,63	105,84	0,79	94,48	,9503	58	,90840	—	48,09	144,33	3,76	69,29	,6969
9	,84858	—	7,46	106,58	0,88	93,82	,9437	59	,90919	—	48,92	145,11	3,81	68,91	,6931
100 + 10	,85042	—	8,29	107,33	0,96	93,17	,9371	100 + 60	,90997	—	49,75	145,90	3,85	68,54	,6894
11	,85222	—	9,12	108,08	1,04	92,53	,9306	61	,91073	—	50,58	146,69	3,89	68,17	,6857
12	,85397	—	9,95	108,83	1,12	91,89	,9242	62	,91149	—	51,41	147,48	3,93	67,81	,6820
13	,85569	—	10,78	109,58	1,20	91,26	,9179	63	,91224	—	52,24	148,26	3,98	67,45	,6784
14	,85738	—	11,61	110,33	1,28	90,64	,9116	64	,91297	—	53,07	149,06	4,01	67,09	,6748
100 + 15	,85902	—	12,44	111,08	1,36	90,02	,9054	100 + 65	,91370	—	53,90	149,85	4,05	66,74	,6712
16	,86063	—	13,27	111,84	1,43	89,41	,8993	66	,91442	—	54,73	150,64	4,09	66,39	,6677
17	,86220	—	14,09	112,60	1,49	88,81	,8933	67	,91514	—	55,56	151,42	4,14	66,04	,6642
18	,86375	—	14,92	113,36	1,56	88,22	,8873	68	,91585	—	56,38	152,21	4,17	65,70	,6608
19	,86527	—	15,75	114,12	1,63	87,63	,8814	69	,91654	—	57,21	153,00	4,21	65,36	,6574
100 + 20	,86676	—	16,58	114,88	1,70	87,05	,8755	100 + 70	,91723	—	58,04	153,79	4,25	65,02	,6540
21	,86823	—	17,41	115,64	1,77	86,48	,8697	71	,91790	—	58,87	154,58	4,29	64,69	,6507
22	,86967	—	18,24	116,40	1,84	85,91	,8640	72	,91856	—	59,70	155,37	4,33	64,36	,6474
23	,87108	—	19,07	117,16	1,91	85,35	,8584	73	,91922	—	60,53	156,17	4,36	64,03	,6441
24	,87247	—	19,90	117,93	1,97	84,80	,8529	74	,91986	—	61,36	156,96	4,40	63,71	,6408
100 + 25	,87384	—	20,73	118,70	2,03	84,25	,8474	100 + 75	,92051	—	62,19	157,75	4,44	63,39	,6376
26	,87518	—	21,56	119,46	2,10	83,71	,8419	76	,92113	—	63,02	158,54	4,48	63,07	,6344
27	,87649	—	22,39	120,23	2,16	83,18	,8365	77	,92175	—	63,85	159,33	4,52	62,76	,6312
28	,87778	—	23,22	121,00	2,22	82,65	,8312	78	,92237	—	64,68	160,13	4,55	62,45	,6281
29	,87905	—	24,05	121,77	2,28	82,12	,8260	79	,92298	—	65,51	160,92	4,59	62,14	,6250
100 + 30	,88030	—	24,88	122,54	2,34	81,61	,8208	100 + 80	,92358	—	66,34	161,72	4,62	61,84	,6219
31	,88153	—	25,70	123,31	2,39	81,10	,8157	81	,92417	—	67,16	162,51	4,65	61,53	,6189
32	,88274	—	26,53	124,08	2,45	80,59	,8106	82	,92476	—	67,99	163,31	4,68	61,23	,6159
33	,88393	—	27,36	124,85	2,51	80,09	,8056	83	,92534	—	68,82	164,10	4,72	60,94	,6129
34	,88510	—	28,19	125,62	2,57	79,60	,8006	84	,92591	—	69,65	164,90	4,75	60,64	,6099
100 + 35	,88626	—	29,02	126,40	2,62	79,12	,7957	100 + 85	,92647	—	70,48	165,69	4,79	60,35	,6070
36	,88739	—	29,85	127,17	2,68	78,64	,7909	86	,92703	—	71,31	166,49	4,82	60,06	,6041
37	,88850	—	30,68	127,94	2,74	78,16	,7861	87	,92758	—	72,14	167,28	4,86	59,78	,6012
38	,88960	—	31,51	128,72	2,79	77,69	,7814	88	,92812	—	72,97	168,08	4,89	59,50	,5984
39	,89068	—	32,34	129,49	2,85	77,22	,7767	89	,92866	—	73,80	168,87	4,93	59,22	,5956
100 + 40	,89174	—	33,17	130,27	2,90	76,76	,7721	100 + 90	,92919	—	74,63	169,67	4,96	58,94	,5928
41	,89279	—	34,00	131,05	2,95	76,31	,7675	91	,92972	—	75,46	170,47	4,99	58,66	,5900
42	,89382	—	34,83	131,82	3,01	75,86	,7630	92	,93024	—	76,29	171,26	5,03	58,39	,5872
43	,89484	—	35,65	132,60	3,05	75,41	,7585	93	,93076	—	77,11	172,06	5,05	58,12	,5845
44	,89585	—	36,48	133,38	3,10	74,97	,7541	94	,93127	—	77,94	172,85	5,09	57,85	,5818
100 + 45	,89684	—	37,31	134,16	3,15	74,54	,7497	100 + 95	,93177	—	78,77	173,65	5,12	57,59	,5792
46	,89782	—	38,14	134,93	3,21	74,11	,7454	96	,93226	—	79,59	174,44	5,15	57,33	,5766
47	,89879	—	38,97	135,71	3,26	73,69	,7411	97	,93275	—	80,42	175,23	5,19	57,07	,5740
48	,89974	—	39,80	136,49	3,31	73,27	,7369	98	,93323	—	81,25	176,03	5,22	56,81	,5714
49	,90068	—	40,63	137,27	3,36	72,85	,7327	99	,93371	—	82,08	176,83	5,25	56,55	,5688

TABLE II. HEAT 50°.

I. Water and spirit by weight.	II. Specific gravity.	III. Spirit by measure.	IV. Water by measure.	V. Bulk of mixture.	VI. Diminution of bulk.	VII. Quantity of spirit per cent.	VIII. Decimal multipliers.
W. + Sp.							
100+100	,93419	100	82,92	177,64	5,28	56,29	,5662
99	,93467	—	83,76	178,46	5,30	56,03	,5636
98	,93515	—	84,62	179,29	5,33	55,77	,5610
97	,93562	—	85,50	180,13	5,37	55,51	,5584
96	,93610	—	86,38	180,98	5,40	55,25	,5558
100 + 95	,93658	—	87,28	181,85	5,43	54,99	,5531
94	,93705	—	88,21	182,75	5,46	54,72	,5504
93	,93753	—	89,16	183,67	5,49	54,44	,5476
92	,93801	—	90,13	184,62	5,51	54,17	,5448
91	,93848	—	91,12	185,58	5,54	53,89	,5420
100 + 90	,93897	—	92,13	186,56	5,57	53,60	,5391
89	,93946	—	93,17	187,56	5,61	53,31	,5362
88	,93996	—	94,23	188,59	5,64	53,02	,5333
87	,94047	—	95,31	189,64	5,67	52,73	,5303
86	,94098	—	96,42	190,72	5,70	52,43	,5273
100 + 85	,94149	—	97,55	191,82	5,73	52,13	,5243
84	,94201	—	98,71	192,95	5,76	51,83	,5213
83	,94254	—	99,90	194,10	5,80	51,52	,5182
82	,94307	—	101,12	195,28	5,84	51,21	,5150
81	,94360	—	102,37	196,50	5,87	50,89	,5118
100 + 80	,94414	—	103,65	197,74	5,91	50,57	,5086
79	,94467	—	104,96	199,02	5,94	50,25	,5054
78	,94521	—	106,31	200,33	5,98	49,92	,5021
77	,94575	—	107,69	201,68	6,01	49,58	,4988
76	,94629	—	109,11	203,06	6,05	49,24	,4954
100 + 75	,94683	—	110,56	204,48	6,08	48,90	,4919
74	,94737	—	112,05	205,94	6,11	48,55	,4884
73	,94792	—	113,58	207,44	6,14	48,20	,4849
72	,94847	—	115,16	208,99	6,17	47,85	,4813
71	,94902	—	116,79	210,58	6,21	47,49	,4776
100 + 70	,94958	—	118,46	212,21	6,25	47,12	,4739
69	,95014	—	120,17	213,89	6,28	46,75	,4702
68	,95071	—	121,94	215,62	6,32	46,38	,4664
67	,95128	—	123,76	217,41	6,35	46,00	,4626
66	,95185	—	125,64	219,25	6,39	45,61	,4587
100 + 65	,95243	—	127,57	221,15	6,42	45,22	,4548
64	,95300	—	129,56	223,11	6,45	44,82	,4508
63	,95358	—	131,61	225,13	6,48	44,42	,4467
62	,95416	—	133,74	227,22	6,52	44,01	,4426
61	,95475	—	135,93	229,38	6,55	43,60	,4384
100 + 60	,95534	—	138,20	231,61	6,59	43,18	,4342
59	,95593	—	140,53	233,92	6,61	42,75	,4299
58	,95653	—	142,96	236,30	6,66	42,32	,4256
57	,95712	—	145,47	238,78	6,69	41,88	,4212
56	,95771	—	148,07	241,35	6,72	41,43	,4167
100 + 55	,95831	—	150,76	244,01	6,75	40,98	,4122
54	,95890	—	153,53	246,77	6,76	40,52	,4076
53	,95949	—	156,42	249,63	6,79	40,05	,4029
52	,96008	—	159,43	252,62	6,81	39,58	,3981
51	,96067	—	162,57	255,73	6,84	39,10	,3933
W. + Sp.							
100 + 50	,96126	100	165,84	258,96	6,88	38,61	,3884
49	,96185	—	169,23	262,32	6,91	38,12	,3834
48	,96244	—	172,75	265,83	6,92	37,62	,3783
47	,96303	—	176,42	269,49	6,93	37,11	,3732
46	,96361	—	180,26	273,31	6,95	36,59	,3680
100 + 45	,96420	—	184,27	277,30	6,97	36,06	,3627
44	,96478	—	188,46	281,48	6,98	35,52	,3573
43	,96536	—	192,84	285,86	6,98	34,98	,3518
42	,96593	—	197,43	290,45	6,98	34,43	,3463
41	,96651	—	202,24	295,26	6,98	33,87	,3406
100 + 40	,96708	—	207,30	300,31	6,99	33,30	,3349
39	,96765	—	212,62	305,62	7,00	32,72	,3291
38	,96823	—	218,21	311,22	6,99	32,13	,3232
37	,96880	—	224,11	317,12	6,99	31,53	,3171
36	,96938	—	230,33	323,35	6,98	30,92	,3110
100 + 35	,96995	—	236,91	329,95	6,96	30,31	,3048
34	,97052	—	243,88	336,96	6,94	29,69	,2985
33	,97110	—	251,27	344,36	6,91	29,04	,2921
32	,97167	—	259,13	352,25	6,88	28,39	,2856
31	,97225	—	267,48	360,64	6,84	27,73	,2789
100 + 30	,97284	—	276,40	369,60	6,80	27,06	,2721
29	,97343	—	285,93	379,18	6,75	26,37	,2652
28	,97403	—	296,14	389,44	6,70	25,68	,2582
27	,97464	—	307,11	400,46	6,65	24,97	,2511
26	,97526	—	318,92	412,34	6,58	24,25	,2439
100 + 25	,97589	—	331,68	425,14	6,54	23,52	,2366
24	,97652	—	345,50	439,02	6,48	22,78	,2291
23	,97717	—	360,52	454,11	6,41	22,02	,2215
22	,97783	—	376,91	470,58	6,33	21,25	,2137
21	,97849	—	394,86	488,61	6,25	20,46	,2058
100 + 20	,97920	—	414,60	508,44	6,16	19,67	,1978
19	,97990	—	436,42	530,36	6,06	18,86	,1896
18	,98062	—	460,67	554,71	5,96	18,03	,1813
17	,98136	—	487,77	581,92	5,85	17,18	,1728
16	,98213	—	518,25	612,53	5,72	16,32	,1642
100 + 15	,98293	—	552,80	647,20	5,60	15,45	,1554
14	,98376	—	592,29	686,83	5,46	14,56	,1464
13	,98462	—	637,85	732,53	5,32	13,65	,1373
12	,98552	—	691,00	785,83	5,17	12,72	,1280
11	,98647	—	753,82	848,80	5,02	11,78	,1185
100 + 10	,98745	—	829,20	924,33	4,87	10,82	,1088
9	,98847	—	921,34	1016,66	4,68	9,84	,0989
8	,98955	—	1036,50	1132,02	4,48	8,83	,0888
7	,99070	—	1184,58	1280,28	4,30	7,81	,0786
6	,99190	—	1382,00	1477,90	4,10	6,77	,0681
100 + 5	,99316	—	1658,41	1754,52	3,89	5,70	,0573
4	,99499	—	2073,01	2169,33	3,66	4,61	,0464
3	,99591	—	2704,02	2860,57	3,45	3,50	,0352
2	,99742	—	4140,03	4242,79	3,24	2,36	,0237
1	,99901	—	8292,06	8388,98	3,08	1,19	,0120

I. Spirit and water by weight.	II. Specific gravity.	III. Spirit by measure.	IV. Water by measure.	V. Bulk of mixture.	VI. Diminution of bulk.	VII. Quantity of spirit per cent.	VIII. Decimal multipliers.	I. Spirit and water by weight.	II. Specific gravity.	III. Spirit by measure.	IV. Water by measure.	V. Bulk of mixture.	VI. Diminution of bulk.	VII. Quantity of spirit per cent.	VIII. Decimal multipliers.
Sp. + W.								Sp. + W.							
100 + 0	,82929	100	—	100,00	—	100,00	1,0052	100 + 50	,90115	100	41,44	138,04	3,40	72,44	,7282
1	,83160	—	0,83	100,72	0,11	99,29	,9981	51	,90205	—	42,27	138,82	3,45	72,03	,7241
2	,83386	—	1,66	101,44	0,22	98,58	,9910	52	,90294	—	43,09	139,60	3,49	71,63	,7201
3	,83606	—	2,49	102,16	0,33	97,88	,9840	53	,90381	—	43,92	140,38	3,54	71,23	,7161
4	,83820	—	3,32	102,89	0,43	97,19	,9770	54	,90467	—	44,75	141,17	3,58	70,84	,7121
100 + 5	,84028	—	4,15	103,63	0,52	96,50	,9700	100 + 55	,90551	—	45,58	141,95	3,63	70,45	,7081
6	,84231	—	4,97	104,36	0,61	95,82	,9632	56	,90634	—	46,40	142,74	3,66	70,06	,7042
7	,84429	—	5,80	105,10	0,70	95,15	,9565	57	,90715	—	47,23	143,52	3,71	69,67	,7004
8	,84622	—	6,63	105,84	0,79	94,48	,9498	58	,90795	—	48,06	144,31	3,75	69,29	,6966
9	,84810	—	7,46	106,58	0,88	93,82	,9432	59	,90874	—	48,89	145,09	3,80	68,92	,6928
100 + 10	,84994	—	8,29	107,33	0,96	93,17	,9366	100 + 60	,90952	—	49,72	145,88	3,84	68,55	,6890
11	,85174	—	9,12	108,08	1,04	92,53	,9301	61	,91028	—	50,55	146,67	3,88	68,18	,6853
12	,85350	—	9,94	108,83	1,11	91,89	,9237	62	,91104	—	51,38	147,46	3,92	67,81	,6817
13	,85522	—	10,78	109,58	1,20	91,26	,9174	63	,91179	—	52,21	148,24	3,97	67,45	,6781
14	,85691	—	11,60	110,33	1,27	90,64	,9111	64	,91252	—	53,04	149,04	4,00	67,09	,6745
100 + 15	,85855	—	12,43	111,08	1,35	90,02	,9049	100 + 65	,91324	—	53,87	149,83	4,04	66,74	,6709
16	,86016	—	13,26	111,84	1,42	89,41	,8988	66	,91397	—	54,70	150,62	4,08	66,39	,6674
17	,86173	—	14,09	112,59	1,50	88,81	,8928	67	,91469	—	55,53	151,40	4,13	66,04	,6639
18	,86328	—	14,92	113,35	1,57	88,22	,8868	68	,91541	—	56,35	152,19	4,16	65,70	,6605
19	,86480	—	15,74	114,11	1,63	87,63	,8809	69	,91610	—	57,18	152,98	4,20	65,36	,6571
100 + 20	,86629	—	16,57	114,88	1,69	87,05	,8750	100 + 70	,91679	—	58,01	153,77	4,24	65,03	,6537
21	,86776	—	17,40	115,64	1,76	86,48	,8692	71	,91747	—	58,84	154,56	4,28	64,70	,6504
22	,86920	—	18,23	116,40	1,83	85,91	,8635	72	,91813	—	59,67	155,35	4,32	64,37	,6471
23	,87062	—	19,06	117,16	1,90	85,35	,8579	73	,91879	—	60,49	156,15	4,34	64,04	,6438
24	,87201	—	19,89	117,93	1,96	84,80	,8524	74	,91944	—	61,32	156,94	4,38	63,72	,6405
100 + 25	,87337	—	20,72	118,69	2,03	84,25	,8469	100 + 75	,92009	—	62,15	157,73	4,42	63,40	,6373
26	,87472	—	21,55	119,46	2,09	83,71	,8415	76	,92078	—	62,98	158,52	4,46	63,08	,6341
27	,87603	—	22,38	120,23	2,15	83,18	,8361	77	,92133	—	63,81	159,31	4,50	62,77	,6309
28	,87732	—	23,21	120,99	2,22	82,65	,8308	78	,92195	—	64,64	160,11	4,53	62,46	,6278
29	,87859	—	24,04	121,76	2,28	82,13	,8256	79	,92256	—	65,48	160,90	4,58	62,15	,6247
100 + 30	,87984	—	24,87	122,54	2,33	81,61	,8204	100 + 80	,92316	—	66,30	161,70	4,60	61,84	,6217
31	,88107	—	25,69	123,30	2,39	81,10	,8153	81	,92375	—	67,12	162,49	4,63	61,54	,6187
32	,88228	—	26,51	124,07	2,44	80,60	,8102	82	,92434	—	67,95	163,29	4,66	61,24	,6157
33	,88347	—	27,35	124,84	2,51	80,10	,8052	83	,92492	—	68,78	164,08	4,70	60,94	,6127
34	,88464	—	28,18	125,62	2,56	79,61	,8002	84	,92549	—	69,61	164,88	4,73	60,65	,6097
100 + 35	,88580	—	29,01	126,39	2,62	79,12	,7953	100 + 85	,92604	—	70,44	165,67	4,77	60,36	,6067
36	,88693	—	29,83	127,16	2,67	78,64	,7905	86	,92661	—	71,27	166,47	4,80	60,07	,6038
37	,88804	—	30,66	127,93	2,73	78,16	,7857	87	,92716	—	72,10	167,26	4,84	59,79	,6010
38	,88914	—	31,49	128,71	2,78	77,69	,7810	88	,92770	—	72,93	168,06	4,87	59,50	,5982
39	,89022	—	32,32	129,48	2,84	77,22	,7763	89	,92824	—	73,76	168,87	4,91	59,22	,5954
100 + 40	,89128	—	33,15	130,26	2,89	76,76	,7717	100 + 90	,92877	—	74,59	169,64	4,95	58,95	,5925
41	,89234	—	33,98	131,04	2,94	76,31	,7671	91	,92930	—	75,42	170,44	4,98	58,67	,5898
42	,89337	—	34,81	131,81	3,00	75,86	,7626	92	,92982	—	76,25	171,24	5,01	58,40	,5870
43	,89439	—	35,63	132,59	3,04	75,42	,7582	93	,93034	—	77,07	172,04	5,03	58,13	,5843
44	,89540	—	36,46	133,37	3,09	74,98	,7538	94	,93085	—	77,90	172,83	5,07	57,86	,5816
100 + 45	,89638	—	37,29	134,15	3,14	74,54	,7494	100 + 95	,93135	—	78,73	173,63	5,10	57,59	,5789
46	,89737	—	38,12	134,92	3,20	74,11	,7450	96	,93184	—	79,55	174,42	5,13	57,33	,5763
47	,89834	—	38,95	135,70	3,25	73,69	,7408	97	,93233	—	80,37	175,21	5,16	57,07	,5737
48	,89929	—	39,78	136,48	3,30	73,27	,7366	98	,93281	—	81,20	176,00	5,20	56,81	,5711
49	,90023	—	40,60	137,26	3,34	72,85	,7324	99	,93329	—	82,03	176,80	5,23	56,56	,5685

TABLE II. HEAT 51°.

I. Water and Spirit by weight.	II. Specific gravity.	III. Spirit by measure.	IV. Water by measure.	V. Bulk of mixture.	VI. Diminution of bulk.	VII. Quantity of spirit per cent.	VIII. Decimal multipliers.
W. + Sp.							
100+100	,93377	100	82,87	177,61	5,26	56,30	,5659
99	,93425	—	83,71	178,43	5,28	56,04	,5633
98	,93474	—	84,57	179,27	5,30	55,78	,5607
97	,93521	—	85,45	180,10	5,35	55,52	,5581
96	,93569	—	86,33	180,95	5,38	55,26	,5555
100 + 95	,93616	—	87,23	181,83	5,40	54,99	,5528
94	,93664	—	88,16	182,73	5,43	54,72	,5502
93	,93713	—	89,11	183,64	5,47	54,45	,5474
92	,93761	—	90,08	184,59	5,49	54,18	,5446
91	,93808	—	91,07	185,55	5,52	53,90	,5418
100 + 90	,93856	—	92,08	186,53	5,55	53,61	,5389
89	,93906	—	93,12	187,54	5,58	53,32	,5360
88	,93956	—	94,18	188,56	5,62	53,03	,5331
87	,94007	—	95,26	189,61	5,65	52,74	,5301
86	,94058	—	96,37	190,69	5,68	52,44	,5271
100 + 85	,94108	—	97,50	191,79	5,71	52,14	,5241
84	,94161	—	98,65	192,92	5,73	51,84	,5211
83	,94214	—	99,84	194,07	5,77	51,53	,5180
82	,94267	—	101,07	195,25	5,82	51,22	,5148
81	,94320	—	102,31	196,47	5,84	50,90	,5116
100 + 80	,94373	—	103,59	197,71	5,88	50,58	,5084
79	,94427	—	104,90	198,99	5,91	50,25	,5052
78	,94481	—	106,25	200,29	5,96	49,92	,5019
77	,94536	—	107,63	201,64	5,99	49,59	,4986
76	,94590	—	109,05	203,03	6,02	49,25	,4952
100 + 75	,94644	—	110,50	204,44	6,06	48,91	,4917
74	,94698	—	111,99	205,91	6,08	48,56	,4882
73	,94754	—	113,52	207,40	6,12	48,21	,4847
72	,94809	—	115,10	208,96	6,14	47,86	,4811
71	,94864	—	116,72	210,54	6,18	47,50	,4774
100 + 70	,94920	—	118,39	212,17	6,22	47,13	,4737
69	,94976	—	120,10	213,85	6,25	46,76	,4700
68	,95034	—	121,87	215,59	6,28	46,38	,4663
67	,95091	—	123,69	217,37	6,32	46,00	,4625
66	,95148	—	125,57	219,21	6,36	45,62	,4586
100 + 65	,95206	—	127,50	221,10	6,40	45,23	,4546
64	,95264	—	129,49	223,07	6,42	44,83	,4506
63	,95322	—	131,54	225,09	6,45	44,43	,4466
62	,95380	—	133,67	227,18	6,49	44,02	,4425
61	,95440	—	135,86	229,34	6,52	43,61	,4383
100 + 60	,95499	—	138,12	231,57	6,55	43,19	,4341
59	,95558	—	140,45	233,87	6,58	42,76	,4298
58	,95619	—	142,88	236,25	6,63	42,32	,4255
57	,95678	—	145,39	238,72	6,67	41,89	,4211
56	,95737	—	147,99	241,30	6,69	41,44	,4166
100 + 55	,95798	—	150,68	243,96	6,72	40,99	,4120
54	,95857	—	153,45	246,72	6,73	40,53	,4075
53	,95916	—	156,33	249,58	6,75	40,06	,4028
52	,95976	—	159,34	252,56	6,78	39,59	,3980
51	,96035	—	162,48	255,66	6,82	39,11	,3932
100 + 50	,96094	100	165,75	258,89	6,86	38,62	,3883
49	,96154	—	169,14	262,26	6,88	38,13	,3833
48	,96213	—	172,66	265,76	6,90	37,63	,3782
47	,96273	—	176,32	269,41	6,91	37,12	,3731
46	,96331	—	180,16	273,23	6,93	36,60	,3679
100 + 45	,96391	—	184,17	277,23	6,94	36,07	,3626
44	,96449	—	188,36	281,40	6,96	35,53	,3572
43	,96508	—	192,73	285,77	6,96	34,99	,3517
42	,96566	—	197,32	290,37	6,95	34,44	,3462
41	,96624	—	202,13	295,17	6,96	33,88	,3406
100 + 40	,96682	—	207,19	300,22	6,97	33,31	,3348
39	,96739	—	212,50	305,52	6,98	32,73	,3290
38	,96798	—	218,09	311,12	6,97	32,14	,3231
37	,96856	—	223,99	317,02	6,97	31,54	,3170
36	,96914	—	230,21	323,25	6,96	30,93	,3109
100 + 35	,96972	—	236,78	329,84	6,94	30,32	,3047
34	,97029	—	243,75	336,82	6,93	29,69	,2984
33	,97088	—	251,14	344,24	6,90	29,05	,2920
32	,97146	—	258,99	352,12	6,87	28,40	,2855
31	,97204	—	267,34	360,50	6,84	27,74	,2789
100 + 30	,97264	—	276,25	369,46	6,79	27,07	,2721
29	,97323	—	285,78	379,03	6,75	26,38	,2652
28	,97384	—	295,98	389,28	6,70	25,69	,2582
27	,97446	—	306,95	400,30	6,65	24,98	,2511
26	,97508	—	318,75	412,17	6,58	24,26	,2439
100 + 25	,97572	—	331,50	424,97	6,53	23,53	,2366
24	,97635	—	345,32	438,84	6,48	22,78	,2291
23	,97701	—	360,33	453,92	6,41	22,03	,2215
22	,97768	—	376,71	470,38	6,33	21,26	,2137
21	,97836	—	394,65	488,40	6,25	20,47	,2058
100 + 20	,97906	—	414,38	508,20	6,16	19,67	,1978
19	,97977	—	436,19	530,12	6,07	18,86	,1896
18	,98049	—	460,43	554,46	5,97	18,03	,1813
17	,98124	—	487,51	581,65	5,86	17,19	,1728
16	,98202	—	517,98	612,24	5,74	16,33	,1642
100 + 15	,98283	—	552,51	646,89	5,62	15,46	,1554
14	,98366	—	591,98	686,50	5,48	14,57	,1464
13	,98453	—	637,51	732,17	5,34	13,66	,1373
12	,98543	—	690,64	785,44	5,20	12,73	,1280
11	,98638	—	753,42	848,38	5,04	11,79	,1185
100 + 10	,98737	—	828,76	923,87	4,89	10,82	,1088
9	,98839	—	920,86	1016,15	4,71	9,84	,0989
8	,98948	—	1035,96	1131,44	4,52	8,84	,0888
7	,99063	—	1183,96	1279,62	4,34	7,81	,0785
6	,99184	—	1381,28	1477,14	4,14	6,77	,0680
100 + 5	,99310	—	1657,54	1753,61	3,93	5,70	,0573
4	,99443	—	2071,91	2168,21	3,70	4,61	,0464
3	,99585	—	2762,56	2859,06	3,50	3,50	,0352
2	,99737	—	4143,84	4240,54	3,30	2,36	,0237
1	,99896	—	8287,67	8384,55	3,12	1,19	,0120

I.	II.	III.	IV.	V.	VI.	VII.	VIII.	I.	II.	III.	IV.	V.	VI.	VII.	VIII.
Spirit and water by weight.	Specific gravity.	Spirit by measure.	Water by measure.	Bulk of mixture.	Diminution of bulk.	Quantity of spirit per cent.	Decimal multipliers.	Spirit and water by weight.	Specific gravity.	Spirit by measure.	Water by measure.	Bulk of mixture.	Diminution of bulk.	Quantity of spirit per cent.	Decimal multipliers.
Sp. + W.								Sp. + W.							
100 + 0	,82881	100	—	100,00	—	100,00	1,0046	100 + 50	,90069	100	41,41	138,03	3,38	72,45	,7278
1	,83112	—	0,83	100,72	0,11	99,29	,9975	51	,90160	—	42,24	138,81	3,43	72,04	,7237
2	,83337	—	1,66	101,44	0,22	98,58	,9904	52	,90248	—	43,07	139,59	3,48	71,64	,7197
3	,83558	—	2,49	102,16	0,33	97,88	,9834	53	,90336	—	43,90	140,37	3,53	71,24	,7157
4	,83771	—	3,32	102,89	0,43	97,19	,9764	54	,90421	—	44,73	141,16	3,57	70,84	,7118
100 + 5	,83980	—	4,15	103,63	0,52	96,50	,9695	100 + 55	,90505	—	45,56	141,94	3,62	70,45	,7078
6	,84183	—	4,97	104,36	0,61	95,82	,9627	56	,90588	—	46,38	142,73	3,65	70,06	,7039
7	,84381	—	5,80	105,10	0,70	95,15	,9560	57	,90669	—	47,21	143,51	3,70	69,68	,7001
8	,84574	—	6,63	105,84	0,79	94,48	,9493	58	,90749	—	48,04	144,30	3,74	69,30	,6962
9	,84762	—	7,45	106,58	0,87	93,82	,9427	59	,90828	—	48,87	145,08	3,79	68,92	,6924
100 + 10	,84946	—	8,28	107,33	0,95	93,17	,9361	100 + 60	,90906	—	49,69	145,87	3,82	68,55	,6887
11	,85126	—	9,11	108,08	1,03	92,53	,9296	61	,90982	—	50,52	146,66	3,86	68,18	,6850
12	,85302	—	9,94	108,82	1,12	91,89	,9232	62	,91059	—	51,36	147,45	3,91	67,82	,6814
13	,85474	—	10,77	109,57	1,20	91,26	,9169	63	,91134	—	52,18	148,23	3,95	67,46	,6778
14	,85643	—	11,60	110,33	1,27	90,64	,9106	64	,91206	—	53,01	149,03	3,98	67,10	,6741
100 + 15	,85808	—	12,43	111,08	1,35	90,02	,9044	100 + 65	,91279	—	53,84	149,82	4,02	66,75	,6706
16	,85969	—	13,25	111,83	1,42	89,42	,8983	66	,91352	—	54,67	150,61	4,06	66,40	,6671
17	,86126	—	14,08	112,59	1,49	88,82	,8923	67	,91424	—	55,50	151,39	4,11	66,05	,6636
18	,86280	—	14,91	113,35	1,56	88,22	,8863	68	,91496	—	56,32	152,18	4,14	65,71	,6602
19	,86433	—	15,73	114,11	1,62	87,63	,8804	69	,91566	—	57,15	152,97	4,18	65,37	,6568
100 + 20	,86582	—	16,56	114,87	1,69	87,05	,8746	100 + 70	,91634	—	57,98	153,76	4,22	65,03	,6534
21	,86729	—	17,39	115,63	1,76	86,48	,8688	71	,91703	—	58,81	154,55	4,26	64,70	,6501
22	,86873	—	18,22	116,39	1,83	85,92	,8631	72	,91770	—	59,64	155,34	4,30	64,37	,6468
23	,87015	—	19,05	117,15	1,90	85,36	,8575	73	,91836	—	60,46	156,13	4,33	64,04	,6435
24	,87154	—	19,88	117,92	1,96	84,81	,8520	74	,91901	—	61,29	156,92	4,37	63,73	,6402
100 + 25	,87291	—	20,71	118,69	2,02	84,26	,8465	100 + 75	,91966	—	62,12	157,71	4,41	63,41	,6370
26	,87425	—	21,54	119,45	2,09	83,72	,8410	76	,92028	—	62,95	158,50	4,45	63,09	,6338
27	,87556	—	22,37	120,22	2,15	83,18	,8357	77	,92090	—	63,78	159,29	4,49	62,78	,6306
28	,87685	—	23,20	120,99	2,21	82,65	,8304	78	,92153	—	64,61	160,09	4,52	62,47	,6275
29	,87812	—	24,02	121,76	2,26	82,13	,8252	79	,92214	—	65,44	160,88	4,56	62,16	,6245
100 + 30	,87937	—	24,86	122,53	2,33	81,62	,8199	100 + 80	,92274	—	66,27	161,68	4,59	61,85	,6214
31	,88060	—	25,68	123,30	2,38	81,11	,8148	81	,92333	—	67,09	162,47	4,62	61,55	,6184
32	,88181	—	26,50	124,07	2,43	80,60	,8097	82	,92392	—	67,92	163,27	4,65	61,25	,6154
33	,88300	—	27,33	124,84	2,49	80,10	,8047	83	,92450	—	68,75	164,06	4,69	60,95	,6124
34	,88418	—	28,16	125,61	2,55	79,61	,7998	84	,92507	—	69,57	164,86	4,71	60,66	,6094
100 + 35	,88533	—	28,99	126,39	2,60	79,13	,7949	100 + 85	,92562	—	70,41	165,65	4,76	60,37	,6065
36	,88647	—	29,82	127,16	2,66	78,65	,7901	86	,92619	—	71,24	166,45	4,79	60,08	,6036
37	,88758	—	30,65	127,92	2,73	78,17	,7853	87	,92674	—	72,06	167,24	4,82	59,80	,6007
38	,88868	—	31,48	128,70	2,78	77,70	,7806	88	,92728	—	72,89	168,03	4,86	59,51	,5979
39	,88976	—	32,30	129,47	2,83	77,23	,7759	89	,92782	—	73,72	168,83	4,89	59,23	,5951
100 + 40	,89083	—	33,13	130,25	2,88	76,77	,7713	100 + 90	,92835	—	74,55	169,62	4,93	58,95	,5922
41	,89188	—	33,96	131,03	2,93	76,32	,7667	91	,92887	—	75,38	170,42	4,96	58,68	,5895
42	,89292	—	34,79	131,80	2,99	75,87	,7622	92	,92939	—	76,21	171,22	4,99	58,40	,5867
43	,89394	—	35,61	132,58	3,03	75,42	,7578	93	,92991	—	77,03	172,02	5,01	58,13	,5840
44	,89495	—	36,44	133,36	3,08	74,98	,7534	94	,93042	—	77,86	172,81	5,05	57,86	,5813
100 + 45	,89593	—	37,27	134,14	3,13	74,55	,7490	100 + 95	,93092	—	78,69	173,61	5,08	57,59	,5787
46	,89692	—	38,10	134,91	3,19	74,12	,7446	96	,93141	—	79,51	174,40	5,11	57,34	,5760
47	,89789	—	38,93	135,69	3,24	73,70	,7404	97	,93191	—	80,33	175,19	5,14	57,08	,5734
48	,89834	—	39,76	136,47	3,29	73,28	,7362	98	,93239	—	81,16	175,98	5,18	56,82	,5708
49	,89978	—	40,58	137,25	3,33	72,86	,7320	99	,93287	—	81,99	176,78	5,21	56,56	,5682

TABLE II. HEAT 52°.

I. Water and spirit by weight.	II. Specific gravity.	III. Spirit by measure.	IV. Water by measure.	V. Bulk of mixture.	VI. Diminution of bulk.	VII. Quantity of spirit per cent.	VIII. Decimal multipliers.	I. Water and spirit by weight.	II. Specific gravity.	III. Spirit by measure.	IV. Water by measure.	V. Bulk of mixture.	VI. Diminution of bulk.	VII. Quantity of spirit per cent.	VIII. Decimal multipliers.
W. + Sp.								W. + Sp.							
100+100	,93334	100	82,83	177,59	5,24	56,30	,5657	100+50	,96062	100	165,66	258,83	6,83	38,63	,3881
99	,93383	—	83,67	178,41	5,26	56,05	,5631	49	,96122	—	169,05	262.20	6,85	38,14	,3832
98	,93432	—	84,53	179,24	5,29	55,79	,5605	48	,96182	—	172,57	265,69	6,88	37,63	,3781
97	,93479	—	85,41	180,08	5,33	55,53	,5579	47	,96242	—	176,23	269,34	6,89	37,13	,3730
96	,93528	—	86,29	180,93	5,36	55,27	,5553	46	,96301	—	180,06	273,16	6,90	36,61	,3678
100+95	,93575	—	87,19	181,80	5,39	55,00	,5526	100+45	,96362	—	184,07	277,16	6,91	36,08	,3625
94	,93623	—	88,12	182,70	5,42	54,73	,5499	44	,96420	—	188,26	281,32	6,94	35,54	,3571
93	,93672	—	89,06	183,62	5,44	54,46	,5471	43	,96479	—	192,63	285,69	6,94	35,00	,3516
92	,93720	—	90,03	184,56	5,47	54,19	,5443	42	,96538	—	197,21	290,28	6,93	34,45	,3461
91	,93768	—	91,02	185,53	5,49	53,91	,5415	41	,96597	—	202,02	295,08	6,94	33,89	,3405
100+90	,93816	—	92,03	186,50	5,53	53,62	,5387	100+40	,96655	—	207,08	300,12	6,96	33,32	,3347
89	,93866	—	93,07	187,51	5,57	53,33	,5358	39	,96713	—	212,39	305,43	6,96	32,74	,3289
88	,93916	—	94,13	188,53	5,60	53,04	,5329	38	,96772	—	217,97	311,02	6,95	32,15	,3230
87	,93967	—	95,21	189,58	5,63	52,75	,5299	37	,96831	—	223,87	316,93	6,94	31,55	,3169
86	,94018	—	96,32	190,66	5,66	52,45	,5269	36	,96890	—	230,09	323,15	6,94	30,94	,3108
100+85	,94068	—	97,45	191,76	5,69	52,15	,5239	100+35	,96949	—	236,66	329,73	6,93	30,33	,3047
84	,94121	—	98,60	192,89	5,71	51,85	,5209	34	,97006	—	243,62	336,71	6,91	29,70	,2984
83	,94173	—	99,79	194,04	5,75	51,54	,5178	33	,97065	—	251,01	344,12	6,89	29,06	,2920
82	,94226	—	101,01	195,22	5,79	51,22	,5146	32	,97124	—	258,85	352,00	6,85	28,41	,2855
81	,94280	—	102,26	196,44	5,82	50,91	,5114	31	,97183	—	267,20	360,37	6,83	27,75	,2788
100+80	,94333	—	103,54	197,68	5,86	50,59	,5082	100+30	,97244	—	276,10	369,32	6,78	27,08	,2720
79	,94387	—	104,85	198,96	5,89	50,26	,5050	29	,97303	—	285,63	378,89	6,74	26,39	,2651
78	,94441	—	106,20	200,26	5,94	49,93	,5017	28	,97364	—	295,82	389,13	6,69	25,70	,2581
77	,94496	—	107,57	201,61	5,96	49,59	,4984	27	,97427	—	306,79	400,14	6,65	24,99	,2510
76	,94550	—	108,99	202,99	6,00	49,26	,4950	26	,97490	—	318,58	412,00	6,58	24,27	,2438
100+75	,94604	—	110,44	204,41	6,03	48,92	,4915	100+25	,97554	—	331,33	424,80	6,53	23,54	,2365
74	,94659	—	111,93	205,87	6,06	48,57	,4880	24	,97618	—	345,14	438,66	6,48	22,79	,2290
73	,94715	—	113,46	207,37	6,09	48,22	,4845	23	,97684	—	360,14	453,73	6,41	22,04	,2214
72	,94770	—	115,04	208,92	6,12	47,86	,4809	22	,97752	—	376,51	470,18	6,33	21,27	,2136
71	,94825	—	116,66	210,51	6,15	47,50	,4773	21	,97821	—	394,44	488,19	6,25	20,48	,2058
100+70	,94881	—	118,33	212,14	6,19	47,14	,4736	100+20	,97892	—	414,16	508,00	6,16	19,68	,1978
69	,94938	—	120,04	213,82	6,22	46,76	,4699	19	,97963	—	435,96	529,88	6,08	18,87	,1896
68	,94996	—	121,80	215,55	6,25	46,39	,4661	18	,98036	—	460,19	554,21	5,98	18,04	,1813
67	,95053	—	123,62	217,33	6,29	46,01	,4623	17	,98112	—	487,25	581,38	5,87	17,20	,1728
66	,95110	—	125,50	219,17	6,33	45,62	,4584	16	,98190	—	517,71	611,95	5,76	16,34	,1612
100+65	,95169	—	127,43	221,06	6,37	45,23	,4544	100+15	,98272	—	552,22	646,58	5,64	15,47	,1554
64	,95227	—	129,42	223,02	6,40	44,83	,4505	14	,98355	—	591,67	686,17	5,50	14,57	,1464
63	,95287	—	131,47	225,04	6,43	44,43	,4464	13	,98443	—	637,18	731,81	5,37	13,66	,1373
62	,95344	—	133,60	227,13	6,47	44,02	,4423	12	,98534	—	698,20	785,06	5,22	12,74	,1280
61	,95404	—	135,79	229,29	6,50	43,61	,4382	11	,98629	—	753,03	847,96	5,07	11,80	,1185
100+60	,95463	—	138,05	231,52	6,53	43,19	,4339	100+10	,98729	—	828,33	923,41	4,92	10,85	,1088
59	,95523	—	140,38	233,82	6,56	42,77	,4297	9	,98831	—	920,38	1015,64	4,74	9,85	,0989
58	,95583	—	142,80	236,20	6,60	42,33	,4253	8	,98940	—	1035,42	1130,87	4,55	8,84	,0888
57	,95643	—	145,31	238,67	6,64	41,89	,4209	7	,99056	—	1183,34	1278,97	4,37	7,82	,0785
56	,95703	—	147,91	241,24	6,67	41,45	,4164	6	,99177	—	1380,56	1476,38	4,18	6,77	,0680
100+55	,95764	—	150,60	243,90	6,70	41,00	,4119	100+5	,99304	—	1657,30	1752,70	3,97	5,70	,0573
54	,95823	—	153,37	246,65	6,71	40,54	,4073	4	,99423	—	2070.83	2167,08	3,75	4,61	,0464
53	,9.883	—	156,25	249,52	6,73	40,07	,4026	3	,99560	—	2761,11	2857,65	3,53	3,50	,0352
52	,95943	—	159,26	252,50	6,76	39,60	,3979	2	,99731	—	4141,66	4238,31	3,35	2,36	,0237
51	,96002	—	162,39	255,60	6,79	33,12	,3930	1	,99890	—	8283,32	8380,15	3,17	1,19	,0120

TABLE I.   HEAT 53°.

I.	II.	III.	IV.	V.	VI.	VII.	VIII.	I.	II.	III.	IV.	V.	VI.	VII.	VIII.
Spirit and water by weight.	Specific gravity.	Spirit by measure.	Water by measure.	Bulk of mixture.	Diminution of bulk.	Quantity of spirit per cent.	Decimal multipliers.	Spirit and water by weight.	Specific gravity.	Spirit by measure.	Water by measure.	Bulk of mixture.	Diminution of bulk.	Quantity of spirit per cent.	Decimal multipliers.
Sp. + W.								Sp. + W.							
100 + 0	,82833	100	—	100,00	—	100,00	1,0040	100 + 50	,90024	100	41,39	138,02	3,37	72,46	,7275
1	,83063	—	0,83	100,72	0,11	99,29	,9969	51	,90114	—	42,22	138,80	3,42	72,05	,7234
2	,83288	—	1,66	101,44	0,22	98,58	,9898	52	,90203	—	43,05	139,58	3,47	71,64	,7194
3	,83509	—	2,49	102,16	0,33	97,88	,9828	53	,90290	—	43,87	140,36	3,51	71,24	,7154
4	,83723	—	3,32	102,89	0,43	97,19	,9758	54	,90375	—	44,70	141,15	3,55	70,85	,7114
100 + 5	,83932	—	4,14	103,63	0,51	96,50	,9689	100 + 55	,90459	—	45,53	141,93	3,60	70,46	,7074
6	,84135	—	4,97	104,36	0,61	95,82	,9622	56	,90542	—	46,35	142,72	3,63	70,07	,7036
7	,84332	—	5,80	105,10	0,70	95,15	,9555	57	,90623	—	47,18	143,50	3,68	69,68	,6997
8	,84526	—	6,63	105,84	0,79	94,48	,9488	58	,90703	—	48,01	144,29	3,72	69,30	,6958
9	,84714	—	7,45	106,58	0,87	93,82	,9422	59	,90782	—	48,84	145,07	3,77	68,93	,6921
100 + 10	,84898	—	8,28	107,33	0,95	93,17	,9355	100 + 60	,90860	—	49,67	145,86	3,81	68,56	,6883
11	,85078	—	9,11	108,08	1,03	92,53	,9291	61	,90937	—	50,50	146,65	3,85	68,19	,6846
12	,85254	—	9,93	108,82	1,11	91,90	,9227	62	,91013	—	51,33	147,44	3,89	67,82	,6811
13	,85426	—	10,76	109,57	1,19	91,27	,9164	63	,91088	—	52,15	148,22	3,93	67,46	,6775
14	,85595	—	11,59	110,33	1,26	90,65	,9101	64	,91161	—	52,98	149,01	3,97	67,11	,6738
100 + 15	,85760	—	12,42	111,08	1,34	90,03	,9039	100 + 65	,91234	—	53,81	149,80	4,01	66,76	,6702
16	,85921	—	13,25	111,83	1,42	89,42	,8978	66	,91307	—	54,64	150,59	4,05	66,41	,6667
17	,86078	—	14,07	112,59	1,48	88,82	,8918	67	,91379	—	55,47	151,37	4,10	66,06	,6633
18	,86233	—	14,90	113,35	1,55	88,22	,8858	68	,91451	—	56,29	152,16	4,13	65,72	,6599
19	,86386	—	15,72	114,10	1,62	87,63	,8799	69	,91522	—	57,12	152,95	4,17	65,38	,6565
100 + 20	,86535	—	16,56	114,87	1,69	87,05	,8741	100 + 70	,91590	—	57,95	153,74	4,21	65,04	,6531
21	,86682	—	17,39	115,63	1,76	86,48	,8684	71	,91659	—	58,78	154,53	4,25	64,71	,6498
22	,86826	—	18,21	116,39	1,82	85,92	,8627	72	,91726	—	59,61	155,32	4,29	64,38	,6465
23	,86968	—	19,04	117,15	1,89	85,36	,8571	73	,91793	—	60,43	156,11	4,32	64,05	,6432
24	,87107	—	19,87	117,91	1,96	84,81	,8516	74	,91858	—	61,26	156,90	4,36	63,73	,6399
100 + 25	,87244	—	20,70	118,68	2,02	84,26	,8460	100 + 75	,91923	—	62,09	157,69	4,40	63,41	,6367
26	,87378	—	21,52	119,44	2,08	83,72	,8406	76	,91986	—	62,91	158,48	4,43	63,09	,6335
27	,87509	—	22,35	120,21	2,14	83,18	,8352	77	,92048	—	63,75	159,27	4,48	62,78	,6303
28	,87638	—	23,19	120,98	2,21	82,66	,8299	78	,92110	—	64,57	160,08	4,49	62,47	,6272
29	,87765	—	24,01	121,75	2,26	82,13	,8247	79	,92171	—	65,41	160,86	4,55	62,16	,6242
100 + 30	,87890	—	24,84	122,52	2,32	81,62	,8195	100 + 80	,92231	—	66,23	161,66	4,57	61,86	,6211
31	,88013	—	25,67	123,29	2,38	81,11	,8144	81	,92290	—	67,05	162,46	4,59	61,55	,6181
32	,88134	—	26,48	124,06	2,42	80,61	,8093	82	,92349	—	67,88	163,25	4,63	61,25	,6151
33	,88254	—	27,32	124,83	2,49	80,11	,8043	83	,92407	—	68,71	164,04	4,67	60,96	,6121
34	,88371	—	28,15	125,60	2,55	79,62	,7994	84	,92464	—	69,53	164,84	4,69	60,66	,6091
100 + 35	,88486	—	28,98	126,38	2,60	79,13	,7945	100 + 85	,92520	—	70,37	165,63	4,74	60,37	,6062
36	,88600	—	29,80	127,15	2,65	78,65	,7897	86	,92577	—	71,20	166,42	4,78	60,09	,6033
37	,88712	—	30,63	127,92	2,71	78,17	,7849	87	,92632	—	72,02	167,22	4,80	59,81	,6004
38	,88822	—	31,46	128,70	2,76	77,70	,7802	88	,92686	—	72,85	168,01	4,84	59,52	,5976
39	,88930	—	32,28	129,47	2,81	77,23	,7755	89	,92740	—	73,68	168,81	4,87	59,24	,5948
100 + 40	,89037	—	33,12	130,25	2,87	76,78	,7709	100 + 90	,92793	—	74,51	169,60	4,91	58,96	,5920
41	,89142	—	33,95	131,02	2,93	76,32	,7663	91	,92844	—	75,34	170,40	4,94	58,68	,5892
42	,89246	—	34,78	131,79	2,99	75,87	,7618	92	,92897	—	76,17	171,20	4,97	58,41	,5865
43	,89349	—	35,60	132,57	3,03	75,43	,7574	93	,92948	—	76,99	172,00	4,99	58,14	,5838
44	,89450	—	36,43	133,35	3,08	74,99	,7530	94	,92999	—	77,82	172,79	5,03	57,87	,5811
100 + 45	,89548	—	37,26	134,13	3,13	74,55	,7486	100 + 95	,93049	—	78,65	173,59	5,06	57,61	,5784
46	,89647	—	38,08	134,90	3,18	74,13	,7443	96	,93098	—	79,47	174,38	5,09	57,35	,5758
47	,89744	—	38,91	135,68	3,23	73,70	,7400	97	,93148	—	80,29	175,17	5,12	57,09	,5732
48	,89839	—	39,74	136,46	3,28	73,28	,7358	98	,93197	—	81,12	175,96	5,16	56,83	,5706
49	,89933	—	40,56	137,24	3,32	72,86	,7316	99	,93244	—	81,95	176,76	5,19	56,57	,5680

I. Water and spirit by weight.	II. Specific gravity.	III. Spirit by measure.	IV. Water by measure.	V. Bulk of mixture.	VI. Diminution of bulk.	VII. Quantity of spirit per cent.	VIII. Decimal multipliers.	I. Water and spirit by weight.	II. Specific gravity.	III. Spirit by measure.	IV. Water by measure.	V. Bulk of mixture.	VI. Diminution of bulk.	VII. Quantity of spirit per cent.	VIII. Decimal multipliers.
W. + Sp.								W. + Sp.							
100+100	,93291	100	82,78	177,57	5,21	56,31	,5654	100 + 50	,96030	100	165,57	258,76	6,81	38,64	,3880
99	,93341	—	83,63	178,38	5,25	56,05	,5628	49	,96091	—	168,96	262,14	6,82	38,14	,3831
98	,93390	—	84,49	179,22	5,27	55,79	,5602	48	,96151	—	172,48	265,62	6,86	37,64	,3780
97	,93438	—	85,36	180,06	5,30	55,53	,5576	47	,96211	—	176,14	269,27	6,87	37,14	,3729
96	,93486	—	86,24	180,91	5,33	55,27	,5550	46	,96271	—	179,97	273,08	6,89	36,62	,3677
100 + 95	,93534	—	87,14	181,78	5,36	55,01	,5523	100 + 45	,96332	—	183,97	277,08	6,89	36,09	,3624
94	,93582	—	88,08	182,68	5,40	54,74	,5496	44	,96391	—	188,16	281,24	6,92	35,55	,3570
93	,93631	—	89,01	183,59	5,42	54,47	,5469	43	,96450	—	192,53	285,60	6,93	35,01	,3515
92	,93679	—	89,98	184,53	5,45	54,19	,5441	42	,96510	—	197,11	290,19	6,92	34,46	,3460
91	,93728	—	90,97	185,50	5,47	53,91	,5413	41	,96570	—	201,92	294,99	6,93	33,90	,3404
100 + 90	,93776	—	91,98	186,47	5,51	53,62	,5384	100 + 40	,96628	—	206,97	300,03	6,94	33,33	,3346
89	,93826	—	93,02	187,48	5,54	53,33	,5356	39	,96687	—	212,28	305,34	6,94	32,75	,3288
88	,93876	—	94,08	188,50	5,58	53,05	,5327	38	,96747	—	217,86	310,93	6,93	32,16	,3229
87	,93927	—	95,16	189,55	5,61	52,75	,5297	37	,96806	—	223,75	316,83	6,92	31,56	,3169
86	,93978	—	96,27	190,63	5,64	52,45	,5267	36	,96866	—	229,97	323,05	6,92	30,95	,3108
100 + 85	,94028	—	97,40	191,73	5,67	52,15	,5237	100 + 35	,96925	—	236,54	329,62	6,92	30,34	,3046
84	,94081	—	98,55	192,86	5,69	51,85	,5207	34	,96983	—	243,50	336,59	6,91	29,71	,2983
83	,94133	—	99,74	194,01	5,73	51,55	,5176	33	,97043	—	250,88	344,00	6,88	29,07	,2919
82	,94186	—	100,96	195,19	5,77	51,23	,5144	32	,97102	—	258,71	351,87	6,84	28,42	,2854
81	,94240	—	102,20	196,41	5,79	50,91	,5112	31	,97162	—	267,06	360,24	6,82	27,76	,2788
100 + 80	,94293	—	103,48	197,65	5,83	50,59	,5080	100 + 30	,97223	—	275,96	369,19	6,77	27,09	,2719
79	,94347	—	104,79	198,92	5,87	50,27	,5048	29	,97283	—	285,48	378,75	6,73	26,40	,2651
78	,94401	—	106,14	200,23	5,91	49,94	,5015	28	,97345	—	295,67	388,98	6,69	25,71	,2581
77	,94456	—	107,51	201,58	5,93	49,60	,4982	27	,97408	—	306,63	399,98	6,65	25,00	,2510
76	,94510	—	108,93	202,96	5,97	49,27	,4948	26	,97472	—	318,41	411,83	6,58	24,28	,2438
100 + 75	,94564	—	110,38	204,38	6,00	48,93	,4913	100 + 25	,97536	—	331,16	424,63	6,53	23,55	,2364
74	,94620	—	111,87	205,83	6,04	48,58	,4878	24	,97601	—	344,96	438,48	6,48	22,80	,2290
73	,94676	—	113,40	207,34	6,06	48,23	,4843	23	,97668	—	359,95	453,54	6,41	22,05	,2214
72	,94731	—	114,98	208,89	6,09	47,87	,4807	22	,97736	—	376,31	469,98	6,33	21,28	,2136
71	,94787	—	116,60	210,47	6,13	47,51	,4771	21	,97806	—	394,23	487,98	6,25	20,49	,2057
100 + 70	,94843	—	118,27	212,11	6,16	47,14	,4734	100 + 20	,97877	—	413,94	507,78	6,18	19,69	,1977
69	,94900	—	119,96	213,78	6,20	46,77	,4697	19	,97949	—	435,73	529,65	6,08	18,88	,1895
68	,94958	—	121,74	215,51	6,24	46,40	,4659	18	,98023	—	459,95	553,96	5,99	18,05	,1812
67	,95015	—	123,56	217,29	6,27	46,02	,4621	17	,98100	—	487,00	581,12	5,88	17,21	,1727
66	,95073	—	125,44	219,13	6,31	45,63	,4582	16	,98179	—	517,44	611,66	5,78	16,35	,1641
100 + 65	,95131	—	127,37	221,02	6,35	45,24	,4543	100 + 15	,98261	—	551,93	646,28	5,65	15,48	,1553
64	,95190	—	129,35	222,98	6,37	44,84	,4503	14	,98345	—	591,36	685,84	5,52	14,58	,1463
63	,95249	—	131,40	225,00	6,40	44,44	,4462	13	,98433	—	636,85	731,46	5,39	13,67	,1372
62	,95308	—	133,53	227,09	6,44	44,03	,4421	12	,98525	—	689,92	784,68	5,24	12,74	,1280
61	,95368	—	135,72	229,25	6,47	43,62	,4380	11	,98620	—	752,64	847,54	5,10	11,80	,1185
100 + 60	,95428	—	137,98	231,47	6,51	43,20	,4338	100 + 10	,98720	—	827,90	922,95	4,95	10,84	,1088
59	,95488	—	140,31	233,77	6,54	42,77	,4295	9	,98823	—	919,90	1015,13	4,77	9,85	,0989
58	,95549	—	142,72	236,15	6,57	42,34	,4251	8	,98933	—	1034,88	1130,30	4,58	8,84	,0888
57	,95609	—	145,24	238,62	6,62	41,90	,4207	7	,99049	—	1182,72	1278,32	4,40	7,82	,0786
56	,95669	—	147,83	241,19	6,64	41,46	,4163	6	,99170	—	1379,84	1475,62	4,22	6,78	,0680
100 + 55	,95728	—	150,52	243,85	6,67	41,01	,4117	100 + 5	,99298	—	1655,81	1751,79	4,20	5,71	,0573
54	,95790	—	153,29	246,60	6,69	40,55	,4071	4	,99431	—	2009,75	2165,96	3,79	4,62	,0464
53	,95850	—	156,17	249,46	6,71	40,08	,4025	3	,99574	—	2759,66	2856,06	3,60	3,50	,0352
52	,95910	—	159,18	252,44	6,74	39,61	,3978	2	,99725	—	4139,50	4236,09	3,41	2,36	,0237
51	,95970	—	162,31	255,54	6,77	39,13	,3929	1	,99884	—	8279,00	8375,77	3,23	1,19	,0120

TABLE I.   HEAT 54°.

I. Spirit and water by weight.	II. Specific gravity.	III. Spirit by measure.	IV. Water by measure.	V. Bulk of mixture.	VI. Diminution of bulk.	VII. Quantity of spirit per cent.	VIII. Decimal multipliers.
Sp. + W.							
100 + 0	,82784	100	—	100,00	—	100,00	1,0034
1	,83015	—	0,83	100,72	0,11	99,29	,9963
2	,83240	—	1,66	101,44	0,22	98,58	,9892
3	,83460	—	2,49	102,16	0,33	97,88	,9822
4	,83674	—	3,31	102,89	0,42	97,19	,9752
100 + 5	,83883	—	4,14	103,63	0,51	96,50	,9683
6	,84086	—	4,96	104,36	0,60	95,82	,9616
7	,84283	—	5,79	105,10	0,69	95,15	,9549
8	,84477	—	6,62	105,84	0,78	94,48	,9482
9	,84666	—	7,45	106,58	0,87	93,83	,9416
100 + 10	,84850	—	8,28	107,32	0,96	93,18	,9350
11	,85030	—	9,11	108,07	1,04	92,54	,9286
12	,85206	—	9,93	108,82	1,11	91,90	,9222
13	,85378	—	10,76	109,57	1,19	91,27	,9159
14	,85547	—	11,59	110,32	1,27	90,65	,9096
100 + 15	,85712	—	12,41	111,08	1,33	90,03	,9034
16	,85873	—	13,24	111,83	1,41	89,42	,8973
17	,86031	—	14,07	112,58	1,49	88,82	,8913
18	,86186	—	14,90	113,34	1,56	88,23	,8853
19	,86339	—	15,71	114,10	1,61	87,64	,8794
100 + 20	,86488	—	16,55	114,86	1,69	87,06	,8736
21	,86635	—	17,38	115,62	1,76	86,49	,8679
22	,86779	—	18,20	116,38	1,82	85,92	,8622
23	,86921	—	19,03	117,14	1,89	85,36	,8566
24	,87060	—	19,86	117,91	1,95	84,81	,8511
100 + 25	,87197	—	20,69	118,68	2,01	84,26	,8455
26	,87331	—	21,51	119,44	2,07	83,72	,8401
27	,87462	—	22,34	120,21	2,13	83,19	,8348
28	,87591	—	23,18	120,98	2,20	82,66	,8295
29	,87718	—	23,99	121,75	2,24	82,14	,8243
100 + 30	,87843	—	24,83	122,52	2,31	81,62	,8190
31	,87966	—	25,65	123,29	2,36	81,11	,8139
32	,88087	—	26,47	124,06	2,41	80,61	,8088
33	,88207	—	27,30	124,83	2,47	80,11	,8039
34	,88324	—	28,14	125,60	2,54	79,62	,7990
100 + 35	,88439	—	28,96	126,37	2,59	79,13	,7941
36	,88554	—	29,79	127,14	2,65	78,65	,7893
37	,88666	—	30,62	127,91	2,71	78,18	,7845
38	,88776	—	31,45	128,69	2,76	77,71	,7798
39	,88884	—	32,26	129,46	2,80	77,24	,7751
100 + 40	,88991	—	33,10	130,24	2,86	76,78	,7705
41	,89097	—	33,93	131,01	2,92	76,33	,7659
42	,89201	—	34,76	131,78	2,98	75,88	,7614
43	,89304	—	35,58	132,56	3,02	75,43	,7570
44	,89405	—	36,41	133,34	3,07	74,99	,7526
100 + 45	,89503	—	37,24	134,12	3,12	74,56	,7482
46	,89602	—	38,06	134,89	3,17	74,13	,7438
47	,89699	—	38,89	135,67	3,22	73,71	,7396
48	,89794	—	39,72	136,45	3,27	73,29	,7354
49	,89887	—	40,54	137,23	3,31	72,87	,7312
100 + 50	,89978	100	41,37	138,01	3,36	72,46	,7271
51	,90068	—	42,20	138,79	3,41	72,05	,7230
52	,90157	—	43,03	139,57	3,46	71,65	,7190
53	,90244	—	43,85	140,35	3,50	71,25	,7150
54	,90329	—	44,68	141,14	3,54	70,85	,7110
100 + 55	,90413	—	45,51	141,92	3,59	70,46	,7071
56	,90496	—	46,33	142,71	3,62	70,07	,7032
57	,90577	—	47,16	143,49	3,67	69,69	,6993
58	,90657	—	47,99	144,28	3,71	69,31	,6955
59	,90736	—	48,81	145,06	3,75	68,93	,6917
100 + 60	,90814	—	49,64	145,85	3,79	68,56	,6880
61	,90891	—	50,47	146,64	3,83	68,20	,6843
62	,90967	—	51,30	147,43	3,87	67,83	,6807
63	,91042	—	52,12	148,21	3,91	67,47	,6771
64	,91115	—	52,95	149,00	3,95	67,11	,6734
100 + 65	,91189	—	53,78	149,79	3,99	66,76	,6699
66	,91262	—	54,61	150,58	4,03	66,41	,6664
67	,91334	—	55,44	151,36	4,08	66,06	,6630
68	,91406	—	56,26	152,15	4,11	65,72	,6596
69	,91477	—	57,09	152,94	4,15	65,38	,6562
100 + 70	,91546	—	57,92	153,73	4,19	65,05	,6528
71	,91615	—	58,75	154,52	4,23	64,72	,6495
72	,91682	—	59,58	155,31	4,27	64,39	,6462
73	,91750	—	60,40	156,09	4,31	64,06	,6429
74	,91815	—	61,23	156,88	4,35	63,74	,6396
100 + 75	,91880	—	62,06	157,68	4,38	63,42	,6364
76	,91944	—	62,88	158,47	4,41	63,10	,6332
77	,92006	—	63,71	159,25	4,46	62,79	,6306
78	,92067	—	64,54	160,06	4,48	62,48	,6269
79	,92128	—	65,37	160,84	4,53	62,17	,6239
100 + 80	,92189	—	66,20	161,64	4,56	61,86	,6208
81	,92247	—	67,02	162,44	4,58	61,56	,6178
82	,92307	—	67,85	163,23	4,62	61,26	,6148
83	,92365	—	68,67	164,02	4,65	60,96	,6118
84	,92421	—	69,50	164,82	4,68	60,67	,6088
100 + 85	,92478	—	70,33	165,61	4,72	60,38	,6059
86	,92535	—	71,17	166,40	4,77	60,09	,6030
87	,92590	—	71,98	167,20	4,78	59,81	,6001
88	,92644	—	72,81	167,99	4,82	59,52	,5973
89	,92797	—	73,64	168,79	4,85	59,24	,5945
100 + 90	,92751	—	74,47	169,58	4,89	58,96	,5917
91	,92802	—	75,30	170,38	4,92	58,69	,5889
92	,92854	—	76,13	171,18	4,95	58,42	,5862
93	,92905	—	76,95	171,98	4,97	58,15	,5835
94	,92956	—	77,78	172,77	5,01	57,88	,5808
100 + 95	,93006	—	78,61	173,57	5,04	57,61	,5781
96	,93055	—	79,43	174,36	5,07	57,35	,5755
97	,93105	—	80,25	175,15	5,10	57,09	,5729
98	,93154	—	81,08	175,94	5,14	56,83	,5703
99	,93202	—	81,91	176,74	5,17	56,58	,5677

TABLE II.   HEAT 54°.

I. Water and spirit by weight.	II. Specific gravity.	III. Spirit by measure.	IV. Water by measure.	V. Bulk of mixture.	VI. Diminution of bulk.	VII. Quantity of spirit per cent.	VIII. Decimal multipliers.	I. Water and spirit by weight.	II. Specific gravity.	III. Spirit by measure.	IV. Water by measure.	V. Bulk of mixture.	VI. Diminution of bulk.	VII. Quantity of spirit per cent.	VIII. Decimal multipliers.
W. + Sp.								W. + Sp.							
100 + 100	,93249	100	82,74	177,55	5,19	56,32	,5652	100 + 50	,95998	100	165,49	258,70	6,79	38,65	,3879
99	,93299	—	83,59	178,36	5,23	56,06	,5626	49	,96059	—	168,87	262,08	6,79	38,15	,3830
98	,93348	—	84,44	179,19	5,25	55,80	,5600	48	,96120	—	172,39	265,55	6,84	37,65	,3779
97	,93396	—	85,32	180,04	5,28	55,54	,5574	47	,96181	—	176,05	269,20	6,85	37,15	,3728
96	,93445	—	86,20	180,89	5,31	55,28	,5548	46	,96241	—	179,88	273,01	6,87	36,63	,3676
100 + 95	,93493	—	87,10	181,76	5,34	55,02	,5521	100 + 45	,96302	—	183,88	277,00	6,88	36,10	,3623
94	,93541	—	88,03	182,66	5,37	54,75	,5494	44	,96362	—	188,06	281,16	6,90	35,56	,3569
93	,93590	—	88,96	183,57	5,39	54,47	,5466	43	,96422	—	192,43	285,52	6,91	35,02	,3514
92	,93639	—	89,93	184,50	5,43	54,20	,5439	42	,96482	—	197,01	290,10	6,91	34,47	,3459
91	,93688	—	90,92	185,47	5,45	53,92	,5410	41	,96543	—	201,82	294,90	6,92	33,91	,3403
100 + 90	,93736	—	91,93	186,44	5,49	53,63	,5382	100 + 40	,96601	—	206,86	299,94	6,92	33,34	,3345
89	,93786	—	92,97	187,45	5,52	53,34	,5353	39	,96661	—	212,17	305,25	6,92	32,76	,3287
88	,93836	—	94,03	188,47	5,56	53,05	,5324	38	,96722	—	217,75	310,83	6,92	32,17	,3228
87	,93887	—	95,11	189,52	5,59	52,76	,5295	37	,96781	—	223,64	316,74	6,90	31,57	,3168
86	,93938	—	96,22	190,60	5,62	52,46	,5265	36	,96842	—	229,85	322,95	6,90	30,96	,3107
100 + 85	,93988	—	97,35	191,70	5,65	52,10	,5235	100 + 35	,96901	—	236,42	329,51	6,91	30,35	,3045
84	,94041	—	98,50	192,83	5,67	51,86	,5205	34	,96960	—	243,37	336,48	6,89	29,72	,2982
83	,94092	—	99,69	193,98	5,71	51,55	,5174	33	,97021	—	250,75	343,89	6,86	29,08	,2918
82	,94146	—	100,91	195,16	5,75	51,24	,5142	32	,97081	—	258,58	351,75	6,83	28,43	,2853
81	,94200	—	102.15	196,38	5,77	50,92	,5110	31	,97141	—	266,92	360,11	6,81	27,77	,2787
100 + 80	,94253	—	103,43	197,02	5,81	50,60	,5078	100 + 30	,97202	—	275,82	369,05	6,77	27,10	,2719
79	,94307	—	104,74	198,89	5,85	50,28	,5046	29	,97263	—	285,33	378,60	6,73	26,41	,2650
78	,94361	—	106,08	200,20	5,88	49,95	,5013	28	,97326	—	295,52	388,83	6,69	25,72	,2581
77	,94416	—	107,46	201,55	5,91	49,61	,4980	27	,97389	—	306,47	399,82	6,65	25,01	,2509
76	,94471	—	108.87	202 93	5,94	49,28	,4946	26	,97454	—	318,25	411,66	6,59	24,29	,2437
100 + 75	,94525	—	110,32	204,35	5,97	48,93	,4910	100 + 25	,97518	—	330,99	424,46	6,53	23,56	,2364
74	,94581	—	111,81	205,80	6,01	48,59	,4876	24	,97584	—	344,78	438,30	6,48	22,81	,2289
73	,94637	—	113,34	207,30	6,04	48,24	,4841	23	,97652	—	359,76	453,36	6,40	22,06	,2213
72	,94693	—	114,92	208,85	6,07	47,88	,4805	22	,97720	—	376,11	469,78	6,33	21,29	,2135
71	,94749	—	116,54	210,44	6,10	47,52	,4769	21	,97791	—	394,03	487,77	6,26	20,50	,2057
100 + 70	,94805	—	118,21	212,07	6,14	47,15	,4732	100 + 20	,97862	—	413,73	507,56	6,17	19,70	,1977
69	,94862	—	119,92	213,74	6,18	46,78	,4695	19	,97935	—	435,51	529,42	6,09	18,89	,1895
68	,94920	—	121,68	215,47	6,21	46,41	,4657	18	,98010	—	459,71	553,71	6,00	18,06	,1812
67	,94978	—	123,50	217,25	6,25	46,03	,4619	17	,98088	—	486,75	580,86	5,89	17,22	,1727
66	,95036	—	125,37	219 09	6,28	45,64	,4580	16	,98168	—	517,17	611,38	5,79	16,36	,1641
100 + 65	,95094	—	127,31	220,98	6,33	45,25	,4541	100 + 15	,98250	—	551,64	645,98	5,66	15,48	,1553
64	,95153	—	129,28	222.94	6,34	44,85	,4501	14	,98335	—	591,05	685,51	5,54	14,59	,1463
63	,95213	—	131,33	224,95	6,38	44,45	,4461	13	,98421	—	636,52	731,11	5,41	13,68	,1372
62	,95272	—	133,46	227,04	6,42	44,04	,4420	12	,98516	—	689,56	784,30	5,26	12,75	,1279
61	,95332	—	135,65	229,20	6,45	43,63	,4378	11	,98611	—	752,25	847,13	5,12	11,81	,1184
100 + 60	,95393	—	137,91	231,42	6,49	43,21	,4336	100 + 10	,98711	—	827,47	922,49	4,98	10,84	,1088
59	,95453	—	140,24	233,72	6,52	42,78	,4293	9	,98815	—	919,42	1014,62	4,80	9,86	,0989
58	,95513	—	142,65	236,10	6,55	42,35	,4250	8	,98920	—	1034,34	1129,73	4,61	8,88	,0888
57	,95575	—	145,16	238,57	6,59	41,91	,4206	7	,99042	—	1182,10	1277,67	4,43	7,83	,0785
56	,95636	—	147,76	241,14	6,62	41,47	,4162	6	,99168	—	1379,12	1474,89	4,12	6,78	,0680
100 + 55	,95696	—	150,44	243,80	6,64	41,02	,4116	100 + 5	,99291	—	1654,95	1750,89	4,06	5,71	,0573
54	,95757	—	153,21	246,54	6,67	40,56	,4070	4	,99425	—	2068,67	2164,83	3,84	4,62	,0463
53	,95817	—	156,09	249,40	6,69	40,09	,4024	3	,99568	—	2758,23	2854,58	3,65	3,50	,0351
52	,95877	—	159,10	252,38	6,72	39,62	,3977	2	,99719	—	4137,35	4233,89	3,46	2,36	,0237
51	,95938	—	162,23	255,48	6,75	39,14	,3928	1	,99878	—	8274,71	8371,43	3,28	1,19	,0120

TABLE I.   HEAT 55°.

I.	II.	III.	IV.	V.	VI.	VII.	VIII.	I.	II.	III.	IV.	V.	VI.	VII.	VIII.
Spirit and water by weight.	Specific gravity.	Spirit by measure.	Water by measure.	Bulk of mixture.	Diminution of bulk.	Quantity of spirit per cent.	Decimal multipliers.	Spirit and water by weight.	Specific gravity.	Spirit by measure.	Water by measure.	Bulk of mixture.	Diminution of Bulk.	Quantity of spirit per cent.	Decimal multipliers.
Sp. + W.								Sp. + W.							
100 + 0	,82736	100	—	100,00	—	100,00	1,0029	100 + 50	,89933	100	41,35	138,00	3,35	72,47	,7267
1	,82967	—	0,83	100,72	0,11	99,29	,9957	51	,90023	—	42,18	138,78	3,40	72,06	,7226
2	,83192	—	1,65	101,44	0,21	98,58	,9886	52	,90111	—	43,01	139,56	3,45	71,65	,7186
3	,83412	—	2,48	102,16	0,32	97,88	,9816	53	,90198	—	43,83	140,34	3,49	71,25	,7146
4	,83626	—	3,31	102,89	0,42	97,19	,9747	54	,90283	—	44,66	141,13	3,53	70,86	,7106
100 + 5	,83834	—	4,13	103,62	0,51	96,50	,9678	100 + 55	,90367	—	45,49	141,91	3,58	70,47	,7067
6	,84037	—	4,96	104,36	0,60	95,82	,9610	56	,90450	—	46,31	142,70	3,61	70,08	,7028
7	,84235	—	5,79	105,10	0,69	95,15	,9543	57	,90531	—	47,14	143,48	3,66	69,69	,6989
8	,84429	—	6,62	105,84	0,78	94,49	,9476	58	,90611	—	47,97	144,27	3,70	69,31	,6951
9	,84618	—	7,44	106,58	0,86	93,83	,9410	59	,90690	—	48,79	145,05	3,74	68,94	,6913
100 + 10	,84802	—	8,27	107,32	0,95	93,18	,9345	100 + 60	,90768	—	49,62	145,84	3,78	68,57	,6876
11	,84982	—	9,10	108,07	1,03	92,54	,9280	61	,90845	—	50,45	146,63	3,82	68,20	,6839
12	,85158	—	9,92	108,81	1,11	91,90	,9216	62	,90921	—	51,28	147,42	3,86	67,83	,6803
13	,85330	—	10,75	109,56	1,19	91,27	,9153	63	,90996	—	52,10	148,20	3,90	67,48	,6767
14	,85499	—	11,58	110,32	1,26	90,65	,9091	64	,91070	—	52,93	148,99	3,94	67,12	,6731
100 + 15	,85664	—	12,41	111,07	1,34	90,03	,9029	100 + 65	,91144	—	53,76	149,78	3,98	66,77	,6696
16	,85825	—	13,23	111,82	1,41	89,43	,8968	66	,91217	—	54,58	150,57	4,01	66,42	,6661
17	,85984	—	14,06	112,58	1,48	88,83	,8908	67	,91289	—	55,41	151,35	4,06	66,07	,6626
18	,86139	—	14,89	113,34	1,55	88,23	,8848	68	,91361	—	56,24	152,14	4,10	65,73	,6592
19	,86292	—	15,71	114,09	1,62	87,64	,8789	69	,91432	—	57,07	152,93	4,14	65,39	,6558
100 + 20	,86441	—	16,54	114,86	1,68	87,06	,8731	100 + 70	,91502	—	57,89	153,71	4,18	65,06	,6524
21	,86588	—	17,37	115,62	1,75	86,49	,8674	71	,91571	—	58,72	154,50	4,22	64,73	,6491
22	,86732	—	18,19	116,38	1,81	85,93	,8617	72	,91639	—	59,55	155,29	4,26	64,40	,6458
23	,86874	—	19,02	117,14	1,88	85,37	,8561	73	,91706	—	60,37	156,08	4,29	64,07	,6425
24	,87013	—	19,85	117,90	1,95	84,82	,8506	74	,91772	—	61,20	156,87	4,33	63,75	,6393
100 + 25	,87150	--	20,68	118,67	2,01	84,27	,8451	100 + 75	,91837	—	62,03	157,66	4,37	63,43	,6361
26	,87284	—	21,50	119,43	2,07	83,73	,8397	76	,91901	—	62,85	158,45	4,40	63,11	,6329
27	,87415	—	22,33	120,20	2,13	83,19	,8343	77	,91963	—	63,68	159,24	4,44	62,80	,6298
28	,87544	—	23,16	120,97	2,19	82,66	,8290	78	,92024	—	64,51	160,04	4,47	62,49	,6267
29	,87671	—	23,98	121,74	2,24	82,14	,8238	79	,92085	—	65,34	160,83	4,51	62,18	,6236
100 + 30	,87796	—	24,81	122,51	2,30	81,63	,8186	100 + 80	,92145	—	66,16	161,62	4,54	61,87	,6205
31	,87919	—	25,64	123,28	2,36	81,12	,8135	81	,92205	—	66,99	162,42	4,57	61,57	,6175
32	,88040	—	26,46	124,05	2,41	80,62	,8084	82	,92264	—	67,82	163,21	4,61	61,27	,6145
33	,88160	—	27,29	124,82	2,47	80,12	,8034	83	,92322	—	68,64	164,00	4,64	60,97	,6115
34	,88277	—	28,12	125,59	2,53	79,63	,7985	84	,92379	—	69,47	164,79	4,68	60,68	,6085
100 + 35	,88393	—	28,95	126,36	2,59	79,14	,7936	100 + 85	,92436	—	70,30	165,58	4,71	60,39	,6056
36	,88507	—	29,77	127,13	2,64	78,66	,7888	86	,92492	—	71,13	166,38	4,75	60,10	,6027
37	,88619	—	30,60	127,90	2,70	78,18	,7840	87	,92547	—	71,95	167,17	4,77	59,82	,5998
38	,88729	—	31,43	128,68	2,75	77,71	,7793	88	,92601	—	72,78	167,97	4,81	59,53	,5970
39	,88838	—	32,25	129,45	2,80	77,25	,7747	89	,92654	—	73,61	168,77	4,84	59,25	,5942
100 + 40	,88945	—	33,08	130,23	2,85	76,79	,7701	100 + 90	,92707	—	74,43	169,56	4,87	58,97	,5914
41	,89051	—	33,91	131,00	2,91	76,33	,7655	91	,92759	—	75,26	170,36	4,90	58,70	,5886
42	,89155	—	34,74	131,77	2,97	75,88	,7610	92	,92811	—	76,09	171,16	4,93	58,43	,5859
43	,89258	—	35,56	132,55	3,01	75,44	,7566	93	,92862	—	76,91	171,96	4,95	58,16	,5832
44	,89359	—	36,39	133,33	3,06	75,00	,7522	94	,92913	—	77,74	172,75	4,99	57,89	,5805
100 + 45	,89458	—	37,22	134,11	3,11	74,57	,7478	100 + 95	,92963	—	78,57	173,55	5,02	57,62	,5779
46	,89556	—	38,04	134,88	3,16	74,14	,7435	96	,93012	—	79,39	174,34	5,05	57,36	,5752
47	,89653	—	38,87	135,66	3,21	73,71	,7392	97	,93061	—	80,21	175,13	5,07	57,10	,5727
48	,89748	—	39,70	136,44	3,26	73,29	,7350	98	,93111	—	81,04	175,92	5,12	56,84	,5701
49	,89841	—	40,52	137,22	3,30	72,88	,7308	99	,93159	—	81,87	176,72	5,15	56,59	,5675

TABLE II.   HEAT 55°.

I. Water and spirit by weight.	II. Specific gravity.	III. Spirit by measure.	IV. Water by measure.	V. Bulk of mixture.	VI. Diminution of bulk.	VII. Quantity of spirit per cent.	VIII. Decimal multipliers.
W. + Sp.							
100 + 100	,93208	100	82,70	177,53	5,17	56,33	,5649
99	,93257	—	83,55	178,34	5,21	56,07	,5623
98	,93306	—	84,40	179,17	5,23	55,81	,5597
97	,93354	—	85,27	180,02	5,25	55,55	,5571
96	,93403	—	86,16	180,87	5,29	55,29	,5545
100 + 95	,93452	—	87,06	181,74	5,32	55,03	,5518
94	,93500	—	87,98	182,64	5,34	54,76	,5491
93	,93549	—	88,92	183,55	5,37	54,48	,5464
92	,93598	—	89,89	184,48	5,41	54,21	,5436
91	,93647	—	90,88	185,44	5,44	53,93	,5408
100 + 90	,93696	—	91,89	186,42	5,47	53,64	,5380
89	,93746	—	92,92	187,42	5,50	53,35	,5351
88	,93796	—	93,98	188,44	5,54	53,06	,5322
87	,93846	—	95,06	189,49	5,57	52,77	,5292
86	,93897	—	96,17	190,57	5,60	52,47	,5262
100 + 85	,93948	—	97,30	191,67	5,63	52,17	,5232
84	,94000	—	98,45	192,80	5,65	51,87	,5202
83	,94052	—	99,64	193,95	5,69	51,56	,5171
82	,94105	—	100,86	195,13	5,73	51,25	,5139
81	,94159	—	102,10	196,35	5,75	50,93	,5107
100 + 80	,94213	—	103,38	197,59	5,79	50,61	,5075
79	,94267	—	104,69	198,86	5,83	50,29	,5043
78	,94321	—	106,03	200,17	5,86	49,96	,5010
77	,94376	—	107,41	201,52	5,89	49,62	,4977
76	,94431	—	108,82	202,90	5,92	49,29	,4943
100 + 75	,94486	—	110,27	204,32	5,95	48,94	,4908
74	,94542	—	111,76	205,77	5,99	48,60	,4874
73	,94598	—	113,29	207,27	6,02	48,25	,4839
72	,94654	—	114,86	208,81	6,05	47,89	,4803
71	,94710	—	116,48	210,40	6,08	47,53	,4767
100 + 70	,94767	—	118,15	212,03	6,12	47,18	,4730
69	,94824	—	119,86	213,70	6,16	46,79	,4693
68	,94882	—	121,62	215,43	6,19	46,42	,4655
67	,94940	—	123,44	217,21	6,23	46,04	,4617
66	,94998	—	125,31	219,05	6,26	45,65	,4578
100 + 65	,95057	—	127,24	220,94	6,30	45,26	,4539
64	,95116	—	129,22	222,89	6,33	44,86	,4499
63	,95176	—	131,27	224,90	6,37	44,46	,4459
62	,95236	—	133,39	226,99	6,40	44,05	,4418
61	,95296	—	135,58	229,15	6,43	43,64	,4376
100 + 60	,95357	—	137,84	231,37	6,47	43,22	,4334
59	,95418	—	140,17	233,67	6,50	42,79	,4292
58	,95479	—	142,58	236,05	6,53	42,36	,4249
57	,95540	—	145,09	238,52	6,57	41,92	,4205
56	,95601	—	147,69	241,09	6,60	41,48	,4160
100 + 55	,95662	—	150,37	243,74	6,63	41,03	,4114
54	,95723	—	153,13	246,48	6,65	40,57	,4068
53	,95784	—	156,01	249,34	6,67	40,10	,4022
52	,95844	—	159,02	252,32	6,70	39,63	,3975
51	,95905	—	162,15	255,42	6,73	39,15	,3926

I. Water and spirit by weight.	II. Specific gravity.	III. Spirit by measure.	IV. Water by measure.	V. Bulk of mixture.	VI. Diminution of bulk.	VII. Quantity of spirit per cent.	VIII. Decimal multipliers.
W. + Sp.							
100 + 50	,95966	100	165,41	258,64	6,77	38,66	,3877
49	,96027	—	168,79	262,03	6,76	38,16	,3828
48	,96088	—	172,30	265,49	6,81	37,66	,3777
47	,96150	—	175,96	269,13	6,83	37,16	,3726
46	,96211	—	179,79	272,94	6,85	36,64	,3674
100 + 45	,96272	—	183,79	276,92	6.87	36,11	,3622
44	,96333	—	187,96	281,08	6,88	35,58	,3568
43	,96393	—	192,33	285,44	6,89	35,03	,3513
42	,96454	—	196,91	290,01	6,90	34,48	,3458
41	,96515	—	201,72	294,80	6,92	33,92	,3402
100 + 40	,96575	—	206,76	299,85	6,91	33,35	,3345
39	,96635	—	212,06	305,16	6,90	32,77	,3286
38	,96696	—	217,64	310,74	6,90	32,18	,3227
37	,96756	—	223,53	316,64	6,89	31,58	,3167
36	,96817	—	229,73	322,86	6,87	30,97	,3106
100 + 35	,96877	—	236,30	329,41	6,89	30,36	,3044
34	,96937	—	243,25	336,37	6,88	29,73	,2981
33	,96998	—	250,62	343,77	6,85	29,09	,2917
32	,97059	—	258,45	351,63	6,82	28,44	,2852
31	,97120	—	266,79	359,98	6,81	27,78	,2786
100 + 30	,97181	—	275,68	368,92	6,76	27,11	,2718
29	,97243	—	285,19	378,46	6,73	26,42	,2650
28	,97306	—	295,37	388,68	6,69	25,73	,2580
27	,97370	—	306,31	399,66	6,65	25,02	,2509
26	,97435	—	318,09	411,49	6,60	24,30	,2437
100 + 25	,97500	—	330,82	424,29	6,53	23,57	,2364
24	,97567	—	344,60	438,13	6,47	22,82	,2289
23	,97635	—	359,58	453,18	6,40	22,06	,2213
22	,97704	—	375,92	469,59	6,33	21,29	,2135
21	,97775	—	393,83	487,57	6,26	20,51	,2057
100 + 20	,97847	—	413,52	507,34	6,18	19,71	,1977
19	,97921	—	435,29	529,19	6,10	18,90	,1895
18	,97997	—	459,47	553,47	6,00	18,07	,1812
17	,98075	—	486,50	580,60	5,90	17,22	,1727
16	,98156	—	516,90	611,10	5,80	16,36	,1641
100 + 15	,98239	—	551,36	645,68	5,68	15,49	,1553
14	,98324	—	590,75	685,19	5,56	14,59	,1463
13	,98413	—	636,19	730,76	5,43	13,68	,1372
12	,98506	—	689,20	783,92	5,28	12,75	,1279
11	,98602	—	751,86	846,72	5,14	11,81	,1184
100 + 10	,98702	—	827,04	922,04	5,00	10,85	,1088
9	,98807	—	918,94	1014,12	4,82	9,86	,0989
8	,98918	—	1033,80	1129,16	4,64	8,85	,0888
7	,99034	—	1181,49	1277,02	4,47	7,83	,0785
6	,99156	—	1378,41	1474,11	4,30	6,78	,0680
100 + 5	,99284	—	1654,09	1749,99	4,11	5,71	,0573
4	,99419	—	2067,61	2163,71	3,90	4,62	,0463
3	,99551	—	2756,81	2853,11	3,70	3,50	,0351
2	,99712	—	4135,22	4231,71	3,51	2,36	,0237
1	,99871	—	8270,45	8367,12	3,33	1,19	,0120

TABLE I.   HEAT 56°.

I. Spirit and water by weight.	II. Specific gravity.	III. Spirit by measure.	IV. Water by measure.	V. Bulk of mixture.	VI. Diminution of bulk.	VII. Quantity of spirit per cent.	VIII. Decimal multipliers.
Sp. + W.							
100 + 0	,82689	100	—	100,00	—	100,00	1,0023
1	,82919	—	0,83	100,72	0,11	99,29	,9952
2	,83145	—	1,65	101,44	0,21	98,58	,9881
3	,83365	—	2,48	102,16	0,32	97,88	,9811
4	,83579	—	3,31	102,89	0,42	97,19	,9741
100 + 5	,83787	—	4,13	103,62	0,51	96,50	,9672
6	,83990	—	4,96	104,36	0,60	95,82	,9605
7	,84188	—	5,79	105,10	0,69	95,15	,9538
8	,84382	—	6,62	105,84	0,78	94,49	,9471
9	,84571	—	7,44	106,58	0,86	93,83	,9405
100 + 10	,84756	—	8,27	107,32	0,95	93,18	,9339
11	,84935	—	9,10	108,07	1,03	92,54	,9275
12	,85111	—	9,92	108,81	1,11	91,90	,9211
13	,85283	—	10,74	109,56	1,18	91,27	,9148
14	,85452	—	11,57	110,32	1,25	90,65	,9086
100 + 15	,85618	—	12,40	111,07	1,33	90,04	,9024
16	,85778	—	13,23	111,82	1,41	89,43	,8963
17	,85938	—	14,05	112,58	1,47	88,83	,8903
18	,86092	—	14,88	113,33	1,55	88,23	,8844
19	,86246	—	15,70	114,09	1,61	87,64	,8785
100 + 20	,86395	—	16,53	114,86	1,67	87,06	,8727
21	,86542	—	17,36	115,62	1,74	86,49	,8670
22	,86686	—	18,19	116,38	1,81	85,93	,8613
23	,86828	—	19,01	117,14	1,87	85,37	,8557
24	,86967	—	19,84	117,90	1,94	84,82	,8502
100 + 25	,87104	—	20,67	118,66	2,01	84,27	,8447
26	,87238	—	21,49	119,43	2,06	83,73	,8393
27	,87369	—	22,32	120,20	2,12	83,19	,8339
28	,87498	—	23,15	120,96	2,19	82,67	,8286
29	,87626	—	23,97	121,74	2,23	82,15	,8234
100 + 30	,87750	—	24,80	122,50	2,30	81,63	,8182
31	,87874	—	25,63	123,27	2,36	81,12	,8131
32	,87994	—	26,45	124,04	2,41	80,62	,8080
33	,88115	—	27,28	124,81	2,47	80,12	,8030
34	,88232	—	28,11	125,58	2,53	79,61	,7981
100 + 35	,88348	—	28,93	126,35	2,58	79,14	,7932
36	,88462	—	29,76	127,12	2,64	78,66	,7884
37	,88574	—	30,58	127,89	2,69	78,19	,7836
38	,88685	—	31,41	128,67	2,74	77,72	,7789
39	,88793	—	32,23	129,44	2,79	77,25	,7743
100 + 40	,88900	—	33,06	130,22	2,84	76,79	,7697
41	,89006	—	33,89	130,99	2,90	76,34	,7652
42	,89110	—	34,72	131,76	2,96	75,89	,7607
43	,89213	—	35,54	132,54	3,00	75,44	,7563
44	,89315	—	36,37	133,32	3,05	75,00	,7519
100 + 45	,89412	—	37,20	134,10	3,10	74,57	,7475
46	,89510	—	38,02	134,87	3,15	74,14	,7432
47	,89608	—	38,85	135,65	3,20	73,72	,7389
48	,89703	—	39,68	136,43	3,25	73,30	,7347
49	,89796	—	40,50	137,21	3,29	72,88	,7305
100 + 50	,89887	100	41,33	137,99	3,34	72,47	,7264
51	,89978	—	42,16	138,77	3,39	72,06	,7223
52	,90066	—	42,98	139,55	3,43	71,66	,7183
53	,90153	—	43,80	140,33	3,47	71,26	,7143
54	,90238	—	44,63	141,11	3,52	70,87	,7103
100 + 55	,90323	—	45,46	141,90	3,56	70,47	,7004
56	,90405	—	46,28	142,68	3,60	70,09	,7025
57	,90487	—	47,11	143,46	3,65	69,70	,6986
58	,90567	—	47,94	144,25	3,69	69,32	,6948
59	,90646	—	48,76	145,03	3,73	68,95	,6910
100 + 60	,90724	—	49,59	145,82	3,77	68,58	,6873
61	,90801	—	50,42	146,61	3,81	68,21	,6836
62	,90877	—	51,25	147,40	3,85	67,84	,6800
63	,90953	—	52,07	148,18	3,89	67,48	,6764
64	,91026	—	52,90	148,97	3,93	67,13	,6728
100 + 65	,91100	—	53,73	149,76	3,97	66,77	,6693
66	,91173	—	54,55	150,55	4,00	66,42	,6658
67	,91246	—	55,38	151,33	4,05	66,07	,6623
68	,91318	—	56,21	152,12	4,09	65,73	,6589
69	,91389	—	57,04	152,91	4,13	65,40	,6555
100 + 70	,91459	—	57,86	153,70	4,10	65,00	,6521
71	,91528	—	58,69	154,48	4,21	64,73	,6488
72	,91596	—	59,52	155,27	4,25	64,40	,6455
73	,91663	—	60,34	156,06	4,28	64,07	,6422
74	,91729	—	61,17	156,85	4,32	63,75	,6390
100 + 75	,91794	—	62,00	157,64	4,36	63,43	,6356
76	,91858	—	62,82	158,43	4,39	63,12	,6326
77	,91920	—	63,64	159,22	4,42	62,80	,6295
78	,91982	—	64,47	160,02	4,45	62,49	,6264
79	,92043	—	65,30	160,81	4,49	62,16	,6233
100 + 80	,92102	—	66,12	161,60	4,52	61,88	,6202
81	,92163	—	66,95	162,40	4,55	61,58	,6172
82	,92222	—	67,78	163,19	4,59	61,28	,6142
83	,92280	—	68,60	163,98	4,62	60,98	,6112
84	,92337	—	69,43	164,77	4,66	60,69	,6083
100 + 85	,92391	—	70,26	165,57	4,69	60,40	,6054
86	,92449	—	71,09	166,36	4,73	60,11	,6025
87	,92505	—	71,91	167,16	4,75	59,82	,5996
88	,92559	—	72,74	167,95	4,79	59,54	,5968
89	,92612	—	73,57	168,75	4,82	59,26	,5940
100 + 90	,92666	—	74,39	169,54	4,85	58,98	,5912
91	,92717	—	75,22	170,33	4,89	58,70	,5884
92	,92769	—	76,05	171,13	4,92	58,43	,5857
93	,92821	—	76,87	171,93	4,94	58,16	,5830
94	,92871	—	77,70	172,73	4,97	57,89	,5803
100 + 95	,92922	—	78,53	173,52	5,01	57,63	,5776
96	,92971	—	79,35	174,31	5,04	57,37	,5750
97	,93020	—	80,17	175,11	5,06	57,11	,5724
98	,93069	—	81,00	175,90	5,10	56,85	,5698
99	,93118	—	81,83	176,70	5,13	56,59	,5672

I. Water and spirit by weight.	II. Specific gravity.	III. Spirit by measure.	IV. Water by measure.	V. Bulk of mixture.	VI. Diminution of bulk.	VII. Quantity of spirit per cent.	VIII. Decimal multipliers.
W. + Sp.							
100+100	,93166	100	82,66	177,50	5,16	56 33	,5646
99	,93215	—	83,50	178,31	5,19	56,08	,5621
98	,93265	—	84,35	179,14	5,21	55,82	,5595
97	,93313	—	85,22	179,99	5,23	55,56	,5569
96	,93362	—	86 11	180 84	5,27	55 30	,5543
100 + 95	93411	—	87,01	181,71	5,30	55,03	,5516
94	,93460	—	87,93	182,62	5,31	54,76	,5489
93	,93509	—	88,87	183,52	5,35	54,49	,5462
92	,93558	—	89,84	184,45	5,39	54,21	,5434
91	,93607	—	90,83	185,41	5,42	53,93	,5406
100 + 90	,93656	—	91,84	186,39	5,45	53 65	,5377
89	,93706	—	92,87	187,39	5,48	53,36	,5349
88	,93756	—	93,93	188,41	5,52	53,07	,5320
87	,93806	—	95,01	189 47	5,54	52,78	,5290
86	,93857	—	96,12	190,54	5,58	52,48	,5260
100 + 85	,93909	—	97,25	191,65	5,60	52 18	,5230
84	,93960	—	98,40	192,77	5,63	51,88	,5200
83	,94013	—	99,59	193,92	5,67	51,57	,5169
82	,94065	—	100,81	195,10	5,71	51,26	,5137
81	,94120	—	102,05	196,32	5,73	50,94	,5105
100 + 80	,94174	—	103,32	197,59	5,70	50,61	,5073
79	,94228	—	104,63	198,83	5,80	50,29	,5040
78	,94282	—	105,97	200,14	5,83	49,96	,5008
77	,94337	—	107 35	201 49	5,86	49,63	,4975
76	,94392	—	108,70	202 80	5,90	49,30	,4941
100 + 75	,94448	—	110 21	204,29	5,92	48,95	,4906
74	,94504	—	111,70	205,74	5 96	48,60	,4872
73	,94560	—	113,23	207,24	5,99	48 25	,4837
72	,94617	—	114,80	208,77	6,03	47,90	,4801
71	,94672	—	116,42	210,37	6,05	47 54	,4765
100 + 70	,94729	—	118,09	212,00	6,09	47,17	,4728
69	,94787	—	119,80	213,66	6,14	46,80	,4691
68	,94845	—	121,56	215,39	6,17	46,43	,4654
67	,94903	—	123,37	217,17	6,20	46,05	,4616
66	,94962	—	125,24	219 01	6,23	45,66	,4577
100 + 65	,95020	—	127,17	220,90	6,27	45,27	,4537
64	,95080	—	129,15	222,85	6,30	44,87	,4497
63	,95140	—	131,20	224,86	6,34	44,46	,4457
62	,95201	—	133,32	226,94	6,38	44,06	,4416
61	,95261	—	135 51	229,10	6,41	43,65	,4375
100 + 60	,95322	—	137,77	231,34	6,43	43,23	,4333
59	,95383	—	140,10	233,63	6,47	42,80	,4291
58	,95445	—	142,51	236,01	6,50	42,37	,4248
57	,95506	—	145,01	238,47	6,54	41,93	,4204
56	,95567	—	147,61	241,03	6 58	41,49	,4159
100 + 55	,95629	—	150,29	243,68	6,61	41,04	,4113
54	,95690	—	153,05	246,42	6,63	40,58	,4067
53	,95751	—	155,93	249,29	6,64	40,11	,4021
52	,95811	—	158,94	252,26	6,68	39,64	,3974
51	,95873	—	162,07	255,37	6,70	39,16	,3925

I. Water and spirit by weight.	II. Specific gravity.	III. Spirit by measure.	IV. Water by measure.	V. Bulk of mixture.	VI. Diminution of bulk.	VII. Quantity of spirit per cent.	VIII. Decimal multipliers.
W + Sp.							
100 + 50	,95933	100	165,32	258,58	6,74	38,67	,3876
49	,95995	—	168,70	261,96	6,74	38,17	,3827
48	,96057	—	172,21	265,42	6,79	37,67	,3776
47	,96119	—	175,87	269,06	0,81	37,17	,3725
46	,96181	—	179 70	272,87	6,83	36,65	,3673
100 + 45	,96242	—	183,70	276,85	6,85	36,12	,3620
44	,96304	—	187,86	281,00	6,86	35,59	,3567
43	,96364	—	192,23	285,36	6,87	35,04	,3512
42	,96426	—	196,81	289,93	6,88	34,49	,3457
41	,96487	—	201,61	294,71	6,90	33,93	,3401
100 + 40	,96548	—	206,05	299,76	6,89	33,36	,3344
39	,96608	—	211,95	305,07	6,88	32,78	,3285
38	,96670	—	217,53	310,64	6,89	32,19	,3226
37	,96730	—	223,41	316,55	6,86	31,59	,3166
36	,96792	—	229,61	322,76	6,85	30,98	,3105
100 + 35	,96852	—	236,18	329,31	6,87	30,37	,3044
34	,96913	—	243,13	336,26	6,87	29,74	,2981
33	,96975	—	250,49	343,65	6,84	29,10	,2917
32	,97036	—	258,32	351,51	6,81	28 45	,2852
31	,97098	—	266 65	359,86	6,79	27,79	,2786
100 + 30	,97160	—	275,54	368,79	6,75	27,11	,2718
29	,97223	—	285,04	378,32	6,72	26,43	,2649
28	,97280	—	295,22	388,54	6,68	25,74	,2580
27	,97351	—	306,16	399,51	6,65	25,03	,2509
26	,97416	—	317,93	411,34	6,59	24,31	,2437
100 + 25	,97482	—	330 05	424,12	6,53	23,58	,2363
24	,97550	—	344,43	437,96	6,47	22,83	,2289
23	,97618	—	359,40	453,00	6,40	22,07	,2213
22	,97688	—	375,73	469,40	6,33	21,30	,2135
21	,97760	—	393,03	487,37	6,26	20,51	,2056
100 + 20	,97831	—	413,31	507,13	6,18	19,72	,1976
19	,97907	—	435,07	528,97	6,10	18,90	,1895
18	,97983	—	459,24	553,20	6,01	18,07	,1812
17	,98062	—	486,25	580,3	5,91	17,23	,1727
16	,98143	—	516,64	610,83	5,81	16,37	,1641
100 + 15	,98227	—	551,08	645,39	5,69	15,49	,1553
14	,98312	—	590,45	684,88	5,57	14,60	,1463
13	,98402	—	635,87	730,42	5,45	13,69	,1372
12	,98496	—	688,86	783,55	5,31	12,76	,1279
11	,98592	—	751,48	846 32	5,16	11,82	,1184
100 + 10	,98693	—	826,63	921,60	5,03	10,85	,1087
9	,98798	—	918,48	1013,63	4,85	9,86	,0989
8	,98900	—	1033,29	1128,61	4,68	8,86	,0888
7	,99026	—	1180,90	1276,40	4,50	7,83	,0735
6	,99148	—	1377,72	1473,28	4,34	6 79	,0680
100 + 5	,99276	—	1653,27	1749,13	4,14	5,72	,0573
4	,99412	—	2066,57	2162,63	3,94	4,62	,0464
3	,99555	—	2755,42	2851,67	3,75	3,50	,0352
2	,99705	—	4133,14	4229,57	3,57	2,36	,0237
1	,99864	—	8266,29	8363,90	3,39	1,20	,0120

I. Spirit and water by weight.	II. Specific gravity.	III. Spirit by measure.	IV. Water by measure.	V. Bulk of mixture.	VI. Diminution of bulk.	VII. Quantity of spirit per cent.	VIII. Decimal multipliers.
Sp. + W.							
100 + 0	,82642	100	—	100,00	—	100,00	1,0017
1	,82872	—	0,83	100,72	0,11	99,29	,9947
2	,83098	—	1,65	101,44	0,21	98,58	,9876
3	,83318	—	2,48	102,16	0,32	97,88	,9806
4	,83532	—	3,31	102,89	0,42	97,19	,9736
100 + 5	,83740	—	4,13	103,62	0,51	96,50	,9667
6	,83943	—	4,96	104,36	0,60	95,82	,9600
7	,84141	—	5,78	105,10	0,68	95,16	,9533
8	,84335	—	6,61	105,84	0,77	94,49	,9466
9	,84524	—	7,43	106,58	0,85	93,83	,9400
100 + 10	,84709	—	8,26	107,32	0,94	93,18	,9334
11	,84888	—	9,09	108,06	1,03	92,54	,9270
12	,85064	—	9,91	108,81	1,10	91,90	,9206
13	,85236	—	10,74	109,56	1,18	91,27	,9143
14	,85405	—	11,57	110,31	1,26	90,65	,9081
100 + 15	,85571	—	12,39	111,07	1,32	90,04	,9019
16	,85732	—	13,22	111,82	1,40	89,43	,8959
17	,85891	—	14,04	112,57	1,47	88,83	,8899
18	,86046	—	14,87	113,33	1,54	88,24	,8839
19	,86199	—	15,69	114,09	1,60	87,65	,8780
100 + 20	,86348	—	16,52	114,85	1,67	87,07	,8722
21	,86496	—	17,35	115,61	1,74	86,50	,8665
22	,86640	—	18,18	116,37	1,81	85,93	,8608
23	,86782	—	19,00	117,13	1,87	85,37	,8552
24	,86920	—	19,83	117,89	1,94	84,82	,8497
100 + 25	,87057	—	20,66	118,66	2,00	84,27	,8442
26	,87191	—	21,48	119,42	2,06	83,73	,8388
27	,87323	—	22,31	120,19	2,12	83,20	,8334
28	,87452	—	23,14	120,96	2,18	82,67	,8281
29	,87580	—	23,96	121,73	2,23	82,15	,8229
100 + 30	,87705	—	24,79	122,50	2,29	81,64	,8177
31	,87829	—	25,61	123,26	2,35	81,13	,8126
32	,87949	—	26,44	124,03	2,41	80,63	,8076
33	,88070	—	27,27	124,80	2,47	80,13	,8026
34	,88187	—	28,09	125,57	2,52	79,64	,7977
100 + 35	,88304	—	28,92	126,35	2,57	79,15	,7928
36	,88417	—	29,74	127,11	2,63	78,67	,7880
37	,88530	—	30,57	127,88	2,69	78,19	,7833
38	,88640	—	31,40	128,66	2,74	77,72	,7786
39	,88748	—	32,22	129,43	2,79	77,26	,7739
100 + 40	,88855	—	33,05	130,21	2,84	76,80	,7693
41	,88960	—	33,87	130,98	2,89	76,34	,7648
42	,89065	—	34,70	131,76	2,94	75,89	,7603
43	,89168	—	35,52	132,53	2,99	75,45	,7559
44	,89269	—	36,35	133,31	3,04	75,01	,7515
100 + 45	,89367	—	37,18	134,09	3,09	74,58	,7471
46	,89465	—	38,00	134,86	3,14	74,15	,7428
47	,89563	—	38,83	135,64	3,19	73,72	,7385
48	,89658	—	39,66	136,42	3,24	73,30	,7343
49	,89751	—	40,48	137,20	3,28	72,88	,7301
100 + 50	,89842	100	41,31	137,98	3,33	72,48	,7260
51	,89933	—	42,13	138,76	3,37	72,07	,7219
52	,90021	—	42,96	139,54	3,42	71,66	,7179
53	,90108	—	43,78	140,32	3,46	71,26	,7139
54	,90194	—	44,61	141,10	3,51	70,87	,7100
100 + 55	,90279	—	45,44	141,89	3,55	70,48	,7060
56	,90361	—	46,26	142,67	3,59	70,09	,7021
57	,90442	—	47,09	143,45	3,64	69,70	,6982
58	,90523	—	47,91	144,24	3,67	69,32	,6945
59	,90602	—	48,74	145,02	3,72	68,95	,6907
100 + 60	,90681	—	49,57	145,81	3,76	68,59	,6870
61	,90758	—	50,40	146,60	3,80	68,22	,6833
62	,90834	—	51,23	147,39	3,84	67,85	,6797
63	,90910	—	52,05	148,17	3,88	67,49	,6761
64	,90983	—	52,88	148,96	3,92	67,13	,6724
100 + 65	,91056	—	53,70	149,75	3,95	66,78	,6689
66	,91130	—	54,53	150,54	3,99	66,43	,6655
67	,91203	—	55,35	151,32	4,03	66,08	,6620
68	,91275	—	56,18	152,11	4,07	65,74	,6586
69	,91346	—	57,01	152,90	4,11	65,40	,6552
100 + 70	,91416	—	57,83	153,68	4,15	65,07	,6518
71	,91485	—	58,66	154,47	4,19	64,74	,6485
72	,91553	—	59,49	155,26	4,23	64,41	,6452
73	,91620	—	60,31	156,05	4,26	64,08	,6419
74	,91686	—	61,14	156,84	4,30	63,76	,6387
100 + 75	,91751	—	61,96	157,63	4,33	63,44	,6355
76	,91815	—	62,79	158,42	4,37	63,13	,6323
77	,91877	—	63,61	159,21	4,40	62,81	,6292
78	,91940	—	64,44	160,00	4,44	62,50	,6261
79	,92001	—	65,27	160,79	4,48	62,19	,6230
100 + 80	,92059	—	66,09	161,58	4,51	61,89	,6199
81	,92120	—	66,92	162,38	4,54	61,59	,6169
82	,92180	—	67,75	163,17	4,58	61,29	,6139
83	,92237	—	68,57	163,96	4,61	60,99	,6110
84	,92294	—	69,40	164,75	4,65	60,70	,6081
100 + 85	,92351	—	70,23	165,55	4,68	60,41	,6051
86	,92408	—	71,06	166,34	4,72	60,12	,6022
87	,92463	—	71,88	167,14	4,74	59,83	,5993
88	,92517	—	72,71	167,93	4,78	59,55	,5965
89	,92571	—	73,53	168,73	4,80	59,27	,5937
100 + 90	,92625	—	74,36	169,52	4,84	58,99	,5909
91	,92676	—	75,18	170,31	4,87	58,71	,5881
92	,92728	—	76,01	171,11	4,90	58,44	,5854
93	,92780	—	76,83	171,91	4,92	58,17	,5827
94	,92830	—	77,66	172,70	4,96	57,90	,5800
100 + 95	,92881	—	78,49	173,50	4,99	57,63	,5774
96	,92930	—	79,31	174,29	5,02	57,38	,5748
97	,92979	—	80,13	175,09	5,04	57,12	,5722
98	,93028	—	80,96	175,88	5,08	56,86	,5696
99	,93077	—	81,79	176,68	5,11	56,60	,5670

TABLE II.   HEAT 57°.

I. Water and spirit by weight.	II. Specific gravity.	III. Spirit by measure.	IV. Water by measure.	V. Bulk of mixture.	VI. Diminution of bulk.	VII. Quantity of spirit per cent.	VIII. Decimal multipliers.
W. + Sp.							
100+100	,93125	100	82.62	177,48	5,14	56,34	,5644
99	,93174	—	83,46	178,29	5,17	56,08	,5618
98	,93224	—	84,31	179,12	5,19	55,83	,5592
97	,93272	—	85,17	179,97	5,20	55,57	,5566
96	,93321	—	86,07	180,82	5,25	55,31	,5540
100+95	,93370	—	86,97	181,69	5,28	55,04	,5514
94	,93419	—	87,89	182,59	5,30	54,77	,5486
93	,93468	—	88,83	183,50	5,33	54,50	,5459
92	,93517	—	89,79	184,43	5,36	54,22	,5432
91	,93566	—	90,78	185,39	5,39	53,94	,5403
100+90	,93616	—	91,79	186,37	5,42	53,65	,5375
89	,93665	—	92,83	187,37	5,46	53,37	,5346
88	,93716	—	93,89	188,39	5,50	53,08	,5317
87	,93766	—	94,97	189,44	5,53	52,78	,5288
86	,93817	—	96,07	190,51	5,56	52,48	,5258
100+85	,93869	—	97,20	191,62	5,58	52,18	,5228
84	,93921	—	98,35	192,74	5,61	51,89	,5198
83	,93973	—	99,54	193,89	5,65	51,58	,5167
82	,94026	—	100,76	195,08	5,68	51,27	,5135
81	,94081	—	102,00	196,29	5,72	50,95	,5103
100+80	,94135	—	103,27	197,53	5,74	50,62	,5071
79	,94190	—	104,58	198,80	5,78	50,30	,5038
78	,94244	—	105,92	200,10	5,82	49,97	,5006
77	,94299	—	107,30	201,45	5,85	49,63	,4973
76	,94354	—	108,71	202,83	5,88	49,30	,4939
100+75	,94410	—	110,16	204,25	5,91	48,96	,4904
74	,94466	—	111,64	205,70	5,94	48,61	,4870
73	,94522	—	113,17	207,20	5,97	48,26	,4835
72	,94579	—	114,75	208,74	6,01	47,90	,4799
71	,94635	—	116,36	210,33	6,03	47,55	,4763
100+70	,94691	—	118,03	211,96	6,07	47,18	,4726
69	,94750	—	119,74	213,62	6,12	46,81	,4689
68	,94808	—	121,50	215,35	6,15	46,43	,4652
67	,94805	—	123,31	217,14	6,17	46,05	,4614
66	,94925	—	125,18	218,97	6,21	45,66	,4575
100+65	,94984	—	127,10	220,86	6,24	45,28	,4537
64	,95044	—	129,09	222.81	6,28	44,88	,4496
63	,95105	—	131,13	224,82	6,31	44,47	,4456
62	,95165	—	133,25	226,90	6,35	44,07	,4415
61	,95225	—	135,45	229,06	6,39	43,66	,4373
100+60	,95287	—	137 70	231,29	6,41	43,24	,4331
59	,95348	—	140,03	233,58	6,45	42,81	,4289
58	,95410	—	142,44	235,96	6,48	42,38	,4246
57	,95471	—	144,94	238,42	6,52	41,94	,4202
56	,95533	—	147,54	240,97	6,57	41,49	,4157
100+55	,95595	—	150,21	243,62	6,59	41,04	,4112
54	,95656	—	1,2,98	246,37	6,61	40,58	4065
53	,95717	—	155,86	249,24	6,63	40,12	,4019
52	,95778	—	158,86	252,20	6,66	39,64	,3972
51	,95840	—	162 00	255,31	6,69	39,16	,3924
W. + Sp.							
100+50	,95900	100	165,24	258,52	6,72	38,68	,3875
49	,95963	—	168,61	261,89	6,72	38,18	,3825
48	,96025	—	172,12	265,35	6,77	37,68	,3775
47	,96088	—	175,78	268,99	6,79	37,17	,3724
46	,96150	—	179,61	272,80	6,81	36,65	,3672
100+45	,96212	—	183,61	276,78	6,83	30,13	,3619
44	,96274	—	187,77	280,92	6,85	35,60	,3566
43	,96335	—	192,13	285,28	6,85	35,05	,3511
42	,96397	—	196,71	289,84	6,87	34,50	,3456
41	,96459	—	201,51	294.63	6,88	33,94	,3400
100+40	,96521	—	206,55	299,68	6,87	33,37	,3343
39	,96581	—	211,84	304,98	6,86	32,79	,3285
38	,96643	—	217,42	310,55	6,87	32,20	,3225
37	,96704	—	223,30	316,45	6,85	31,60	,3165
36	,96766	—	229,50	322,66	6,84	30,99	,3104
100+35	,96827	—	236,06	329,21	6,85	30,38	,3043
34	,96889	—	243,01	336,15	6,86	29,75	,2980
33	,96951	—	250,36	343,54	6,82	29,11	,2916
32	,97013	—	258,19	351,39	6,80	28,46	,2851
31	,97076	—	266,52	359,24	6,78	27,80	,2785
100+30	,97138	—	275,40	368,66	6,74	27,12	,2717
29	,97202	—	284,90	378,18	6,72	26,44	,2649
28	,97266	—	295,07	388,40	6,67	25,75	,2579
27	,97331	—	306,01	399.37	6,64	25,04	,2508
26	,97397	—	317,77	411,19	6,58	24,32	,2436
100+25	,97464	—	330,48	423,96	6,52	23,59	,2363
24	,97532	—	344,26	437,79	6,47	22,84	,2288
23	,97601	—	359,22	452,82	6,40	22,08	,2212
22	,97671	—	375,54	469,21	6,33	21,31	,2135
21	,97744	—	393,43	487,17	6,26	20,52	,2056
100+20	,97816	—	413,10	500.92	6,18	19,72	,1976
19	,97892	—	434,85	528,75	6,10	18,91	,1895
18	,97909	—	459,01	552,99	6,02	18,08	,1812
17	,98048	—	486,01	580,09	5,92	17,24	,1727
16	,98130	—	516,38	610,56	5,82	16,38	,1641
100+15	,98215	—	550,81	645,10	5,71	15,50	,1553
14	,98300	—	590,16	684,57	5,59	14,61	,1463
13	,98390	—	635,55	730,09	5 46	13,69	,1372
12	,98485	—	688,52	783,19	5 33	12,77	,1279
11	,98582	—	751,11	845.92	5.19	11,82	,1184
100+10	,98684	—	826,22	921,16	5,06	10,85	,1087
9	,98789	—	918,02	1013,14	4,88	9,87	,0989
8	,98904	—	1032,78	1128,06	4,72	8,86	,0888
7	,99017	—	1180,31	1275,78	4,53	7,84	,0785
6	,99140	—	1377,04	1472,66	4.38	6,79	,0680
100+5	,99266	—	1652,45	1748,27	4,18	5,72	,0573
4	,99404	—	2005,54	2161,56	3,98	4,63	,0463
3	,99548	—	2754,04	2850,24	3,80	3,51	,0354
2	,99698	—	4131,07	4227,44	3.03	2,37	,0237
1	,99857	—	8262.16	8358,71	3.45	1,20	,0120

I. Spirit and water by weight.	II. Specific gravity.	III. Spirit by measure.	IV. Water by measure.	V. Bulk of mixture.	VI. Diminution of bulk.	VII. Quantity of Spirit per cent.	VIII. Decimal multipliers.
Sp. + W.							
100 + 0	,82594	100	—	100,00	—	100,00	1,0011
1	,82825	—	0,83	100,72	0,11	99,29	,9941
2	,83051	—	1,65	101,44	0,21	98,58	,9871
3	,83271	—	2,48	102,16	0,32	97,88	,9801
4	,83485	—	3,31	102,89	0,42	97,19	,9731
100 + 5	,83693	—	4,13	103,63	0,51	96,50	,9662
6	,83896	—	4,96	104,36	0,60	95,82	,9595
7	,84094	—	5,78	105,10	0,68	95,16	,9528
8	,84289	—	6,61	105,84	0,77	94,49	,9461
9	,84477	—	7,43	106,58	0,85	93,83	,9395
100 + 10	,84662	—	8,26	107,32	0,94	93,16	,9329
11	,84842	—	9,09	108,06	1,03	92,54	,9265
12	,85017	—	9,91	108,81	1,10	91,90	,9201
13	,85189	—	10,73	109,56	1,17	91,27	,9138
14	,85359	—	11,56	110,31	1,25	90,65	,9076
100 + 15	,85524	—	12,39	111,06	1,33	90,04	,9014
16	,85685	—	13,21	111,82	1,39	89,43	,8954
17	,85844	—	14,04	112,57	1,47	88,83	,8894
18	,85999	—	14,87	113,33	1,54	88,24	,8835
19	,86152	—	15,68	114,09	1,59	87,65	,8776
100 + 20	,86302	—	16,51	114,85	1,66	87,07	,8717
21	,86449	—	17,34	115,61	1,73	86,50	,8660
22	,86593	—	18,17	116,37	1,80	85,93	,8604
23	,86735	—	18,99	117,13	1,86	85,37	,8548
24	,86874	—	19,82	117,89	1,93	84,82	,8493
100 + 25	,87011	—	20,64	118,65	1,99	84,27	,8438
26	,87145	—	21,47	119,42	2,05	83,73	,8384
27	,87276	—	22,30	120,19	2,11	83,20	,8330
28	,87406	—	23,13	120,95	2,18	82,67	,8277
29	,87534	—	23,94	121,73	2,21	82,15	,8225
100 + 30	,87659	—	24,78	122,49	2,29	81,64	,8173
31	,87784	—	25,60	123,26	2,34	81,13	,8122
32	,87904	—	26,42	124,03	2,39	80,63	,8072
33	,88025	—	27,25	124,80	2,45	80,13	,8022
34	,88143	—	28,08	125,57	2,51	79,64	,7973
100 + 35	,88259	—	28,91	126,34	2,57	79,15	,7924
36	,88373	—	29,73	127,10	2,63	78,67	,7876
37	,88485	—	30,55	127,88	2,67	78,20	,7829
38	,88595	—	31,38	128,66	2,72	77,73	,7782
39	,88703	—	32,20	129,42	2,78	77,26	,7735
100 + 40	,88810	—	33,03	130,21	2,82	76,80	,7689
41	,88915	—	33,86	130,97	2,89	76,35	,7644
42	,89020	—	34,69	131,75	2,94	75,90	,7599
43	,89123	—	35,51	132,52	2,99	75,45	,7555
44	,89224	—	36,34	133,31	3,03	75,01	,7511
100 + 45	,89322	—	37,16	134,08	3,08	74,58	,7467
46	,89420	—	37,98	134,85	3,13	74,15	,7424
47	,89518	—	38,81	135,63	3,18	73,73	,7382
48	,89613	—	39,64	136,41	3,23	73,31	,7340
49	,89700	—	40,46	137,19	3,27	72,89	,7297
100 + 50	,89797	100	41,29	137,97	3,32	72,48	,7256
51	,89888	—	42,11	138,75	3,36	72,07	,7215
52	,89976	—	42,94	139,53	3,41	71,67	,7176
53	,90063	—	43,76	140,31	3,45	71,27	,7136
54	,90149	—	44,59	141,09	3,50	70,88	,7096
100 + 55	,90234	—	45,42	141,88	3,54	70,48	,7056
56	,90317	—	46,24	142,66	3,58	70,10	,7018
57	,90398	—	47,06	143,44	3,62	69,71	,6979
58	,90479	—	47,89	144,23	3,66	69,33	,6941
59	,90558	—	48,71	145,01	3,70	68,96	,6904
100 + 60	,90638	—	49,54	145,80	3,74	68,59	,6860
61	,90714	—	50,37	146,58	3,79	68,22	,6830
62	,90791	—	51,20	147,37	3,83	67,85	,6794
63	,90866	—	52,02	148,15	3,87	67,50	,6758
64	,90940	—	52,85	148,94	3,91	67,14	,6721
100 + 65	,91013	—	53,67	149,73	3,94	66,79	,6686
66	,91087	—	54,50	150,52	3,98	66,44	,6652
67	,91160	—	55,32	151,30	4,02	66,09	,6617
68	,91232	—	56,15	152,09	4,06	65,75	,6583
69	,91303	—	56,98	152,88	4,10	65,41	,6549
100 + 70	,91373	—	57,80	153,67	4,13	65,08	,6515
71	,91442	—	58,63	154,45	4,18	64,75	,6482
72	,91510	—	59,46	155,24	4,22	64,42	,6449
73	,91577	—	60,28	156,03	4,25	64,09	,6416
74	,91643	—	61,11	156,82	4,29	63,77	,6384
100 + 75	,91708	—	61,93	157,61	4,32	63,45	,6352
76	,91772	—	62,76	158,40	4,36	63,13	,6320
77	,91834	—	63,58	159,19	4,39	62,82	,6289
78	,91897	—	64,41	159,99	4,42	62,51	,6258
79	,91958	—	65,23	160,77	4,46	62,20	,6227
100 + 80	,92017	—	66,06	161,56	4,50	61,90	,6196
81	,92078	—	66,88	162,36	4,52	61,59	,6167
82	,92137	—	67,71	163,15	4,56	61,29	,6136
83	,92195	—	68,53	163,94	4,59	60,99	,6107
84	,92252	—	69,36	164,74	4,62	60,71	,6078
100 + 85	,92309	—	70,19	165,53	4,66	60,41	,6048
86	,92366	—	71,02	166,32	4,70	60,12	,6019
87	,92421	—	71,84	167,12	4,72	59,84	,5990
88	,92475	—	72,67	167,91	4,76	59,56	,5962
89	,92529	—	73,49	168,71	4,78	59,27	,5935
100 + 90	,92583	—	74,32	169,50	4,82	58,99	,5906
91	,92635	—	75,14	170,29	4,85	58,72	,5878
92	,92687	—	75,97	171,09	4,88	58,45	,5851
93	,92739	—	76,79	171,88	4,91	58,18	,5825
94	,92789	—	77,62	172,68	4,94	57,91	,5798
100 + 95	,92840	—	78,45	173,47	4,98	57,04	,5771
96	,92889	—	79,27	174,27	5,00	57,38	,5745
97	,92938	—	80,09	175,07	5,02	57,12	,5719
98	,92987	—	80,92	175,86	5,06	56,86	,5693
99	,93036	—	81,75	176,66	5,09	56,61	,5667

TABLE II. HEAT 58°.

I. Water and Spirit by weight.	II. Specific gravity.	III. Spirit by measure.	IV. Water by measure.	V. Bulk of mixture.	VI. Diminution of bulk.	VII. Quantity of spirit per cent.	VIII. Decimal multipliers.
W. + Sp.							
100+100	,93084	100	82,58	177,45	5,13	56,35	,5641
99	,93133	—	83,42	178,26	5,16	56,09	,5616
98	,93183	—	84,27	179,09	5,18	55,84	,5590
97	,93231	—	85,12	179,94	5,18	55,57	,5564
96	,93280	—	86,02	180,79	5,23	55,31	,5538
100+95	,93329	—	86,92	181,66	5,26	55,05	,5512
94	,93378	—	87,84	182,56	5,28	54,78	,5484
93	,93427	—	88,79	183,47	5,32	54,50	,5457
92	,93476	—	89,75	184,40	5,35	54,22	,5429
91	,93525	—	90,74	185,36	5,38	53,95	,5401
100+90	,93575	—	91,75	186,34	5,41	53,66	,5373
89	,93625	—	92,78	187,34	5,44	53,38	,5344
88	,93676	—	93,84	188,36	5,48	53,09	,5315
87	,93726	—	94,92	189,41	5,51	52,79	,5286
86	,93777	—	96,02	190,48	5,54	52,49	,5256
100+85	,93829	—	97,15	191,59	5,56	52,19	,5226
84	,93881	—	98,30	192,71	5,59	51,89	,5196
83	,93934	—	99,49	193,86	5,63	51,59	,5165
82	,93987	—	100,71	195,05	5,66	51,28	,5133
81	,94042	—	101,95	196,26	5,69	50,95	,5101
100+80	,94090	—	103,22	197,50	5,72	50,63	,5069
79	,94151	—	104,53	198,77	5,76	50,31	,5036
78	,94205	—	105,87	200,07	5,80	49,98	,5004
77	,94261	—	107,24	201,42	5,82	49,64	,4971
76	,94316	—	108,65	202,79	5,86	49,31	,4937
100+75	,94372	—	110,10	204,22	5,88	48,97	,4902
74	,94428	—	111,59	205,67	5,92	48,62	,4868
73	,94484	—	113,11	207,17	5,94	48,27	,4833
72	,94541	—	114,69	208,71	5,98	47,91	,4797
71	,94597	—	116,30	210,30	6,00	47,56	,4761
100+70	,94653	—	117,97	211,92	6,05	47,19	,4724
69	,94712	—	119,68	213,59	6,09	46,81	,4687
68	,94771	—	121,44	215,31	6,13	46,44	,4650
67	,94828	—	123,25	217,10	6,15	46,07	,4612
66	,94889	—	125,12	218,93	6,19	45,67	,4573
100+65	,94948	—	127,04	220,82	6,22	45,28	,4534
64	,95008	—	129,02	222,77	6,25	44,89	,4494
63	,95069	—	131,07	224,78	6,29	44,48	,4454
62	,95129	—	133,18	226,86	6,32	44,08	,4413
61	,95190	—	135,38	229,02	6,36	43,66	,4371
100+60	,95252	—	137,63	231,24	6,39	43,25	,4330
59	,95313	—	139,96	233,54	6,42	42,82	,4287
58	,95376	—	142,37	235,91	6,46	42,39	,4244
57	,95437	—	144,87	238,38	6,49	41,95	,4200
56	,95499	—	147,46	240,92	6,54	41,50	,4155
100+55	,95561	—	150,14	243,57	6,57	41,05	,4110
54	,95622	—	152,91	246,32	6,59	40,59	,4064
53	,95684	—	155,79	249,19	6,60	40,13	,4018
52	,95745	—	158,79	252,15	6,64	39,65	,3970
51	,95807	—	161,92	255,25	6,67	39,17	,3922
100+50	,95868	100	165,16	258,46	6,70	38,69	,3873
49	,95931	—	168,53	261,82	6,71	38,19	,3824
48	,95994	—	172,03	265,28	6,75	37,69	,3774
47	,96057	—	175,69	268,92	6,77	37,18	,3723
46	,96120	—	179,52	272,73	6,79	36,66	,3671
100+45	,96182	—	183,52	276,71	6,81	36,14	,3618
44	,96244	—	187,68	280,85	6,83	35,61	,3565
43	,96306	—	192,04	285,20	6,84	35,06	,3510
42	,96368	—	196,61	289,76	6,85	34,51	,3455
41	,96431	—	201,41	294,54	6,87	33,95	,3399
100+40	,96493	—	206,45	299,59	6,86	33,38	,3342
39	,96554	—	211,74	304,89	6,85	32,80	,3284
38	,96616	—	217,31	310,46	6,85	32,21	,3225
37	,96678	—	223,19	316,35	6,84	31,61	,3165
36	,96740	—	229,39	322,56	6,83	31,00	,3104
100+35	,96802	—	235,94	329,10	6,84	30,39	,3042
34	,96865	—	242,89	336,04	6,85	29,76	,2979
33	,96927	—	250,24	343,43	6,81	29,12	,2915
32	,96990	—	258,06	351,22	6,79	28,47	,2850
31	,97054	—	266,39	359,62	6,77	27,81	,2784
100+30	,97116	—	275,26	368,54	6,72	27,13	,2716
29	,97181	—	284,76	378,05	6,71	26,45	,2648
28	,97246	—	294,93	388,26	6,63	25,76	,2578
27	,97312	—	305,86	399,23	6,63	25,05	,2508
26	,97378	—	317,62	411,04	6,58	24,33	,2436
100+25	,97446	—	330,32	423,80	6,52	23,60	,2362
24	,97514	—	344,09	437,62	6,47	22,85	,2287
23	,97584	—	359,04	452,65	6,39	22,09	,2212
22	,97655	—	375,36	469,02	6,34	21,32	,2134
21	,97728	—	393,24	486,97	6,27	20,53	,2056
100+20	,97801	—	412,90	506,71	6,19	19,73	,1976
19	,97877	—	434,63	528,53	6,10	18,92	,1894
18	,97955	—	458,78	552,75	6,03	18,09	,1811
17	,98034	—	485,77	579,84	5,93	17,25	,1726
16	,98117	—	516,12	610,29	5,83	16,39	,1640
100+15	,98202	—	550,54	644,81	5,73	15,51	,1553
14	,98288	—	589,87	684,26	5,61	14,62	,1463
13	,98379	—	635,22	729,76	5,46	13,70	,1372
12	,98474	—	688,18	782,83	5,35	12,77	,1279
11	,98572	—	750,74	845,52	5,22	11,83	,1184
100+10	,98670	—	825,81	920,73	5,08	10,86	,1087
9	,98780	—	917,57	1012,66	4,91	9,87	,0989
8	,98891	—	1032,27	1127,52	4,75	8,87	,0888
7	,99008	—	1179,73	1275,16	4,57	7,84	,0785
6	,99132	—	1376,36	1471,94	4,42	6,79	,0680
100+5	,99260	—	1651,63	1747,41	4,22	5,72	,0573
4	,99396	—	2064,51	2160,49	4,02	4,63	,0463
3	,99540	—	2752,68	2848,83	3,85	3,51	,0351
2	,99691	—	4129,03	4225,35	3,68	2,37	,0237
1	,99850	—	8258,07	8354,57	3,50	1,20	,0120

Sp. + W.

I. Spirit and water by weight.	II. Specific gravity.	III. Spirit by measure.	IV. Water by measure.	V. Bulk of mixture.	VI. Diminution of bulk.	VII. Quantity of spirit per cent.	VIII. Decimal multipliers.
100 + 0	,82547	100	—	100,00	—	100,00	1,0006
1	,82778	—	0,83	100,72	0,11	99,29	,9935
2	,83004	—	1,65	101,44	0,21	98,58	,9865
3	,83124	—	2,48	102,16	0,32	97,88	,9795
4	,83438	—	3,30	102,89	0,41	97,19	,9725
100 + 5	,83646	—	4,12	103,62	0,50	96,50	,9656
6	,83849	—	4,95	104,35	0,60	95,82	,9589
7	,84048	—	5,78	105,10	0,68	95,16	,9522
8	,84242	—	6,61	105,83	0,78	94,49	,9456
9	,84430	—	7,43	106,57	0,86	93,83	,9390
100 + 10	,84615	—	8,26	107,31	0,95	93,19	,9324
11	,84795	—	9,08	108,06	1,02	92,55	,9260
12	,84970	—	9,90	108,80	1,10	91,91	,9196
13	,85143	—	10,73	109,55	1,18	91,27	,9133
14	,85312	—	11,55	110,31	1,24	90,66	,9071
100 + 15	,85477	—	12,38	111,06	1,32	90,04	,9009
16	,85639	—	13,21	111,81	1,40	89,44	,8949
17	,85797	—	14,03	112,57	1,46	88,84	,8889
18	,85953	—	14,86	113,32	1,54	88,24	,8830
19	,86106	—	15,68	114,08	1,60	87,65	,8771
100 + 20	,86256	—	16,51	114,85	1,66	87,07	,8713
21	,86402	—	17,33	115,61	1,72	86,50	,8656
22	,86547	—	18,16	116,36	1,80	85,94	,8599
23	,86688	—	18,98	117,12	1,86	85,38	,8543
24	,86827	—	19,81	117,88	1,93	84,83	,8488
100 + 25	,86964	—	20,63	118,65	1,98	84,28	,8433
26	,87098	—	21,46	119,41	2,05	83,74	,8379
27	,87229	—	22,28	120,18	2,10	83,21	,8326
28	,87360	—	23,12	120,95	2,17	82,68	,8273
29	,87488	—	23,93	121,72	2,21	82,16	,8221
100 + 30	,87614	—	24,76	122,49	2,27	81,65	,8169
31	,87738	—	25,58	123,25	2,33	81,14	,8118
32	,87859	—	26,41	124,02	2,39	80,64	,8068
33	,87980	—	27,24	124,79	2,45	80,14	,8018
34	,88098	—	28,06	125,56	2,50	79,64	,7969
100 + 35	,88214	—	28,89	126,33	2,56	79,16	,7920
36	,88328	—	29,71	127,10	2,61	78,68	,7872
37	,88440	—	30,54	127,87	2,67	78,20	,7825
38	,88550	—	31,37	128,65	2,72	77,73	,7778
39	,88658	—	32,19	129,42	2,77	77,27	,7731
100 + 40	,88765	—	33,02	130,20	2,82	76,81	,7685
41	,88870	—	33,84	130,97	2,87	76,35	,7640
42	,88973	—	34,67	131,75	2,92	75,90	,7595
43	,89078	—	35,49	132,52	2,97	75,46	,7551
44	,89179	—	36,32	133,30	3,02	75,02	,7507
100 + 45	,89277	—	37,14	134,07	3,07	74,59	,7463
46	,89375	—	37,96	134,85	3,11	74,16	,7420
47	,89473	—	38,79	135,62	3,17	73,73	,7378
48	,89568	—	39,62	136,40	3,22	73,31	,7336
49	,89661	—	40,44	137,18	3,26	72,90	,7294
100 + 50	,89752	100	41,27	137,96	3,31	72,49	,7253
51	,89843	—	42,09	138,74	3,35	72,08	,7212
52	,89931	—	42,92	139,52	3,40	71,67	,7172
53	,90018	—	43,74	140,30	3,44	71,27	,7132
54	,90104	—	44,57	141,08	3,49	70,88	,7093
100 + 55	,90189	—	45,40	141,87	3,53	70,49	,7053
56	,90272	—	46,22	142,65	3,57	70,10	,7014
57	,90354	—	47,04	143,43	3,61	69,71	,6976
58	,90435	—	47,87	144,22	3,65	69,33	,6938
59	,90514	—	48,69	145,00	3,69	68,96	,6901
100 + 60	,90595	—	49,52	145,79	3,73	68,60	,6803
61	,90670	—	50,34	146,57	3,77	68,23	,6827
62	,90747	—	51,17	147,36	3,81	67,86	,6791
63	,90822	—	51,99	148,14	3,85	67,50	,6755
64	,90897	—	52,82	148,93	3,89	67,14	,6718
100 + 65	,90970	—	53,64	149,72	3,92	66,79	,6683
66	,91044	—	54,47	150,51	3,96	66,44	,6649
67	,91117	—	55,29	151,29	4,00	66,09	,6614
68	,91189	—	56,12	152,08	4,04	65,75	,6580
69	,91260	—	56,95	152,86	4,09	65,41	,6546
100 + 70	,91330	—	57,77	153,65	4,12	65,08	,6512
71	,91399	—	58,60	154,44	4,16	64,75	,6479
72	,91467	—	59,43	155,22	4,21	64,42	,6446
73	,91534	—	60,25	156,02	4,23	64,09	,6413
74	,91600	—	61,08	156,80	4,28	63,77	,6381
100 + 75	,91665	—	61,90	157,59	4,31	63,45	,6349
76	,91729	—	62,73	158,38	4,35	63,14	,6317
77	,91791	—	63,55	159,18	4,37	62,82	,6286
78	,91854	—	64,38	159,97	4,41	62,51	,6255
79	,91915	—	65,20	160,75	4,45	62,20	,6224
100 + 80	,91975	—	66,03	161,54	4,49	61,90	,6194
81	,92036	—	66,85	162,34	4,51	61,60	,6164
82	,92095	—	67,68	163,13	4,55	61,30	,6134
83	,92152	—	68,50	163,92	4,58	61,00	,6104
84	,92210	—	69,33	164,72	4,61	60,71	,6075
100 + 85	,92267	—	70,15	165,51	4,64	60,42	,6045
86	,92324	—	70,98	166,30	4,68	60,13	,6017
87	,92379	—	71,81	167,10	4,71	59,84	,5988
88	,92433	—	72,63	167,89	4,74	59,56	,5960
89	,92487	—	73,45	168,69	4,76	59,28	,5932
100 + 90	,92541	—	74,28	169,48	4,80	59,00	,5904
91	,92594	—	75,10	170,27	4,83	58,72	,5876
92	,92646	—	75,93	171,07	4,86	58,45	,5850
93	,92698	—	76,75	171,86	4,89	58,18	,5822
94	,92748	—	77,58	172,66	4,92	57,91	,5795
100 + 95	,92799	—	78,41	173,45	4,96	57,65	,5768
96	,92848	—	79,23	174,25	4,98	57,39	,5742
97	,92897	—	80,05	175,05	5,00	57,13	,5716
98	,92946	—	80,88	175,84	5,04	56,87	,5691
99	,92995	—	81,71	176,64	5,07	56,61	,5665

TABLE II. HEAT 59°.

I. Water and spirit by weight.	II. Specific gravity.	III. Spirit by measure.	IV. Water by measure.	V. Bulk of mixture.	VI. Diminution of bulk.	VII. Quantity by spirit per cent.	VIII. Decimal multipliers.
W. + Sp.							
100+100	,93043	100	82,54	177,43	5,11	56,35	,5639
99	,93092	—	83,38	178,24	5,14	56,10	,5613
98	,93142	—	84,23	179,07	5,16	55,84	,5587
97	,93190	—	85,07	179,91	5,16	55,58	,5562
96	,93239	—	85,98	180,77	5,21	55,32	,5535
100+95	,93288	—	86,88	181,64	5,24	55,05	,5509
94	,93337	—	87,80	182,53	5,27	54,78	,5481
93	,93386	—	88,75	183,44	5,31	54,51	,5454
92	,93435	—	89,71	184,37	5,34	54,23	,5427
91	,93484	—	90,70	185,34	5,36	53,95	,5398
100+90	,93534	—	91,71	186,32	5,39	53,67	,5370
89	,93585	—	92,74	187,31	5,43	53,39	,5342
88	,93636	—	93,79	188,34	5,45	53,10	,5313
87	,93686	—	94,87	189,38	5,49	52,80	,5284
86	,93737	—	95,97	190,45	5,52	52,50	,5254
100+85	,93789	—	97,10	191,56	5,54	52,20	,5223
84	,93842	—	98,25	192,68	5,57	51,90	,5193
83	,93895	—	99,44	193,83	5,61	51,60	,5163
82	,93949	—	100,66	195,02	5,64	51,29	,5131
81	,94003	—	101,90	196,23	5,67	50,96	,5099
100+80	,94052	—	103,17	197,47	5,70	50,64	,5067
79	,94112	—	104,48	198,74	5,74	50,32	,5034
78	,94167	—	105,82	200,04	5,78	49,99	,5002
77	,94223	—	107,19	201,39	5,80	49,65	,4969
76	,94278	—	108,60	202,76	5,84	49,32	,4935
100+75	,94334	—	110,05	204,19	5,86	48,98	,4900
74	,94390	—	111,53	205,64	5,89	48,63	,4866
73	,94446	—	113,05	207,13	5,92	48,28	,4831
72	,94503	—	114,64	208,68	5,96	47,92	,4795
71	,94560	—	116,25	210,26	5,99	47,56	,4760
100+70	,94616	—	117,91	211,88	6,03	47,19	,4722
69	,94674	—	119,62	213,55	6,07	46,82	,4685
68	,94734	—	121,38	215,27	6,11	46,45	,4648
67	,94791	—	123,19	217,06	6,13	46,08	,4610
66	,94853	—	125,06	218,89	6,17	45,68	,4571
100+65	,94912	—	126,98	220,78	6,20	45,30	,4532
64	,94971	—	128,96	222,73	6,23	44,90	,4492
63	,95033	—	131,01	224,74	6,27	44,49	,4452
62	,95094	—	133,12	226,82	6,30	44,09	,4412
61	,95155	—	135,31	228,97	6,34	43,67	,4370
100+60	,95215	—	137,56	231,19	6,37	43,26	,4328
59	,95278	—	139,89	233,49	6,40	42,83	,4285
58	,95341	—	142,30	235,85	6,44	42,40	,4243
57	,95403	—	144,80	238,33	6,47	41,96	,4199
56	,95465	—	147,39	240,87	6,52	41,51	,4154
100+55	,95527	—	150,07	243,52	6,55	41,06	,4109
54	,95589	—	152,84	246,27	6,57	40,60	,4063
53	,95651	—	155,72	249,13	6,59	40,14	,4017
52	,95712	—	158,72	252,10	6,62	39,66	,3969
51	,95774	—	161,84	255,19	6,65	39,18	,3921

I. Water and spirit by weight.	II. Specific gravity.	III. Spirit by measure.	IV. Water by measure.	V. Bulk of mixture.	VI. Diminution of bulk.	VII. Quantity by spirit per cent.	VIII. Decimal multipliers.
W. + Sp.							
100+50	,95836	100	165,08	258,40	6,68	38,70	,3872
49	,95899	—	168,45	261,75	6,70	38,20	,3823
48	,95963	—	171,95	265,22	6,73	37,70	,3773
47	,96026	—	175,61	268,86	6,75	37,19	,3722
46	,96089	—	179,43	272,66	6,77	36,67	,3670
100+45	,96152	—	183,43	276,64	6,79	36,15	,3617
44	,96215	—	187,59	280,77	6,82	35,62	,3564
43	,96277	—	191,95	285,12	6,83	35,07	,3509
42	,96340	—	196,52	289,68	6,84	34,52	,3454
41	,96403	—	201,31	294,46	6,85	33,96	,3398
100+40	,96465	—	206,35	299,51	6,84	33,39	,3341
39	,96527	—	211,64	304,80	6,84	32,81	,3283
38	,96590	—	217,20	310,37	6,83	32,22	,3224
37	,96652	—	223,08	316,25	6,83	31,62	,3164
36	,96715	—	229,28	322,46	6,82	31,01	,3103
100+35	,96777	—	235,82	329,00	6,82	30,40	,3041
34	,96841	—	242,77	335,94	6,83	29,77	,2978
33	,96904	—	250,12	343,32	6,80	29,13	,2914
32	,96967	—	257,93	351,15	6,78	28,48	,2850
31	,97032	—	266,26	359,50	6,76	27,82	,2784
100+30	,97095	—	275,13	368,41	6,72	27,14	,2716
29	,97160	—	284,62	377,92	6,70	26,46	,2648
28	,97226	—	294,79	388,12	6,67	25,77	,2578
27	,97293	—	305,71	399,09	6,62	25,06	,2507
26	,97359	—	317,47	410,89	6,58	24,34	,2435
100+25	,97429	—	330,16	423,64	6,52	23,61	,2362
24	,97497	—	343,92	437,45	6,47	22,86	,2287
23	,97567	—	358,87	452,48	6,39	22,10	,2211
22	,97639	—	375,18	468,77	6,36	21,33	,2134
21	,97712	—	393,05	486,77	6,28	20,54	,2055
100+20	,97786	—	412,70	506,50	6,20	19,74	,1975
19	,97863	—	434,42	528,31	6,11	18,93	,1894
18	,97941	—	458,55	552,52	6,03	18,10	,1811
17	,98020	—	485,53	579,59	5,94	17,26	,1726
16	,98104	—	515,87	610,02	5,85	16,40	,1640
100+15	,98189	—	550,27	644,53	5,74	15,52	,1552
14	,98276	—	589,58	683,96	5,62	14,62	,1463
13	,98368	—	634,91	729,43	5,48	13,71	,1372
12	,98463	—	687,84	782,47	5,37	12,78	,1279
11	,98562	—	750,37	845,13	5,24	11,84	,1184
100+10	,98664	—	825,40	920,30	5,10	10,86	,1087
9	,98771	—	917,12	1012,18	4,94	9,88	,0988
8	,98882	—	1031,76	1126,98	4,78	8,87	,0888
7	,99000	—	1179,15	1274,24	4,61	7,84	,0785
6	,99124	—	1375,68	1471,23	4,45	6,80	,0680
100+5	,99252	—	1650,81	1746,55	4,26	5,72	,0573
4	,99388	—	2063,50	2159,43	4,07	4,63	,0463
3	,99532	—	2751,34	2847,43	3,91	3,51	,0351
2	,99683	—	4127,00	4223,27	3,73	2,37	,0237
1	,99842	—	8254,02	8350,45	3,57	1,20	,0120

I. Spirit and water by weight.	II. Specific gravity.	III. Spirit by measure.	IV. Water by measure.	V. Bulk of mixture.	VI. Diminution of bulk.	VII. Quantity of spirit per cent.	VIII. Decimal multipliers.
Sp. + W.							
100 + 0	,82500	100	—	100,00	—	100,00	1,0000
1	,82731	—	0,83	100,72	0,11	99,29	,9929
2	,82957	—	1,65	101,44	0,21	98,58	,9858
3	,83177	—	2,47	102,16	0,31	97,88	,9789
4	,83391	—	3,30	102,89	0,41	97,19	,9719
100 + 5	,83599	—	4,12	103,62	0,50	96,51	,9651
6	,83802	—	4,95	104,35	0,60	95,83	,9583
7	,84001	—	5,77	105,09	0,68	95,16	,9516
8	,84195	—	6,60	105,83	0,77	94,50	,9450
9	,84384	—	7,42	106,57	0,85	93,84	,9384
100 + 10	,84568	—	8,25	107,31	0,94	93,19	,9319
11	,84748	—	9,07	108,05	1,02	92,55	,9255
12	,84924	—	9,90	108,80	1,10	91,91	,9191
13	,85096	—	10,72	109,55	1,17	91,28	,9128
14	,85265	—	11,55	110,30	1,25	90,66	,9066
100 + 15	,85430	—	12,37	111,05	1,32	90,04	,9005
16	,85592	—	13,20	111,81	1,39	89,44	,8944
17	,85750	—	14,02	112,56	1,46	88,84	,8884
18	,85906	—	14,85	113,32	1,53	88,25	,8825
19	,86058	—	15,67	114,08	1,59	87,66	,8766
100 + 20	,86208	—	16,50	114,84	1,66	87,08	,8708
21	,86355	—	17,32	115,60	1,72	86,51	,8651
22	,86500	—	18,15	116,36	1,79	85,94	,8594
23	,86642	—	18,97	117,12	1,85	85,38	,8538
24	,86781	—	19,80	117,88	1,92	84,83	,8483
100 + 25	,86918	—	20,62	118,64	1,98	84,28	,8428
26	,87052	—	21,45	119,41	2,04	83,74	,8374
27	,87183	—	22,27	120,18	2,09	83,21	,8321
28	,87314	—	23,10	120,94	2,16	82,68	,8268
29	,87442	—	23,92	121,71	2,21	82,16	,8216
100 + 30	,87569	—	24,75	122,48	2,27	81,65	,8165
31	,87692	—	25,57	123,24	2,33	81,14	,8114
32	,87814	—	26,40	124,01	2,39	80,64	,8064
33	,87935	—	27,22	124,78	2,44	80,14	,8014
34	,88053	—	28,05	125,55	2,50	79,65	,7965
100 + 35	,88169	—	28,87	126,32	2,55	79,16	,7916
36	,88283	—	29,70	127,09	2,61	78,68	,7868
37	,88395	—	30,52	127,86	2,66	78,21	,7821
38	,88505	—	31,35	128,64	2,71	77,74	,7774
39	,88613	—	32,17	129,41	2,76	77,27	,7727
100 + 40	,88720	—	33,00	130,19	2,81	76,81	,7681
41	,88825	—	33,82	130,96	2,86	76,36	,7636
42	,88929	—	34,65	131,74	2,91	75,91	,7591
43	,89032	—	35,47	132,51	2,96	75,47	,7547
44	,89133	—	36,30	133,29	3,01	75,03	,7503
100 + 45	,89232	—	37,12	134,06	3,06	74,59	,7459
46	,89330	—	37,95	134,84	3,11	74,16	,7416
47	,89427	—	38,77	135,61	3,16	73,74	,7374
48	,89522	—	39,60	136,39	3,21	73,32	,7332
49	,89615	—	40,42	137,17	3,25	72,90	,7290

I. Spirit and water by weight.	II. Specific gravity.	III. Spirit by measure.	IV. Water by measure.	V. Bulk of mixture.	VI. Diminution of bulk.	VII. Quantity of spirit per cent.	VIII. Decimal multipliers.
Sp. + W.							
100 + 50	,89707	100	41,25	137,95	3,30	72,49	,7249
51	,89797	—	42,07	138,73	3,34	72,08	,7208
52	,89886	—	42,90	139,51	3,39	71,68	,7108
53	,89973	—	43,72	140,29	3,43	71,28	,7128
54	,90059	—	44,55	141,07	3,48	70,89	,7089
100 + 55	,90144	—	45,38	141,86	3,52	70,49	,7049
56	,90227	—	46,20	142,64	3,56	70,11	,7011
57	,90309	—	47,02	143,42	3,60	69,72	,6972
58	,90391	—	47,85	144,21	3,64	69,34	,6934
59	,90470	—	48,67	144,99	3,68	68,97	,6897
100 + 60	,90549	—	49,50	145,78	3,72	68,60	,6860
61	,90626	—	50,32	146,56	3,76	68,23	,6823
62	,90703	—	51,15	147,35	3,80	67,87	,6787
63	,90778	—	51,97	148,13	3,84	67,51	,6751
64	,90853	—	52,80	148,92	3,88	67,15	,6715
100 + 65	,90927	—	53,62	149,71	3,91	66,80	,6680
66	,91001	—	54,45	150,50	3,95	66,45	,6645
67	,91074	—	55,27	151,28	3,99	66,10	,6610
68	,91146	—	56,10	152,07	4,03	65,76	,6576
69	,91217	—	56,92	152,85	4,07	65,42	,6542
100 + 70	,91287	—	57,75	153,64	4,11	65,09	,6509
71	,91356	—	58,57	154,42	4,15	64,76	,6476
72	,91424	—	59,40	155,21	4,19	64,43	,6443
73	,91491	—	60,22	156,00	4,22	64,10	,6410
74	,91557	—	61,05	156,79	4,26	63,78	,6378
100 + 75	,91622	—	61,87	157,58	4,29	63,46	,6346
76	,91686	—	62,70	158,37	4,33	63,14	,6314
77	,91748	—	63,52	159,16	4,36	62,83	,6283
78	,91811	—	64,35	159,95	4,40	62,52	,6252
79	,91872	—	65,17	160,74	4,43	62,21	,6221
100 + 80	,91933	—	66,00	161,53	4,47	61,91	,6191
81	,91993	—	66,82	162,32	4,50	61,61	,6161
82	,92052	—	67,65	163,11	4,54	61,31	,6131
83	,92110	—	68,47	163,90	4,57	61,01	,6101
84	,92168	—	69,30	164,70	4,60	60,72	,6072
100 + 85	,92225	—	70,12	165,49	4,63	60,43	,6043
86	,92281	—	70,95	166,28	4,66	60,14	,6014
87	,92336	—	71,77	167,08	4,69	59,85	,5985
88	,92391	—	72,60	167,87	4,73	59,57	,5957
89	,92445	—	73,42	168,66	4,76	59,29	,5929
100 + 90	,92499	—	74,25	169,46	4,79	59,01	,5901
91	,92552	—	75,07	170,25	4,82	58,73	,5873
92	,92604	—	75,90	171,05	4,85	58,46	,5846
93	,92656	—	76,72	171,84	4,88	58,19	,5819
94	,92707	—	77,55	172,64	4,91	57,92	,5792
100 + 95	,92758	—	78,37	173,43	4,94	57,66	,5766
96	,92807	—	79,20	174,23	4,97	57,40	,5740
97	,92856	—	80,02	175,02	5,00	57,14	,5714
98	,92905	—	80,85	175,82	5,03	56,88	,5688
99	,92954	—	81,68	176,62	5,06	56,62	,5662

TABLE II.   HEAT 60°.

I. Water and spirit by weight.	II. Specific gravity.	III. Spirit by measure.	IV. Water by measure.	V. Bulk of mixture.	VI. Diminution of bulk.	VII. Quantity of spirit per cent.	VIII. Decimal multipliers.	I. Water and spirit by weight.	II. Specific gravity.	III. Spirit by measure.	IV. Water by measure.	V. Bulk of mixture.	VI. Diminution of bulk.	VII. Quantity of spirit per cent.	VIII. Decimal multipliers.
W. + Sp.								W. + Sp.							
100+100	,93002	100	82,50	177,41	5,09	56,36	,5636	100 + 50	,95804	100	165,00	258,34	6,66	38,71	,3871
99	,93051	—	83,34	178,22	5,12	56,11	,5611	49	,95867	—	168,37	261,68	6,69	38,21	,3821
98	,93100	—	84,19	179,05	5,14	55,85	,5585	48	,95931	—	171,87	265,16	6,71	37,71	,3771
97	,93149	—	85,02	179,89	5,13	55,59	,5559	47	,95995	—	175,53	268,80	6,73	37,20	,3720
96	,93198	—	85,94	180,74	5,20	55,33	,5533	46	,96058	—	179,35	272,59	6,76	36,68	,3668
100 + 95	,93247	—	86,84	181,61	5,23	55,06	,5506	100 + 45	,96122	—	183,34	276,56	6,78	36,10	,3616
94	,93290	—	87,76	182,50	5,26	54,79	,5479	44	,96185	—	187,50	280,70	6,80	35,63	,3563
93	,93345	—	88,71	183,42	5,29	54,52	,5452	43	,96248	—	191,86	285,05	6,81	35,08	,3508
92	,93394	—	89,67	184,35	5,32	54,24	,5424	42	,96311	—	196,43	289,60	6,83	34,53	,3453
91	,93443	—	90,66	185,31	5,35	53,96	,5396	41	,96374	—	201,21	294,38	6,83	33,97	,3397
100 + 90	,93493	—	91,67	186,29	5,38	53,68	,5368	100 + 40	,96437	—	206,25	299,42	6,83	33,40	,3340
89	,93544	—	92,70	187,29	5,41	53,39	,5339	39	,96500	—	211,54	304,71	6,83	32,82	,3282
88	,93595	—	93,75	188,31	5,44	53,10	,5310	38	,96563	—	217,10	310,28	6,82	32,23	,3223
87	,93646	—	94,83	189,35	5,48	52,81	,5281	37	,96626	—	222,97	316,15	6,82	31,63	,3163
86	,93697	—	95,93	190,42	5,51	52,51	,5251	36	,96689	—	229,17	322,36	6,81	31,02	,3102
100 + 85	,93749	—	97,06	191,53	5,53	52,21	,5221	100 + 35	,96752	—	235,71	328,90	6,81	30,40	,3040
84	,93802	—	98,21	192,65	5,56	51,91	,5191	34	,96816	—	242,65	335,84	6,81	29,78	,2978
83	,93855	—	99,39	193,80	5,59	51,60	,5160	33	,96880	—	250,00	343,21	6,79	29,14	,2914
82	,93909	—	100,61	194,99	5,62	51,29	,5129	32	,96944	—	257,81	351,04	6,77	28,49	,2849
81	,93963	—	101,85	196,20	5,65	50,97	,5097	31	,97009	—	266,13	359,38	6,75	27,83	,2783
100 + 80	,94018	—	103,12	197,44	5,68	50,65	,5065	100 + 30	,97074	—	275,00	368,28	6,72	27,15	,2715
79	,94073	—	104,43	198,71	5,72	50,32	,5032	29	,97139	—	284,48	377,79	6,69	26,47	,2647
78	,94128	—	105,77	200,01	5,76	50,00	,5000	28	,97206	—	294,64	387,99	6,65	25,77	,2577
77	,94184	—	107,14	201,35	5,79	49,66	,4966	27	,97273	—	305,56	398,95	6,61	25,07	,2507
76	,94240	—	108,55	202,73	5,82	49,33	,4933	26	,97340	—	317,31	410,74	6,57	24,35	,2435
100 + 75	,94296	·	110,00	204,15	5,85	48,98	,4898	100 + 25	,97410	—	330,00	423,48	6,52	23,61	,2361
74	,94352	—	111,48	205,60	5,88	48,64	,4864	24	,97479	—	343,75	437,29	6,46	22,87	,2287
73	,94408	—	113,01	207,10	5,91	48,29	,4829	23	,97550	—	358,70	452,31	6,39	22,11	,2211
72	,94465	—	114,58	208,64	5,94	47,93	,4793	22	,97621	—	375,00	468,64	6,36	21,34	,2134
71	,94522	—	116,20	210,22	5,98	47,57	,4757	21	,97696	—	392,86	486,58	6,28	20,55	,2055
100 + 70	,94579	—	117,86	211,84	6,02	47,20	,4720	100 + 20	,97771	—	412,50	506,29	6,21	19,75	,1975
69	,94637	—	119,56	213,51	6,05	46,83	,4683	19	,97848	—	434,21	528,08	6,13	18,94	,1894
68	,94696	—	121,32	215,24	6,08	46,46	,4646	18	,97926	—	458,33	552,29	6,04	18,11	,1811
67	,94756	—	123,13	217,02	6,11	46,08	,4608	17	,98006	—	485,29	579,34	5,95	17,26	,1726
66	,94816	—	125,00	218,85	6,15	45,69	,4569	16	,98090	—	515,62	609,76	5,86	16,40	,1640
100 + 65	,94876	—	126,92	220,74	6,18	45,30	,4530	100 + 15	,98176	—	550,00	644,25	5,75	15,52	,1552
64	,94936	—	128,90	222,69	6,21	44,91	,4491	14	,98264	—	589,29	683,66	5,63	14,63	,1463
63	,94997	—	130,95	224,70	6,24	44,50	,4450	13	,98356	—	634,61	729,10	5,51	13,72	,1372
62	,95058	—	133,06	226,78	6,28	44,10	,4410	12	,98452	—	687,50	782,11	5,39	12,79	,1279
61	,95119	—	135,25	228,93	6,32	43,68	,4368	11	,98551	—	750,00	844,74	5,26	11,84	,1184
100 + 60	,95181	—	137,50	231,14	6,36	43,26	,4326	100 + 10	,98654	—	825,00	919,87	5,13	10,87	,1087
59	,95243	—	139,82	233,44	6,38	42,84	,4284	9	,98761	—	916,67	1011,70	4,97	9,88	,0988
58	,95305	—	142,23	235,82	6,41	42,41	,4241	8	,98873	—	1031,25	1126,44	4,81	8,88	,0888
57	,95368	—	144,73	238,28	6,45	41,97	,4197	7	,98991	—	1178,57	1273,92	4,65	7,85	,0785
56	,95430	—	147,32	240,82	6,50	41,52	,4152	6	,99115	—	1375,00	1470,52	4,48	6,80	,0680
100 + 55	,95493	—	150,00	243,47	6,53	41,07	,4107	100 + 5	,99244	—	1650,00	1745,70	4,30	5,73	,0573
54	,95555	—	152,77	246,22	6,55	40,61	,4061	4	,99380	—	2062,50	2158,37	4,13	4,63	,0463
53	,95617	—	155,65	249,08	6,57	40,15	,4015	3	,99524	—	2750,00	2846,04	3,96	3,51	,0351
52	,95679	—	158,65	252,05	6,60	39,67	,3967	2	,99675	—	4125,00	4221,21	3,79	2,37	,0237
51	,95741	—	161,77	255,14	6,63	39,19	,3919	1	,99834	—	8250,00	8346,38	3,62	1,20	,0120

TABLE I.   HEAT 61°.

I. Spirit and water by weight.	II. Specific gravity.	III. Spirit by measure.	IV. Water by measure.	V. Bulk of mixture.	VI. Diminution of bulk.	VII. Quantity of spirit per cent.	VIII. Decimal multipliers.
Sp. + W.							
100 + 0	,82453	100	—	100,00	—	100,00	,9994
1	,82684	—	0,82	100,72	0,10	99,29	,9924
2	,82910	—	1,65	101,44	0,21	98,58	,9853
3	,83130	—	2,47	102,16	0,31	97,88	,9784
4	,83344	—	3,30	102,89	0,41	97,19	,9714
100 + 5	,83552	—	4,12	103,62	0,50	96,51	,9645
6	,83755	—	4,95	104,35	0,60	95,83	,9578
7	,83954	—	5,77	105,09	0,68	95,16	,9511
8	,84148	—	6,60	105,83	0,77	94,50	,9445
9	,84337	—	7,42	106,57	0,85	93,84	,9379
100 + 10	,84520	—	8,25	107,31	0,94	93,19	,9314
11	,84701	—	9,07	108,05	1,02	92,55	,9250
12	,84877	—	9,90	108,80	1,10	91,91	,9186
13	,85049	—	10,72	109,55	1,17	91,28	,9123
14	,85218	—	11,54	110,30	1,24	90,66	,9061
100 + 15	,85382	—	12,36	111,05	1,31	90,04	,9000
16	,85545	—	13,19	111,81	1,38	89,44	,8939
17	,85703	—	14,02	112,56	1,46	88,84	,8879
18	,85860	—	14,84	113,32	1,52	88,25	,8820
19	,86012	—	15,66	114,08	1,58	87,66	,8762
100 + 20	,86162	—	16,49	114,84	1,65	87,08	,8703
21	,86309	—	17,31	115,60	1,71	86,51	,8646
22	,86454	—	18,14	116,35	1,79	85,94	,8589
23	,86596	—	18,96	117,11	1,85	85,38	,8534
24	,86735	—	19,79	117,87	1,92	84,83	,8479
100 + 25	,86872	—	20,61	118,64	1,97	84,28	,8424
26	,87006	—	21,44	119,40	2,04	83,74	,8370
27	,87137	—	22,26	120,17	2,09	83,21	,8316
28	,87268	—	23,09	120,94	2,15	82,68	,8264
29	,87396	—	23,91	121,71	2,20	82,16	,8212
100 + 30	,87523	—	24,74	122,48	2,26	81,65	,8161
31	,87646	—	25,56	123,23	2,33	81,15	,8110
32	,87768	—	26,39	124,00	2,39	80,64	,8060
33	,87889	—	27,21	124,78	2,43	80,14	,8010
34	,88007	—	28,04	125,55	2,49	79,65	,7961
100 + 35	,88123	—	28,86	126,32	2,54	79,16	,7912
36	,88237	—	29,69	127,09	2,60	78,68	,7864
37	,88349	—	30,51	127,85	2,66	78,21	,7817
38	,88459	—	31,33	128,63	2,70	77,74	,7770
39	,88567	—	32,16	129,40	2,76	77,28	,7723
100 + 40	,88674	—	32,98	130,18	2,80	76,82	,7677
41	,88780	—	33,80	130,95	2,85	76,36	,7632
42	,88884	—	34,63	131,73	2,90	75,91	,7588
43	,88987	—	35,45	132,50	2,95	75,47	,7544
44	,89089	—	36,28	133,28	3,00	75,03	,7500
100 + 45	,89186	—	37,10	134,05	3,05	74,60	,7456
46	,89285	—	37,93	134,83	3,10	74,17	,7413
47	,89382	—	38,75	135,60	3,15	73,74	,7371
48	,89477	—	39,58	136,38	3,20	73,32	,7329
49	,89570	—	40,40	137,16	3,24	72,91	,7286

I. Spirit and water by weight.	II. Specific gravity.	III. Spirit by measure.	IV. Water by measure.	V. Bulk of mixture.	VI. Diminution of bulk.	VII. Quantity of spirit per cent.	VIII. Decimal multipliers.
Sp. + W.							
100 + 50	,89662	100	41,23	137,94	3,29	72,50	,7245
51	,89752	—	42,05	138,72	3,33	72,09	,7205
52	,89841	—	42,88	139,50	3,38	71,69	,7165
53	,89928	—	43,70	140,28	3,42	71,28	,7125
54	,90014	—	44,52	141,06	3,46	70,89	,7086
100 + 55	,90100	—	45,35	141,85	3,50	70,50	,7046
56	,90183	—	46,17	142,62	3,55	70,12	,7007
57	,90265	—	46,99	143,40	3,59	69,73	,6968
58	,90347	—	47,82	144,19	3,63	69,35	,6931
59	,90426	—	48,64	144,97	3,67	68,97	,6893
100 + 60	,90504	—	49,47	145,76	3,71	68,60	,6856
61	,90582	—	50,29	146,54	3,75	68,24	,6820
62	,90659	—	51,12	147,33	3,79	67,88	,6783
63	,90734	—	51,94	148,11	3,83	67,52	,6748
64	,90809	—	52,77	148,90	3,87	67,16	,6712
100 + 65	,90883	—	53,59	149,69	3,90	66,80	,6676
66	,90957	—	54,42	150,48	3,94	66,45	,6642
67	,91030	—	55,24	151,26	3,98	66,11	,6607
68	,91102	—	56,07	152,05	4,02	65,76	,6573
69	,91173	—	56,89	152,83	4,06	65,42	,6539
100 + 70	,91242	—	57,72	153,62	4,10	65,09	,6506
71	,91312	—	58,54	154,41	4,13	64,76	,6473
72	,91380	—	59,37	155,19	4,18	64,43	,6440
73	,91447	—	60,19	155,98	4,21	64,10	,6407
74	,91513	—	61,02	156,77	4,25	63,78	,6375
100 + 75	,91578	—	61,84	157,56	4,28	63,47	,6343
76	,91642	—	62,67	158,35	4,32	63,15	,6311
77	,91704	—	63,49	159,14	4,35	62,84	,6280
78	,91767	—	64,32	159,93	4,39	62,52	,6249
79	,91829	—	65,14	160,72	4,42	62,21	,6218
100 + 80	,91890	—	65,90	161,51	4,45	61,91	,6188
81	,91950	—	66,78	162,30	4,48	61,61	,6158
82	,92009	—	67,61	163,09	4,52	61,31	,6128
83	,92067	—	68,43	163,88	4,55	61,01	,6098
84	,92125	—	69,26	164,68	4,58	60,72	,6069
100 + 85	,92182	—	70,08	165,48	4,60	60,43	,6040
86	,92238	—	70,91	166,27	4,64	60,15	,6011
87	,92293	—	71,73	167,07	4,66	59,86	,5983
88	,92348	—	72,56	167,86	4,70	59,58	,5954
89	,92402	—	73,38	168,64	4,74	59,29	,5926
100 + 90	,92455	—	74,21	169,44	4,77	59,01	,5898
91	,92509	—	75,03	170,24	4,79	58,74	,5871
92	,92562	—	75,86	171,03	4,83	58 47	,5844
93	,92614	—	76,68	171,82	4,86	58,20	,5816
94	,92665	—	77,51	172,62	4,89	57,93	,5789
100 + 95	,92716	—	78,33	173,41	4,92	57,66	,5763
96	,92765	—	79,16	174,20	4,96	57,40	,5738
97	,92814	—	79,98	175,00	4,98	57,14	,5712
98	,92864	—	80,81	175,80	5,01	56,88	,5686
99	,92913	—	81,64	176,59	5,05	56,63	,5660

TABLE II. HEAT 61°.

I. Water and spirit by weight.	II. Specific gravity.	III. Spirit by measure.	IV. Water by measure.	V. Bulk of mixture.	VI. Diminution of bulk.	VII. Quantity of spirit per cent.	VIII. Decimal multipliers.
W. + Sp.							
100+100	,92961	100	82,46	177,38	5,08	56,37	,5634
99	,93010	—	83,29	178,19	5,10	56,11	,5609
98	,93059	—	84,14	179.02	5,12	55,86	,5583
97	,93108	—	84.98	179,86	5,12	55,60	,5557
96	,93157	—	85.89	180,71	5,18	55,33	,5531
100+95	,93206	—	86,79	181.59	5,20	55,07	,5504
94	,93255	—	87,72	182.47	5,25	54,80	,5477
93	,93304	—	88,66	183,40	5,26	54,53	,5450
92	,93353	—	89,62	184,32	5,30	54,25	,5422
91	,93402	—	90.61	185,29	5,32	53,97	,5394
100+90	,93452	—	91,62	186,26	5,36	53,69	,5366
89	,93503	—	92,65	187.26	5,39	53,40	,5337
88	,93554	—	93,70	188,28	5,42	53,11	,5308
87	,93605	—	94,78	189,13	5,45	52,81	,5279
86	,93657	—	95.88	190.40	5,48	52,52	,5249
100+85	,93709	—	97,01	191.50	5,51	52,22	,5219
84	,93762	—	98,16	192,63	5,53	51,91	,5189
83	,93815	—	99,34	193,78	5,56	51,61	,5158
82	,93870	—	100.56	194,96	5,60	51,30	,5127
81	,93924	—	101,80	196.17	5,63	50,98	,5095
100+80	,93979	—	103,07	197.41	5,66	50,66	,5063
79	,94034	—	104,38	198.68	5,70	50,33	,5030
78	,94089	—	105,71	199.97	5,74	50,00	,4998
77	,94145	—	107 09	201,32	5,77	49,67	,4964
76	,94201	—	108,49	202,69	5 80	49,33	,4931
100+75	,94257	—	109,94	204,11	5,83	48,99	,4897
74	,94313	—	111,42	205.56	5,86	48,65	,4862
73	,94369	—	112,95	207.06	5,89	48,30	,4827
72	,94427	—	114,52	208.60	5,92	47,94	,4791
71	,94484	—	116,13	210.18	5,95	47,58	,4756
100+70	,94540	—	117,80	211 80	6,00	47,21	,4719
69	,94559	—	119,50	213,48	6,02	46,84	,4682
68	,94659	—	121,26	215.20	6,06	46,47	,4645
67	,94719	—	123,07	216 98	6,09	46,09	,4607
66	,94779	—	124,94	218 81	6,13	45,70	,4508
100+65	,94839	—	126 86	220,70	6,16	45,31	,4526
64	,94899	—	128 83	222,65	6,18	44,91	,4489
63	,94951	—	130,88	224 66	6,22	44,51	,4449
62	,95022	—	132,99	226,74	6,25	44,10	,4409
61	,95083	—	135,18	228.88	6,30	43,69	,4367
100+60	,95145	—	137,43	231·10	6,33	43,27	,4325
59	,95207	—	139 75	233,40	6,35	42,85	,4283
58	,95270	—	142,16	235.77	6,39	42,41	,4240
57	,95333	—	144,66	238.23	6,43	41,97	,4196
56	,95395	—	147,25	240,77	6,48	41,53	,4151
100+55	,95458	—	149,92	243,42	6,50	41,08	,4106
54	,95521	—	152,69	246.16	6,53	40,62	,4060
53	,95583	—	155,57	249,02	6,55	40,15	,4014
52	,95645	—	158,57	252,00	6,57	39,68	,3966
51	,95707	—	161,69	255,08	6,61	39,20	,3918
W. + Sp.							
100+50	,95771	100	164 92	258,28	6,64	38,71	,3869
49	,95834	—	168,29	261,62	6,67	38,22	,3820
48	,95898	—	171,78	265,10	6,68	37,72	,3770
47	,95902	—	175,44	268,73	6,71	37,21	,3719
46	,96026	—	179,26	272,52	6,74	36,69	,3667
100+45	,96090	—	183,25	276,49	6,76	36,16	,3615
44	,96154	—	187,41	280,63	6,78	35,63	,3562
43	,96217	—	191,76	284,97	6,79	35,08	,3507
42	,96281	—	196,33	289.52	6,81	34,53	,3452
41	,96344	—	201,11	294.30	6,81	33,97	,3396
100+40	,96408	—	206,15	299,34	6,81	33,40	,3339
39	,96471	—	211,43	304,62	6,81	32,83	,3281
38	,96535	—	216,99	310,19	6,80	32,24	,3222
37	,96598	—	222.86	316,06	6,80	31,64	,3162
36	,96662	—	229,06	322,25	6,81	31,03	,3101
100+35	,96725	—	235,59	328,79	6,80	30,41	,3040
34	,96791	—	242,53	335,73	6,80	29,78	,2977
33	,96855	—	249,88	343,10	6 78	29,14	,2913
32	,96920	—	257,68	350,93	6,75	28,49	,2848
31	,96986	—	266,00	359,26	6,74	27,83	,2782
100+30	,97051	—	274,86	368,16	6,70	27,16	,2715
29	,97117	—	284,34	377,65	6,69	26,48	,2647
28	,97185	—	294,50	387,85	6,65	25,78	,2577
27	,97252	—	305,41	398,80	6,61	25,08	,2507
26	,97320	—	317,15	410,58	6,57	24,36	,2435
100+25	,97389	—	329,84	423,32	6,52	23,62	,2361
24	,97460	—	343,58	437,12	6 46	22,87	,2287
23	,97532	—	358,52	452,13	6,39	22,12	,2211
22	,97604	—	374,82	468,45	6,37	21,35	,2134
21	,97679	—	392,67	486.38	6,29	20,56	,2055
100+20	,97755	—	412,30	506,08	6,22	19,76	,1975
19	,97832	—	434,00	527,86	6,14	18,95	,1894
18	,97911	—	458,11	552,05	6,06	18,11	,1811
17	,97992	—	485,06	579,09	5,97	17,27	,1726
16	,98076	—	515,37	609.50	5,87	16,41	,1640
100+15	,98162	—	549,73	643,97	5,76	15,53	,1552
14	,98251	—	589,00	683,36	5,64	14,63	,1463
13	,98343	—	634,30	728,77	5,53	13,72	,1372
12	,98440	—	687,17	781,76	5,41	12.79	,1279
11	,98539	—	749,64	844.36	5.28	11,84	,1184
100+10	,98642	—	824,60	919.45	5,15	10,87	,1087
9	,98750	—	916,23	1011,23	5,00	9,89	,0988
8	,98862	—	1030,75	1125,91	4,84	8,88	,0888
7	,98981	—	1178,00	1273.32	4,68	7,85	,0785
6	,99105	—	1374 34	1469,82	4.52	6,80	,0680
100+5	,99234	—	1649,21	1744,86	4.25	5,73	,0573
4	,99371	—	2061 51	2157,33	4,18	4,63	,0463
3	,99515	—	2748.67	2844 60	4,01	3,51	,0351
2	,99666	—	4123,01	4219,16	3,85	2,37	,0237
1	,99825	—	8246 02	8342,33	3.69	1,20	,0120

TABLE I. HEAT 62°.

I. Spirit and water by weight.	II. Specific gravity.	III. Spirit by measure.	IV. Water by measure.	V. Bulk of mixture.	VI. Diminution of bulk.	VII. Quantity of spirit per cent.	VIII. Decimal multipliers.	I. Spirit and water by weight.	II. Specific gravity.	III. Spirit by measure.	IV. Water by measure.	V. Bulk of mixture.	VI. Diminution of bulk.	VII. Quantity of spirit per cent.	VIII. Decimal multipliers.
Sp. + W.								Sp. + W.							
100 + 0	,82405	100	—	100,00	—	100,00	,9988	100 + 50	,89617	100	41,21	137,93	3,28	72,50	,7242
1	,82637	—	0,82	100,72	0,10	99,29	,9918	51	,89707	—	42,03	138,71	3,32	72,09	,7201
2	,82863	—	1,65	101,44	0,21	98,58	,9848	52	,89796	—	42,86	139,49	3,37	71,69	,7161
3	,83083	—	2,47	102,16	0,31	97,88	,9778	53	,89883	—	43,68	140,27	3,41	71,29	,7121
4	,83297	—	3,30	102,89	0,41	97,19	,9709	54	,89969	—	44,50	141,05	3,45	70,90	,7082
100 + 5	,83504	—	4,12	103,62	0,50	96,51	,9640	100 + 55	,90055	—	45,33	141,83	3,50	70,51	,7042
6	,83708	—	4,95	104,35	0,60	95,83	,9573	56	,90138	—	46,15	142,61	3,54	70,12	,7004
7	,83907	—	5,77	105,09	0,68	95,16	,9506	57	,90221	—	46,97	143,39	3,58	69,73	,6965
8	,84101	—	6,59	105,83	0,76	94,50	,9440	58	,90302	—	47,80	144,18	3,62	69,35	,6928
9	,84290	—	7,42	106,57	0,85	93,84	,9374	59	,90382	—	48,62	144,96	3,66	68,98	,6890
100 + 10	,84473	—	8,24	107,31	0,93	93,19	,9309	100 + 60	,90460	—	49,45	145,75	3,70	68,61	,6853
11	,84654	—	9,06	108,05	1,01	92,55	,9245	61	,90538	—	50,27	146,53	3,74	68,24	,6816
12	,84830	—	9,89	108,80	1,09	91,91	,9181	62	,90615	—	51,10	147,32	3,78	67,88	,6780
13	,85002	—	10,71	109,55	1,16	91,28	,9118	63	,90690	—	51,92	148,10	3,82	67,52	,6745
14	,85171	—	11,54	110,30	1,24	90,66	,9056	64	,90765	—	52,75	148,89	3,86	67,16	,6709
100 + 15	,85335	—	12,36	111,05	1,31	90,04	,8995	100 + 65	,90839	—	53,56	149,68	3,88	66,81	,6673
16	,85498	—	13,18	111,80	1,38	89,44	,8934	66	,90913	—	54,40	150,47	3,93	66,46	,6639
17	,85656	—	14,01	112,56	1,45	88,84	,8874	67	,90986	—	55,21	151,25	3,96	66,11	,6604
18	,85813	—	14,83	113,31	1,52	88,25	,8815	68	,91058	—	56,05	152,04	4,01	65,77	,6570
19	,85965	—	15,65	114,07	1,58	87,66	,8757	69	,91129	—	56,87	152,82	4,05	65,43	,6536
100 + 20	,86115	—	16,48	114,83	1,65	87,08	,8698	100 + 70	,91198	—	57,69	153,61	4,08	65,10	,6503
21	,86263	—	17,30	115,59	1,71	86,51	,8641	71	,91268	—	58,51	154,39	4,12	64,77	,6470
22	,86407	—	18,13	116,35	1,78	85,95	,8585	72	,91336	—	59,34	155,18	4,16	64,44	,6437
23	,86549	—	18,95	117,11	1,84	85,39	,8529	73	,91403	—	60,16	155,97	4,19	64,11	,6404
24	,86689	—	19,78	117,87	1,91	84,84	,8474	74	,91469	—	60,99	156,76	4,23	63,79	,6372
100 + 25	,86825	—	20,60	118,63	1,97	84,29	,8419	100 + 75	,91533	—	61,81	157,55	4,26	63,47	,6340
26	,86960	—	21,43	119,40	2,03	83,75	,8365	76	,91598	—	62,64	158,34	4,30	63,16	,6308
27	,87091	—	22,25	120,17	2,08	83,22	,8312	77	,91660	—	63,46	159,13	4,33	62,84	,6277
28	,87222	—	23,08	120,93	2,15	82,69	,8259	78	,91723	—	64,29	159,92	4,37	62,53	,6246
29	,87350	—	23,90	121,70	2,20	82,17	,8208	79	,91786	—	65,11	160,71	4,40	62,22	,6215
100 + 30	,87476	—	24,73	122,47	2,26	81,65	,8156	100 + 80	,91847	—	65,93	161,50	4,43	61,92	,6185
31	,87600	—	25,55	123,23	2,32	81,15	,8106	81	,91907	—	66,75	162,29	4,40	61,62	,6155
32	,87722	—	26,38	124,00	2,38	80,65	,8055	82	,91966	—	67,58	163,07	4,51	61,32	,6125
33	,87843	—	27,19	124,77	2,42	80,15	,8006	83	,92024	—	68,40	163,87	4,53	61,02	,6095
34	,87961	—	28,02	125,54	2,48	79,65	,7957	84	,92082	—	69,23	164,67	4,56	60,73	,6066
100 + 35	,88076	—	28,85	126,31	2,54	79,17	,7908	100 + 85	,92139	—	70,05	165,46	4,59	60,44	,6037
36	,88191	—	29,67	127,08	2,59	78,69	,7860	86	,92195	—	70,88	166,25	4,63	60,15	,6008
37	,88303	—	30,49	127,85	2,64	78,22	,7813	87	,92250	—	71,70	167,05	4,65	59,87	,5980
38	,88413	—	31,32	128,62	2,70	77,75	,7766	88	,92305	—	72,52	167,84	4,68	59,58	,5951
39	,88521	—	32,14	129,39	2,75	77,28	,7719	89	,92359	—	73,35	168,63	4,72	59,30	,5924
100 + 40	,88628	—	32,97	130,17	2,80	76,82	,7674	100 + 90	,92412	—	74,18	169,43	4,75	59,00	,5896
41	,88734	—	33,79	130,94	2,85	76,37	,7628	91	,92466	—	75,00	170,22	4,78	58,74	,5868
42	,88839	—	34,62	131,72	2,90	75,92	,7584	92	,92519	—	75,83	171,01	4,82	58,47	,5841
43	,88942	—	35,44	132,49	2,95	75,48	,7540	93	,92571	—	76,65	171,81	4,84	58,20	,5814
44	,89043	—	36,26	133,27	2,99	75,04	,7496	94	,92623	—	77,47	172,60	4,87	57,93	,5787
100 + 45	,89141	—	37,09	134,04	3,05	74,60	,7452	100 + 95	,92674	—	78,29	173,39	4,90	57,67	,5761
46	,89240	—	37,91	134,82	3,09	74,17	,7409	96	,92723	—	79,12	174,18	4,94	57,41	,5735
47	,89337	—	38,73	135,59	3,14	73,75	,7367	97	,92772	—	79,94	174,98	4,96	57,15	,5709
48	,89432	—	39,56	136,37	3,19	73,33	,7325	98	,92822	—	80,76	175,77	4,99	56,89	,5683
49	,89525	—	40,38	137,15	3,23	72,91	,7283	99	,92871	—	81,59	176,57	5,02	56,64	,5657

I. Water and spirit by weight.	II. Specific gravity.	III. Spirit by measure.	IV. Water by measure.	V. Bulk of mixture.	VI. Diminution of bulk.	VII. Quantity of spirit per cent.	VIII. Decimal multipliers.
W. + Sp.							
100+100	,92920	100	82,42	177,36	5,06	56,38	,5631
99	,92969	—	83,25	178,17	5,08	56,12	,5606
98	,93018	—	84,10	179,00	5,10	55,87	,5580
97	,93066	—	84,94	179,84	5,10	55,60	,5554
96	,93116	—	85,85	180,69	5,16	55,34	,5528
100+95	,93164	—	86,75	181,56	5,19	55,08	,5501
94	,93213	—	87,68	182,45	5,23	54,81	,5474
93	,93262	—	88,62	183,37	5,25	54,54	,5447
92	,93311	—	89,58	184,30	5,28	54,26	,5419
91	,93360	—	90,57	185,26	5,31	53,98	,5391
100+90	,93411	—	91,58	186,24	5,34	53,70	,5363
89	,93461	—	92,61	187,24	5,37	53,41	,5335
88	,93512	—	93,66	188,26	5,40	53,12	,5306
87	,93564	—	94,74	189,30	5,44	52,82	,5277
86	,93616	—	95,84	190,37	5,47	52,52	,5247
100+85	,93669	—	96,97	191,47	5,50	52,22	,5217
84	,93721	—	98,12	192,60	5,52	51,92	,5187
83	,93775	—	99,30	193,75	5,55	51,62	,5156
82	,93830	—	100,51	194,93	5,58	51,31	,5125
81	,93884	—	101,75	196,13	5,61	50,99	,5093
100+80	,93940	—	103,02	197,37	5,65	50,67	,5061
79	,93995	—	104,33	198,64	5,69	50,34	,5028
78	,94050	—	105,66	199,94	5,72	50,01	,4996
77	,94105	—	107,03	201,29	5,74	49,68	,4962
76	,94162	—	108,44	202,66	5,78	49,34	,4929
100+75	,94218	—	109,89	204,08	5,81	49,00	,4894
74	,94274	—	111,37	205,53	5,84	48,66	,4860
73	,94330	—	112,89	207,03	5,86	48,30	,4825
72	,94388	—	114,46	208,57	5,89	47,95	,4789
71	,94445	—	116,07	210,14	5,93	47,59	,4754
100+70	,94502	—	117,74	211,77	5,97	47,22	,4717
69	,94561	—	119,44	213,44	6,00	46,85	,4680
68	,94621	—	121,20	215,16	6,04	46,47	,4643
67	,94681	—	123,01	216,94	6,07	46,10	,4605
66	,94741	—	124,88	218,77	6,11	45,71	,4566
100+65	,94801	—	126,80	220,65	6,15	45,32	,4527
64	,94862	—	128,77	222,60	6,17	44,92	,4488
63	,94924	—	130,81	224,61	6,20	44,52	,4447
62	,94985	—	132,92	226,70	6,22	44,11	,4407
61	,95046	—	135,11	228,84	6,27	43,70	,4365
100+60	,95109	—	137,37	231,05	6,32	43,28	,4323
59	,95171	—	139,69	233,35	6,34	42,86	,4281
58	,95234	—	142,09	235,72	6,37	42,42	,4238
57	,95297	—	144,59	238,18	6,41	41,98	,4194
56	,95360	—	147,18	240,72	6,46	41,54	,4149
100+55	,95423	—	149,85	243,37	6,48	41,09	,4104
54	,95486	—	152,61	246,10	6,51	40,63	,4058
53	,95548	—	155,49	248,96	6,53	40,16	,4012
52	,95610	—	158,48	251,94	6,54	39,69	,3965
51	,95673	—	161,60	255,02	6,58	39,21	,3917

I. Water and spirit by weight.	II. Specific gravity.	III. Spirit by measure.	IV. Water by measure.	V. Bulk of mixture.	VI. Diminution of bulk.	VII. Quantity of spirit per cent.	VIII. Decimal multipliers.
W. + Sp.							
100+50	,95737	100	164,84	258,23	6,61	38,72	,3868
49	,95800	—	168,21	261,56	6,65	38,23	,3819
48	,95865	—	171,70	265,04	6,66	37,73	,3769
47	,95929	—	175,35	268,67	6,68	37,22	,3718
46	,95993	—	179,17	272,46	6,71	36,70	,3666
100+45	,96058	—	183,16	276,42	6,74	36,17	,3613
44	,96122	—	187,32	280,56	6,76	35,64	,3561
43	,96186	—	191,67	284,90	6,77	35,09	,3506
42	,96250	—	196,23	289,45	6,78	34,54	,3451
41	,96313	—	201,01	294,23	6,78	33,98	,3395
10+40	,96378	—	206,05	299,26	6,79	33,41	,3338
39	,96442	—	211,33	304,53	6,80	32,84	,3280
38	,96506	—	216,89	310,10	6,79	32,25	,3221
37	,96570	—	222,75	315,96	6,79	31,65	,3161
36	,96635	—	228,95	322,15	6,80	31,04	,3100
100+35	,96698	—	235,48	328,69	6,79	30,42	,3039
34	,96765	—	242,41	335,62	6,79	29,79	,2977
33	,96830	—	249,76	343,00	6,78	29,15	,2913
32	,96895	—	257,56	350,82	6,74	28,50	,2847
31	,96962	—	265,87	359,15	6,72	27,84	,2781
100+30	,97028	—	274,73	368,03	6,70	27,17	,2714
29	,97095	—	284,20	377,52	6,68	26,49	,2646
28	,97163	—	294,36	387,70	6,66	25,79	,2576
27	,97231	—	305,26	398,65	6,61	25,09	,2506
26	,97299	—	317,00	410,42	6,58	24,37	,2434
100+25	,97369	—	329,68	423,16	6,52	23,63	,2360
24	,97440	—	343,41	436,95	6,46	22,88	,2286
23	,97513	—	358,35	451,95	6,40	22,13	,2210
22	,97586	—	374,64	468,26	6,38	21,35	,2133
21	,97662	—	392,48	486,18	6,30	20,57	,2054
100+20	,97740	—	412,10	505,87	6,23	19,77	,1974
19	,97816	—	433,79	527,64	6,15	18,95	,1893
18	,97895	—	457,89	551,82	6,07	18,12	,1810
17	,97977	—	484,83	578,84	5,99	17,28	,1726
16	,98061	—	515,12	609,24	5,88	16,42	,1640
100+15	,98148	—	549,47	643,69	5,78	15,54	,1552
14	,98237	—	588,72	683,06	5,66	14,64	,1463
13	,98330	—	634,00	728,44	5,56	13,73	,1371
12	,98427	—	686,84	781,41	5,43	12,80	,1278
11	,98526	—	749,28	843,98	5,30	11,85	,1184
100+10	,98630	—	824,20	919,03	5,17	10,88	,1087
9	,98738	—	915,79	1010,76	5,03	9,89	,0987
8	,98851	—	1030,26	1125,38	4,88	8,88	,0888
7	,98970	—	1177,44	1272,72	4,72	7,86	,0785
6	,99094	—	1373,68	1469,12	4,56	6,81	,0680
100+5	,99224	—	1648,42	1744,00	4,39	5,73	,0573
4	,99361	—	2060,52	2156,29	4,23	4,64	,0463
3	,99505	—	2747,35	2843,29	4,06	3,52	,0351
2	,99656	—	4121,03	4217,12	3,91	2,37	,0237
1	,99815	—	8242,06	8338,30	3,76	1,20	,0120

TABLE I.   HEAT 63°.

I. Spirit and water by weight.	II. Specific gravity.	III. Spirit by measure.	IV. Water by measure.	V. Bulk of mixture.	VI. Diminution of bulk.	VII. Quantity of spirit per cent.	VIII. Decimal multipliers.	I. Spirit and water by weight.	II. Specific gravity.	III. Spirit by measure.	IV. Water by measure.	V. Bulk of mixture.	VI. Diminution of bulk.	VII. Quantity of spirit per cent.	VIII. Decimal multipliers.
Sp. + W.								Sp. + W.							
100 + 0	,82357	100	—	100,00	—	100,00	,9983	100 + 50	,89571	100	41,19	137,92	3,27	72,51	,7238
1	,82589	—	0,82	100,72	0,10	99,29	,9912	51	,89661	—	42,01	138,70	3,31	72,10	,7198
2	,82815	—	1,65	101,44	0,21	98,58	,9842	52	,89750	—	42,84	139,48	3,35	71,69	,7158
3	,83035	—	2,47	102,16	0,31	97,88	,9772	53	,89838	—	43,66	140,26	3,40	71,29	,7117
4	,83250	—	3,30	102,89	0,41	97,19	,9703	54	,89924	—	44,48	141,04	3,44	70,90	,7078
100 + 5	,83457	—	4,12	103,62	0,50	96,51	,9634	100 + 55	,90010	—	45,31	141,82	3,49	70,51	,7039
6	,83661	—	4,94	104,35	0,59	95,83	,9567	56	,90094	—	46,13	142,60	3,53	70,13	,7001
7	,83860	—	5,77	105,09	0,68	95,16	,9501	57	,90177	—	46,95	143,38	3,57	69,74	,6962
8	,84054	—	6,59	105,83	0,76	94,50	,9435	58	,90258	—	47,78	144,17	3,61	69,36	,6924
9	,84243	—	7,42	106,57	0,85	93,84	,9369	59	,90337	—	48,60	144,95	3,65	68,99	,6885
100 + 10	,84426	—	8,24	107,31	0,93	93,19	,9304	100 + 60	,90416	—	49,42	145,73	3,69	68,62	,6850
11	,84607	—	9,06	108,05	1,01	92,55	,9240	61	,90494	—	50,24	146,52	3,72	68,25	,6813
12	,84783	—	9,89	108,80	1,09	91,91	,9176	62	,90571	—	51,07	147,31	3,76	67,89	,6777
13	,84955	—	10,71	109,55	1,16	91,28	,9113	63	,90646	—	51,89	148,09	3,80	67,53	,6741
14	,85124	—	11,53	110,30	1,23	90,66	,9051	64	,90721	—	52,72	148,87	3,83	67,17	,6705
100 + 15	,85288	—	12,35	111,05	1,30	90,05	,8990	100 + 65	,90795	—	53,54	149,66	3,88	66,82	,6670
16	,85451	—	13,18	111,80	1,38	89,44	,8929	66	,90869	—	54,37	150,45	3,92	66,47	,6636
17	,85609	—	14,00	112,56	1,44	88,85	,8869	67	,90941	—	55,19	151,23	3,96	66,12	,6601
18	,85760	—	14,82	113,31	1,51	88,25	,8810	68	,91014	—	56,02	152,02	4,00	65,78	,6567
19	,85919	—	15,65	114,07	1,58	87,66	,8752	69	,91085	—	56,84	152,80	4,04	65,44	,6533
100 + 20	,86069	—	16,47	114,83	1,64	87,08	,8694	100 + 70	,91154	—	57,66	153,59	4,07	65,11	,6500
21	,86217	—	17,29	115,58	1,71	86,51	,8637	71	,91224	—	58,48	154,38	4,10	64,78	,6467
22	,86361	—	18,12	116,34	1,78	85,95	,8580	72	,91291	—	59,31	155,17	4,14	64,45	,6434
23	,86503	—	18,95	117,10	1,85	85,39	,8525	73	,91358	—	60,13	155,95	4,18	64,12	,6401
24	,86642	—	19,77	117,87	1,90	84,84	,8470	74	,91425	—	60,96	156,74	4,22	63,80	,6369
100 + 25	,86779	—	20,60	118,63	1,97	84,29	,8415	100 + 75	,91488	—	61,78	157,53	4,25	63,48	,6337
26	,86914	—	21,42	119,39	2,03	83,75	,8361	76	,91554	—	62,61	158,32	4,29	63,16	,6305
27	,87045	—	22,24	120,16	2,08	83,22	,8307	77	,91616	—	63,43	159,11	4,32	62,85	,6274
28	,87176	—	23,07	120,92	2,15	82,69	,8255	78	,91679	—	64,26	159,90	4,36	62,54	,6243
29	,87304	—	23,89	121,69	2,20	82,17	,8203	79	,91742	—	65,08	160,69	4,39	62,23	,6212
100 + 30	,87430	—	24,71	122,46	2,25	81,66	,8152	100 + 80	,91803	—	65,90	161,48	4,42	61,93	,6182
31	,87553	—	25,53	123,22	2,31	81,15	,8102	81	,91863	—	66,72	162,27	4,45	61,63	,6152
32	,87676	—	26,36	123,99	2,37	80,65	,8051	82	,91922	—	67,55	163,06	4,49	61,33	,6122
33	,87797	—	27,18	124,76	2,42	80,15	,8002	83	,91981	—	68,37	163,85	4,52	61,03	,6092
34	,87915	—	28,01	125,53	2,48	79,66	,7953	84	,92039	—	69,19	164,65	4,54	60,74	,6063
100 + 35	,88030	—	28,83	126,30	2,53	79,17	,7904	100 + 85	,92090	—	70,02	165,44	4,58	60,45	,6034
36	,88145	—	29,66	127,07	2,59	78,69	,7856	86	,92152	—	70,84	166,23	4,61	60,16	,6005
37	,88257	—	30,48	127,84	2,64	78,22	,7809	87	,92207	—	71,67	167,03	4,64	59,87	,5977
38	,88367	—	31,30	128,62	2,68	77,75	,7762	88	,92262	—	72,49	167,82	4,67	59,59	,5949
39	,88475	—	32,13	129,39	2,74	77,29	,7715	89	,92316	—	73,32	168,61	4,71	59,31	,5921
100 + 40	,88582	—	32,95	130,17	2,78	76,83	,7669	100 + 90	,92369	—	74,14	169,41	4,73	59,03	,5893
41	,88686	—	33,77	130,94	2,83	76,37	,7624	91	,92423	—	74,96	170,20	4,76	58,75	,5866
42	,88793	—	34,60	131,71	2,89	75,92	,7580	92	,92476	—	75,79	170,99	4,80	58,48	,5838
43	,88896	—	35,42	132,48	2,94	75,48	,7536	93	,92528	—	76,61	171,79	4,82	58,21	,5811
44	,88998	—	36,25	133,26	2,99	75,04	,7492	94	,92580	—	77,43	172,58	4,85	57,94	,5785
100 + 45	,89096	—	37,07	134,03	3,04	74,61	,7448	100 + 95	,92632	—	78,25	173,37	4,88	57,68	,5758
46	,89195	—	37,89	134,81	3,08	74,18	,7405	96	,92681	—	79,08	174,15	4,93	57,42	,5732
47	,89292	—	38,71	135,58	3,13	73,75	,7363	97	,92730	—	79,90	174,95	4,95	57,16	,5706
48	,89386	—	39,54	136,36	3,18	73,33	,7321	98	,92780	—	80,72	175,75	4,97	56,90	,5681
49	,89479	—	40,36	137,14	3,22	72,92	,7279	99	,92829	—	81,55	176,55	5,00	56,64	,5655

TABLE II.　　　　HEAT 63°.

I. Water and spirit by weight.	II. Specific gravity.	III. Spirit by measure.	IV. Water by measure.	V. Bulk of mixture.	VI. Diminution of bulk.	VII. Quantity of spirit per cent.	VIII. Decimal multipliers.
W. + Sp.							
100+100	,92878	100	82,38	177,34	5,04	56,38	,5629
99	,92928	—	83,21	178,15	5,06	56,13	,5604
98	,92976	—	84,06	178,97	5,09	55,87	,5578
97	,93025	—	84,90	179,81	5,09	55,61	,5552
96	,93074	—	85,81	180,66	5,15	55,35	,5526
100+95	,93122	—	86,71	181,54	5,17	55,08	,5499
94	,93172	—	87,64	182,42	5,22	54,81	,5472
93	,93220	—	88,58	183,34	5,24	54,54	,5445
92	,93269	—	89,54	184,27	5,27	54,26	,5417
91	,93318	—	90,52	185,24	5,28	53,99	,5389
100+90	,93369	—	91,53	186,21	5,32	53,70	,5361
89	,93419	—	92,56	187,21	5,35	53,41	,5333
88	,93471	—	93,61	188,23	5,38	53,12	,5304
87	,93523	—	94,69	189,27	5,42	52,83	,5275
86	,93575	—	95,79	190,35	5,44	52,53	,5245
100+85	,93628	—	96,92	191,45	5,47	52,23	,5215
84	,93681	—	98,07	192,57	5,50	51,93	,5185
83	,93735	—	99,25	193,72	5,53	51,62	,5154
82	,93790	—	100,46	194,90	5,56	51,31	,5123
81	,93845	—	101,70	196,10	5,60	50,99	,5091
100+80	,93900	—	102,97	197,34	5,63	50,67	,5059
79	,93956	—	104,28	198,61	5,67	50,35	,5026
78	,94011	—	105,61	199,91	5,70	50,02	,4994
77	,94066	—	106,98	201,26	5,72	49,69	,4960
76	,94123	—	108,39	202,63	5,76	49,35	,4927
100+75	,94178	—	109,83	204,05	5,78	49,01	,4892
74	,94235	—	111,31	205,50	5,81	48,66	,4858
73	,94291	—	112,83	206,99	5,84	48,31	,4823
72	,94349	—	114,41	208,54	5,87	47,95	,4787
71	,94407	—	116,01	210,11	5,90	47,60	,4752
100+70	,94364	—	117,68	211,73	5,95	47,23	,4715
69	,94523	—	119,38	213,40	5,98	46,86	,4678
68	,94583	—	121,14	215,12	6,02	46,48	,4641
67	,94643	—	122,95	216,90	6,05	46,10	,4603
66	,94703	—	124,82	218,73	6,09	45,72	,4564
100+65	,94763	—	126,74	220,61	6,13	45,33	,4525
64	,94825	—	128,71	222,56	6,15	44,93	,4486
63	,94887	—	130,75	224,57	6,18	44,53	,4445
62	,94948	—	132,86	226,65	6,21	44,12	,4405
61	,95010	—	135,05	228,79	6,26	43,71	,4363
100+60	,95073	—	137,30	231,00	6,30	43,29	,4321
59	,95135	—	139,62	233,30	6,32	42,87	,4279
58	,95198	—	142,02	235,67	6,35	42,43	,4236
57	,95261	—	144,52	238,13	6,39	41,99	,4192
56	,95325	—	147,11	240,67	6,44	41,55	,4148
100+55	,95388	—	149,78	243,32	6,46	41,09	,4103
54	,95451	—	152,53	246,05	6,48	40,64	,4057
53	,95513	—	155,41	248,90	6,51	40,17	,4011
52	,95570	—	158,10	251,88	6,52	39,70	,3964
51	,95639	—	161,52	254,96	6,56	39,22	,3916

I. Water and spirit by weight.	II. Specific gravity.	III. Spirit by measure.	IV. Water by measure.	V. Bulk of mixture.	VI. Diminution of bulk.	VII. Quantity of spirit per cent.	VIII. Decimal multipliers.
W. + Sp.							
100+50	,95703	100	164,76	258,17	6,59	38,73	,3867
49	,95767	—	168,13	261,50	6,63	38,24	,3818
48	,95832	—	171,62	264,98	6,64	37,74	,3768
47	,95896	—	175,27	268,61	6,66	37,23	,3717
46	,95961	—	179,09	272,40	6,69	36,71	,3665
100+45	,96026	—	183,07	276,36	6,71	36,18	,3612
44	,96090	—	187,23	280,49	6,74	35,65	,3560
43	,96155	—	191,58	284,83	6,75	35,11	,3505
42	,96219	—	196,14	289,38	6,76	34,55	,3450
41	,96283	—	200,92	294,15	6,77	33,99	,3394
100+40	,96348	—	205,95	299,18	6,77	33,42	,3337
39	,96413	—	211,23	304,45	6,78	32,85	,3279
38	,96477	—	216,79	310,01	6,78	32,26	,3220
37	,96542	—	222,64	315,87	6,77	31,66	,3161
36	,96608	—	228,84	322,05	6,79	31,05	,3100
100+35	,96672	—	235,37	328,59	6,78	30,43	,3038
34	,96739	—	242,30	335,52	6,78	29,80	,2976
33	,96806	—	249,64	342,90	6,74	29,16	,2912
32	,96871	—	257,44	350,72	6,72	28,51	,2847
31	,96938	—	265,74	359,03	6,71	27,85	,2781
100+30	,97005	—	274,60	367,90	6,70	27,18	,2714
29	,97073	—	284,06	377,39	6,67	26,50	,2645
28	,97141	—	294,22	387,56	6,66	25,80	,2575
27	,97210	—	305,11	398,50	6,61	25,10	,2505
26	,97278	—	316,85	410,27	6,58	24,38	,2433
100+25	,97349	—	329,52	423,00	6,52	23,64	,2360
24	,97421	—	343,25	436,78	6,47	22,89	,2286
23	,97494	—	358,18	451,77	6,41	22,13	,2210
22	,97568	—	374,46	468,08	6,38	21,36	,2133
21	,97645	—	392,29	485,98	6,31	20,58	,2054
100+20	,97723	—	411,90	505,66	6,24	19,77	,1974
19	,97800	—	433,58	527,42	6,16	18,96	,1893
18	,97879	—	457,67	551,59	6,08	18,13	,1810
17	,97962	—	484,60	578,59	6,01	17,29	,1725
16	,98047	—	514,88	608,98	5,90	16,43	,1639
100+15	,98134	—	549,21	643,41	5,80	15,55	,1552
14	,98223	—	588,44	682,76	5,68	14,65	,1463
13	,98317	—	633,70	728,12	5,58	13,74	,1371
12	,98414	—	686,51	781,06	5,45	12,80	,1278
11	,98514	—	748,92	843,60	5,32	11,85	,1183
100+10	,98618	—	823,81	918,61	5,20	10,88	,1087
9	,98726	—	915,35	1010,29	5,06	9,90	,0988
8	,98840	—	1029,77	1124,86	4,91	8,89	,0888
7	,98960	—	1176,88	1272,12	4,76	7,86	,0785
6	,99084	—	1373,02	1468,42	4,60	6,81	,0680
100+5	,99214	—	1647,63	1743,20	4,43	5,73	,0573
4	,99351	—	2059,53	2155,26	4,27	4,64	,0463
3	,99495	—	2746,03	2841,93	4,10	3,52	,0351
2	,99646	—	4119,06	4215,10	3,96	2,37	,0237
1	,99805	—	8238,12	8334,30	3,82	1,20	,0120

I.	II.	III.	IV.	V.	VI.	VII.	VIII.	I.	II.	III.	IV.	V.	VI.	VII.	VIII.
Spirit and water by weight.	Specific gravity.	Spirit by measure.	Water by measure.	Bulk of mixture.	Diminution of bulk.	Quantity of spirit per cent.	Decimal multipliers.	Spirit and water by weight.	Specific gravity.	Spirit by measure.	Water by measure.	Bulk of mixture.	Diminution of bulk.	Quantity of spirit per cent.	Decimal multipliers.
Sp. + W.								Sp. + W.							
100 + 0	,82310	100	—	100,00	—	100,00	,9977	100 + 50	,89525	100	41,17	137,91	3,26	72,51	,7235
1	,82541	—	0,82	100,72	0,10	99,29	,9906	51	,89616	—	41,99	138,69	3,30	72,10	,7194
2	,82767	—	1,65	101,44	0,21	98,58	,9836	52	,89705	—	42,82	139,47	3,35	71,70	,7154
3	,82987	—	2,47	102,16	0,31	97,88	,9766	53	,89793	—	43,64	140,25	3,39	71,30	,7114
4	,83202	—	3,29	102,89	0,40	97,19	,9698	54	,89879	—	44,46	141,03	3,43	70,91	,7075
100 + 5	,83409	—	4,11	103,62	0,49	96,51	,9629	100 + 55	,89965	—	45,29	141,81	3,48	70,51	,7035
6	,83614	—	4,94	104,35	0,59	95,83	,9562	56	,90049	—	46,11	142,59	3,52	70,13	,6997
7	,83813	—	5,77	105,09	0,68	95,17	,9495	57	,90132	—	46,93	143,37	3,56	69,74	,6958
8	,84007	—	6,59	105,82	0,77	94,50	,9429	58	,90214	—	47,76	144,16	3,60	69,36	,6921
9	,84196	—	7,41	106,56	0,85	93,84	,9363	59	,90293	—	48,58	144,94	3,64	68,99	,6883
100 + 10	,84379	—	8,24	107,30	0,94	93,20	,9298	100 + 60	,90372	—	49,40	145,72	3,68	68,63	,6846
11	,84560	—	9,06	108,05	1,01	92,56	,9234	61	,90450	—	50,22	146,51	3,71	68,26	,6809
12	,84736	—	9.88	108,79	1,09	91,92	,9171	62	,90527	—	51,05	147,29	3,76	67,89	,6773
13	,84908	—	10,70	109,55	1,15	91,29	,9108	63	,90602	—	51,87	148,08	3,79	67,54	,6738
14	,85076	—	11,52	110,29	1,23	90,67	,9046	64	,90677	—	52,69	148,86	3,83	67,18	,6702
100 + 15	,85241	—	12,35	111,05	1,30	90,05	,8985	100 + 65	,90751	—	53,52	149,65	3,87	66,82	,6667
16	,85403	—	13,17	111,80	1,37	89,45	,8924	66	,90825	—	54,34	150,43	3,91	66,47	,6632
17	,85562	—	14.00	112,56	1,44	88,85	,8864	67	,90897	—	55,16	151,22	3,94	66,12	,6598
18	,85719	—	14,82	113,31	1,51	88,26	,8805	68	,90970	—	55,99	152,00	3,99	65,78	,6564
19	,85872	—	15.64	114,07	1,57	87,67	,8747	69	,91041	—	56,81	152,79	4,02	65,45	,6530
100 + 20	,86022	—	16,46	114,82	1,64	87,09	,8689	100 + 70	,91110	—	57,63	153,58	4,05	65,11	,6496
21	,86170	—	17,29	115,58	1,70	86,52	,8632	71	,91180	—	58,45	154,36	4,09	64,78	,6464
22	,86314	—	18,11	116,34	1,77	85,95	,8576	72	,91247	—	59,28	155,15	4,13	64,45	,6431
23	,86456	—	18,94	117,10	1,84	85,39	,8521	73	,91314	—	60,10	155,94	4,16	64,12	,6398
24	,86596	—	19,76	117,86	1,90	84,84	,8465	74	,91380	—	60,93	156,73	4,20	63,80	,6366
100 + 25	,86732	—	20,59	118,63	1,96	84,29	,8410	100 + 75	,91444	—	61,75	157,51	4,24	63,48	,6334
26	,86868	—	21,41	119,39	2,02	83,76	,8356	76	,91510	—	62,58	158,31	4,27	63,17	,6302
27	,86999	—	22,23	120,16	2,07	83,23	,8303	77	,91572	—	63,40	159,09	4,31	62,85	,6271
28	,87130	—	23,06	120,92	2,14	82,70	,8250	78	,91635	—	64,23	159,88	4,35	62,54	,6240
29	,87258	—	23,88	121,69	2,19	82,17	,8198	79	,91698	—	65,05	160,68	4,37	62,23	,6209
100 + 30	,87384	—	24,70	122,46	2,24	81,66	,8148	100 + 80	,91759	—	65,87	161,46	4,41	61,93	,6179
31	,87507	—	25,52	123,22	2,30	81,16	,8097	81	,91819	—	66,69	162,25	4,44	61,63	,6149
32	,87630	—	26,35	123,99	2,36	80,66	,8047	82	,91879	—	67,52	163,04	4,48	61,33	,6119
33	,87751	—	27,17	124,76	2,41	80,16	,7998	83	,91938	—	68,34	163,84	4,50	61,03	,6089
34	,87869	—	27,99	125,53	2,46	79,66	,7949	84	,91996	—	69,16	164,63	4,53	60,74	,6060
100 + 35	,87984	—	28,82	126,30	2,52	79,17	,7900	100 + 85	,92053	—	69,99	165,42	4,57	60,45	,6031
36	,88099	—	29,64	127,07	2,57	78,69	,7852	86	,92109	—	70,81	166,21	4,60	60,16	,6002
37	,88211	—	30,46	127,83	2,63	78,22	,7805	87	,92164	—	71,64	167,01	4,63	59,88	,5974
38	,88321	—	31,29	128,61	2,68	77,75	,7758	88	,92219	—	72,46	167,80	4,66	59,60	,5946
39	,88429	—	32.11	129,38	2,73	77,29	,7711	89	,92273	—	73,28	168,60	4,69	59,31	,5918
100 + 40	,88536	—	32,93	130,16	2,77	76,83	,7666	100 + 90	,92326	—	74,11	169,39	4,72	59,03	,5890
41	,88642	—	33,76	130,93	2,83	76,37	,7620	91	,92380	—	74,93	170,18	4,75	58,76	,5863
42	,88747	—	34,59	131,70	2,89	75,92	,7576	92	,92433	—	75,76	170,97	4,79	58,49	,5836
43	,88850	—	35,41	132,47	2,94	75,49	,7532	93	,92485	—	76,57	171,77	4,80	58,22	,5808
44	,88952	—	36,23	133,25	2,98	75,04	,7488	94	,92537	—	77,39	172,56	4,83	57,95	,5782
100 + 45	,89051	—	37,06	134,02	3,04	74,61	,7444	100 + 95	,92589	—	78,21	173,35	4,86	57,68	,5755
46	,89150	—	37,87	134,80	3,08	74,18	,7401	96	,92639	—	79,04	174,13	4,90	57,42	,5730
47	,89246	—	38,69	135,57	3,12	73,76	,7359	97	,92688	—	79,86	174,93	4,93	57,16	,5704
48	,89340	—	39,52	136,35	3,17	73,34	,7317	98	,92738	—	80,68	175,72	4,96	56,90	,5678
49	,89433	—	40,34	137,13	3,21	72,92	,7275	99	,92787	—	81,51	176,52	4,99	56,65	,5652

TABLE II. HEAT 64°.

I. Water and spirit by weight.	II. Specific gravity.	III. Spirit by measure.	IV. Water by measure.	V. Bulk of mixture.	VI. Diminution of bulk.	VII. Quantity of spirit per cent.	VIII. Decimal multipliers.
W. + Sp.							
100+100	,92836	100	82,34	177,32	5,02	56,39	,5626
99	,92886	—	83,17	178,13	5,04	56,13	,5601
98	,92935	—	84,02	178,95	5,07	55,88	,5576
97	,92983	—	84,87	179,79	5,08	55,62	,5549
96	,93033	—	85,77	180,64	5,13	55,35	,5523
100+95	,93081	—	86,67	181,51	5,16	55,09	,5497
94	,93130	—	87,60	182,40	5,20	54,82	,5469
93	,93179	—	88,54	183,32	5,22	54,55	,5442
92	,93228	—	89,50	184,25	5,25	54,27	,5414
91	,93277	—	90,48	185,21	5,27	53,99	,5386
100+90	,93327	—	91,49	186,19	5,30	53,71	,5358
89	,93378	—	92,51	187,19	5,32	53,42	,5331
88	,93430	—	93,56	188,20	5,36	53,13	,5302
87	,93482	—	94,64	189,25	5,39	52,84	,5273
86	,93534	—	95,75	190,32	5,43	52,54	,5243
100+85	,93587	—	96,87	191,42	5,45	52,24	,5212
84	,93641	—	98,03	192,54	5,49	51,94	,5182
83	,93695	—	99,21	193,69	5,52	51,63	,5152
82	,93750	—	100,41	194,87	5,54	51,32	,5121
81	,93806	—	101,65	196,07	5,58	51,00	,5089
100+80	,93861	—	102,92	197,31	5,61	50,68	,5056
79	,93917	—	104,23	198,58	5,65	50,36	,5024
78	,93972	—	105,56	199,88	5,68	50,03	,4992
77	,94027	—	106,93	201,23	5,70	49,70	,4958
76	,94183	—	108,34	202,60	5,74	49,35	,4925
100+75	,94139	—	109,78	204,01	5,77	49,01	,4890
74	,94196	—	111,26	205,47	5,79	48,67	,4856
73	,94252	—	112,78	206,95	5,83	48,32	,4821
72	,94310	—	114,35	208,50	5,85	47,96	,4785
71	,94368	—	115,96	210,08	5,88	47,61	,4750
100+70	,94326	—	117,62	211,70	5,92	47,24	,4713
69	,94485	—	119,32	213,36	5,96	46,87	,4676
68	,94545	—	121,08	215,08	6,00	46,49	,4639
67	,94605	—	122,89	216,86	6,03	46,11	,4601
66	,94666	—	124,76	218,69	6,07	45,73	,4562
100+65	,94725	—	126,68	220,57	6,11	45,34	,4523
64	,94788	—	128,65	222,51	6,14	44,94	,4484
63	,94850	—	130,69	224,52	6,17	44,54	,4444
62	,94912	—	132,80	226,61	6,19	44,13	,4403
61	,94974	—	134,98	228,74	6,24	43,72	,4362
100+60	,95037	—	137,23	230,96	6,27	43,30	,4320
59	,95099	—	139,55	233,25	6,30	42,87	,4277
58	,95162	—	141,95	235,62	6,33	42,44	,4234
57	,95226	—	144,45	238,08	6,37	42,00	,4190
56	,95290	—	147,04	240,62	6,42	41,56	,4146
100+55	,95353	—	149,71	243,27	6,44	41,10	,4101
54	,95416	—	152,45	246,00	6,45	40,65	,4055
53	,95479	—	155,33	248,85	6,48	40,18	,4010
52	,95542	—	158,32	251,82	6,50	39,71	,3963
51	,95605	—	161,44	254,90	6,54	39,23	,3915

I. Water and spirit by weight.	II. Specific gravity.	III. Spirit by measure.	IV. Water by measure.	V. Bulk of mixture.	VI. Diminution of bulk.	VII. Quantity of spirit per cent.	VIII. Decimal multipliers.
W. + Sp.							
100+50	,95669	100	164,68	258,11	6,57	38,74	,3865
49	,95734	—	168,05	261,44	6,61	38,25	,3816
48	,95799	—	171,54	264,92	6,62	37,75	,3767
47	,95863	—	175,19	268,55	6,64	37,24	,3716
46	,95929	—	179,00	272,33	6,67	36,72	,3664
100+45	,95994	—	182,98	276,29	6,69	36,19	,3611
44	,96059	—	187,14	280,42	6,72	35,66	,3558
43	,96124	—	191,49	284,76	6,73	35,11	,3504
42	,96188	—	196,15	289,31	6,74	34,56	,3449
41	,96253	—	200,83	294,08	6,75	34,00	,3393
100+40	,96318	—	205,85	299,10	6,75	33,43	,3336
39	,96384	—	211,13	304,37	6,76	32,86	,3278
38	,96449	—	216,69	309,92	6,77	32,27	,3219
37	,96514	—	222,54	315,77	6,77	31,67	,3160
36	,96581	—	228,73	321,95	6,78	31,06	,3099
100+35	,96646	—	235,26	328,49	6,77	30,44	,3037
34	,96713	—	242,18	335,42	6,76	29,81	,2975
33	,96780	—	249,52	342,80	6,72	29,17	,2911
32	,96847	—	257,32	350,61	6,71	28,52	,2846
31	,96914	—	265,61	358,92	6,69	27,86	,2780
100+30	,96982	—	274,47	367,78	6,69	27,19	,2713
29	,97051	—	283,93	377,26	6,67	26,51	,2645
28	,97119	—	294,08	387,42	6,66	25,81	,2575
27	,97189	—	304,97	398,36	6,61	25,11	,2505
26	,97258	—	316,70	410,12	6,58	24,39	,2433
100+25	,97329	—	329,36	422,84	6,52	23,65	,2360
24	,97402	—	343,09	436,61	6,48	22,90	,2286
23	,97475	—	358,01	451,59	6,42	22,14	,2210
22	,97550	—	374,28	467,90	6,38	21,37	,2132
21	,97628	—	392,10	485,79	6,31	20,58	,2054
100+20	,97706	—	411,71	505,45	6,26	19,78	,1974
19	,97784	—	433,37	527,20	6,17	18,97	,1892
18	,97864	—	457,45	551,36	6,09	18,14	,1809
17	,97947	—	484,37	578,35	6,02	17,29	,1725
16	,98033	—	514,64	608,72	5,92	16,43	,1639
100+15	,98120	—	548,95	643,13	5,82	15,55	,1551
14	,98210	—	588,16	682,46	5,71	14,65	,1462
13	,98304	—	633,40	727,80	5,60	13,74	,1371
12	,98401	—	686,18	780,71	5,47	12,81	,1278
11	,98502	—	748,56	843,22	5,34	11,86	,1183
100+10	,98606	—	823,42	918,19	5,23	10,89	,1087
9	,98715	—	914,91	1009,83	5,08	9,90	,0988
8	,98829	—	1029,28	1124,34	4,94	8,89	,0887
7	,98950	—	1176,32	1271,52	4,80	7,86	,0785
6	,99074	—	1372,37	1467,73	4,64	6,81	,0680
100+5	,99204	—	1646,84	1742,37	4,47	5,74	,0573
4	,99341	—	2058,65	2154,24	4,31	4,64	,0463
3	,99485	—	2744,72	2840,58	4,14	3,52	,0351
2	,99636	—	4117,09	4213,09	4,00	2,37	,0237
1	,99795	—	8234,20	8330,33	3,87	1,20	,0120

TABLE I.   HEAT 65°.

I. Spirit and water by weight.	II. Specific gravity.	III. Spirit by measure.	IV. Water by measure.	V. Bulk of mixture.	VI. Diminution of bulk.	VII. Quantity of spirit per cent.	VIII. Decimal multipliers.
Sp. + W.							
100 + 0	,82262	100	—	100,00	—	100,00	,9971
1	,82493	—	0,82	100,72	0,10	99,29	,9900
2	,82719	—	1,65	101,44	0,21	98,58	,9830
3	,82939	—	2,47	102,16	0,31	97,88	,9760
4	,83154	—	3,29	102,89	0,40	97,19	,9691
100 + 5	,83362	—	4,11	103,02	0,49	96,51	,9623
6	,83566	—	4,94	104,35	0,59	95,84	,9556
7	,83765	—	5,76	105,08	0,68	95,17	,9489
8	,83960	—	6,58	105,82	0,76	94,50	,9423
9	,84150	—	7,41	106,56	0,85	93,85	,9358
100 + 10	,84334	—	8,23	107,30	0,93	93,20	,9293
11	,84513	—	9,05	108,04	1,01	92,56	,9229
12	,84689	—	9,88	108,79	1,09	91,92	,9166
13	,84861	—	10,70	109,54	1,16	91,29	,9103
14	,85029	—	11,52	110,29	1,23	90,67	,9041
100 + 15	,85193	—	12,34	111,04	1,30	90,05	,8980
16	,85355	—	13,17	111,79	1,38	89,45	,8919
17	,85515	—	13,99	112,55	1,44	88,85	,8859
18	,85672	—	14,81	113,30	1,51	88,26	,8800
19	,85826	—	15,64	114,06	1,58	87,67	,8742
100 + 20	,85976	—	16,46	114,82	1,64	87,09	,8684
21	,86123	—	17,28	115,57	1,71	86,52	,8627
22	,86268	—	18,10	116,33	1,77	85,96	,8571
23	,86410	—	18,93	117,09	1,84	85,40	,8515
24	,86549	·	19,75	117,86	1,89	84,85	,8460
100 + 25	,86680	—	20,58	118,62	1,96	84,30	,8406
26	,86821	—	21,40	119,38	2,02	83,76	,8352
27	,86953	—	22,22	120,15	2,07	83,23	,8299
28	,87083	—	23,05	120,91	2,14	82,70	,8246
29	,87211	—	23,87	121,68	2,19	82,18	,8194
100 + 30	,87337	—	24,69	122,45	2,24	81,67	,8143
31	,87461	—	25,51	123,21	2,30	81,16	,8092
32	,87583	—	26,34	123,98	2,36	80,66	,8042
33	,87704	—	27,16	124,75	2,41	80,16	,7993
34	,87822	—	27,98	125,52	2,46	79,67	,7944
100 + 35	,87938	—	28,81	126,29	2,52	79,18	,7896
36	,88052	—	29,63	127,06	2,57	78,70	,7848
37	,88164	—	30,45	127,83	2,62	78,23	,7800
38	,88274	—	31,27	128,60	2,67	77,76	,7753
39	,88382	—	32,10	129,37	2,73	77,30	,7707
100 + 40	,88490	—	32,92	130,15	2,77	76,84	,7662
41	,88596	—	33,74	130,92	2,82	76,38	,7616
42	,88701	—	34,57	131,69	2,88	75,93	,7572
43	,88804	—	35,39	132,46	2,93	75,49	,7528
44	,88906	··	36,21	133,24	2,97	75,05	,7484
100 + 45	,89006	—	37,04	134,01	3,03	74,62	,7440
46	,89104	—	37,86	134,79	3,07	74,19	,7397
47	,89201	—	38,68	135,57	3,11	73,76	,7355
48	,89294	—	39,50	136,34	3,16	73,34	,7313
49	,89387	—	40,33	137,12	3,21	72,93	,7272
100 + 50	,89479	100	41,15	137,90	3,25	72,52	,7231
51	,89570	—	41,97	138,68	3,29	72,11	,7190
52	,89659	—	42,80	139,46	3,34	71,70	,7150
53	,89747	—	43,62	140,24	3,38	71,30	,7110
54	,89834	—	44,44	141,02	3,42	70,91	,7071
100 + 55	,89920	—	45,27	141,80	3,47	70,52	,7032
56	,90004	—	46,09	142,58	3,51	70,13	,6993
57	,90087	—	46,91	143,36	3,55	69,75	,6955
58	,90168	—	47,74	144,15	3,59	69,37	,6917
59	,90248	—	48,56	144,93	3,63	69,00	,6880
100 + 60	,90328	—	49,38	145,71	3,67	68,63	,6843
61	,90406	—	50,20	146,50	3,70	68,26	,6806
62	,90483	—	51,03	147,28	3,75	67,90	,6770
63	,90558	—	51,85	148,07	3,78	67,54	,6734
64	,90633	—	52,67	148,85	3,82	67,18	,6698
100 + 65	,90707	—	53,50	149,64	3,86	66,83	,6663
66	,90781	—	54,32	150,42	3,90	66,48	,6628
67	,90853	—	55,14	151,21	3,93	66,13	,6594
68	,90925	—	55,97	151,99	3,98	65,79	,6560
69	,90996	—	56,79	152,78	4,01	65,45	,6526
100 + 70	,91066	—	57,61	153,56	4,05	65,12	,6493
71	,91135	—	58,43	154,35	4,08	64,79	,6460
72	,91202	—	59,26	155,14	4,12	64,46	,6427
73	,91269	—	60,08	155,93	4,15	64,13	,6395
74	,91335	—	60,90	156,72	4,18	63,81	,6363
100 + 75	,91400	—	61,73	157,50	4,23	63,49	,6331
76	,91465	—	62,55	158,29	4,26	63,17	,6299
77	,91528	—	63,37	159,08	4,29	62,86	,6268
78	,91591	—	64,20	159,87	4,33	62,55	,6237
79	,91654	—	65,02	160,66	4,36	62,24	,6206
100 + 80	,91715	—	65,84	161,45	4,39	61,94	,6170
81	,91776	—	66,66	162,24	4,42	61,64	,6146
82	,91835	—	67,49	163,03	4,46	61,34	,6116
83	,91894	—	68,31	163,82	4,49	61,04	,6086
84	,91953	—	69,13	164,61	4,52	60,75	,6057
100 + 85	,92010	—	69,96	165,40	4,56	60,46	,6028
86	,92066	—	70,78	166,19	4,59	60,17	,5999
87	,92121	—	71,61	166,99	4,62	59,89	,5971
88	,92176	—	72,43	167,78	4,65	59,60	,5943
89	,92230	—	73,25	168,58	4,67	59,32	,5915
100 + 90	,92283	—	74,07	169,37	4,70	59,04	,5887
91	,92337	—	74,90	170,16	4,74	58,77	,5860
92	,92390	—	75,72	170,95	4,77	58,50	,5833
93	,92442	—	76,54	171,75	4,79	58,23	,5806
94	,92494	—	77,36	172,54	4,82	57,96	,5779
100 + 95	,92546	—	78,18	173,33	4,85	57,69	,5753
96	,92596	—	79,00	174,11	4,89	57,43	,5726
97	,92646	—	79,82	174,90	4,92	57,17	,5701
98	,92696	—	80,64	175,70	4,94	56,91	,5675
99	,92745	—	81,47	176,50	4,97	56,66	,5649

TABLE II.   HEAT 65°.

I.	II.	III.	IV.	V.	VI.	VII.	VIII.	I.	II.	III.	IV.	V.	VI.	VII.	VIII
Water and spirit by weight.	Specific gravity.	Spirit by measure.	Water by measure.	Bulk of mixture.	Diminution of bulk.	Quantity of spirit per cent.	Decimal multipliers.	Water and spirit by weight.	Specific gravity.	Spirit by measure.	Water by measure.	Bulk of mixture.	Diminution of bulk.	Quantity of spirit per cent.	Decimal multipliers.
W. + Sp.								**W. + Sp.**							
100+100	,92794	100	82,30	177,30	5,00	56.40	,5624	100 + 50	,95635	100	164,61	258,05	6,56	38,75	,3864
99	,92844	—	83,13	178,11	5,01	56,14	,5599	49	,95700	—	167,97	261,38	6,59	38,26	,3815
98	,92893	—	83,98	178,93	5,05	55,89	,5574	48	,95765	—	171,46	264,86	6,60	37,76	,3765
97	,92942	—	84,84	179,77	5,07	55.63	,5547	47	,95830	—	175,11	268,49	6,62	37,25	,3714
96	,92991	—	85,72	180,62	5,10	55,36	,5521	46	,95896	—	178,92	272,27	6,65	36,73	,3662
100 + 95	,93040	—	86,63	181,49	5,14	55,10	,5494	100 + 45	,95962	—	182,90	276,22	6,68	36,20	,3610
94	,93088	—	87,56	182,38	5,18	54,83	,5467	44	,96027	—	187,05	280,35	6,70	35,67	,3557
93	,93137	—	88,50	183,29	5,21	54,56	,5440	43	,96092	—	191,40	284,69	6,71	35,12	,3502
92	,93186	—	89,46	184,23	5,23	54,28	,5412	42	,96157	—	195,96	289,24	6,72	34,57	,3447
91	,93235	—	90,44	185,19	5,25	54,00	,5384	41	,96222	—	200,74	294,01	6,73	34,01	,3392
100 + 90	,93285	—	91,45	186,16	5,29	53,72	,5356	100 + 40	,96288	—	205,76	299,02	6,74	33,44	,3335
89	,93336	—	92,47	187,16	5,31	53,43	,5328	39	,96354	—	211,03	304,29	6,74	32,86	,3277
88	,93388	—	93,52	188,18	5,34	53,14	,5299	38	,96420	—	216,59	309,83	6,76	32,28	,3218
87	,93440	—	94,60	189,23	5,37	52,84	,5270	37	,96486	—	222,44	315,68	6,76	31,68	,3159
86	,93493	—	95,70	190,30	5,40	52,55	,5240	36	,96553	—	228,62	321,85	6,77	31,07	,3098
100 + 85	,93546	—	96,83	191,39	5,44	52,25	,5210	100 + 35	,96620	—	235,15	328,39	6,76	30,45	,3036
84	,93600	—	97,98	192,51	5,47	51,94	,5180	34	,96687	—	242,07	335,32	6,75	29,82	,2974
83	,93655	—	99,16	193,66	5,50	51,63	,5149	33	,96754	—	249,40	342,70	6,70	29,18	,2910
82	,93710	—	100,37	194,84	5,53	51,32	,5118	32	,96822	—	257,20	350,51	6,69	28,53	,2845
81	,93766	—	101,61	196,04	5,57	51,01	,5086	31	,96890	—	265,49	358,81	6,68	27,87	,2779
100 + 80	,93822	—	102,88	197,28	5,60	50,70	,5054	100 + 30	,96959	—	274,34	367,65	6,69	27,20	,2712
79	,93877	—	104,18	198,55	5,63	50,37	,5022	29	,97028	—	283,80	377,13	6,67	26,52	,2644
78	,93932	—	105,51	199,85	5,66	50,04	,4989	28	,97097	—	293,94	387,28	6,66	25,82	,2574
77	,93987	—	106,88	201,19	5,69	49,70	,4956	27	,97167	—	304,83	398,22	6,61	25,11	,2504
76	,94043	—	108,29	202,57	5,72	49,36	,4922	26	,97237	—	316,55	409,97	6,58	24,39	,2432
100 + 75	,94099	—	109,73	203,98	5,75	49,02	,4888	100 + 25	,97309	—	329,21	422,68	6,53	23,66	,2359
74	,94156	—	111,21	205,43	5,78	48,68	,4854	24	,97382	—	342,93	430,45	6,48	22,91	,2285
73	,94213	—	112,73	206,92	5,81	48,33	,4819	23	,97456	—	357,84	451,41	6,43	22,15	,2209
72	,94271	—	114,30	208,46	5,84	47,97	,4783	22	,97532	—	374,10	467,72	6,38	21,38	,2132
71	,94329	—	115,91	210,04	5,87	47,61	,4747	21	,97610	—	391,92	485,60	6,32	20,59	,2053
100 + 70	,94388	—	117,57	211,66	5,91	47,25	,4711	100 + 20	,97688	—	411,52	505,25	6,27	19,79	,1973
69	,94447	—	119,27	213,32	5,95	46,88	,4674	19	,97767	—	433,17	526,99	6,17	18,97	,1892
68	,94507	—	121,02	215,04	5,98	46,50	,4637	18	,97848	—	457,24	551,13	6,11	18,14	,1809
67	,94567	—	122,83	216,82	6,01	46,12	,4599	17	,97932	—	484,14	578,11	6,03	17,30	,1725
66	,94628	—	124,70	218,65	6,05	45,74	,4560	16	,98018	—	514,40	608,46	5,94	16,44	,1639
100 + 65	,94689	—	126,62	220,53	6,09	45,35	,4521	100 + 15	,98106	—	548,69	642,85	5,84	15,56	,1551
64	,94751	—	128,59	222,47	6,12	44,95	,4482	14	,98196	—	587,88	682,16	5,72	14,66	,1462
63	,94813	—	130,63	224,48	6,15	44,55	,4442	13	,98290	—	633,10	727,48	5,62	13,75	,1371
62	,94875	—	132,74	226,56	6,18	44,14	,4401	12	,98388	—	685,86	780,36	5,50	12,81	,1278
61	,94937	—	134,92	228,70	6,22	43,73	,4360	11	,98489	—	748,21	842,84	5,37	11,86	,1183
100 + 60	,95000	—	137,17	230,91	6,26	43,31	,4318	100 + 10	,98594	—	823,03	917,77	5,26	10,89	,1086
59	,95063	—	139,49	233,20	6,29	42,88	,4276	9	,98703	—	914,48	1009,37	5,11	9,91	,0988
58	,95126	—	141,89	235,57	6,32	42,45	,4233	8	,98818	—	1028,79	1123,84	4,97	8,90	,0887
57	,95190	—	144,38	238,03	6,35	42,01	,4189	7	,98939	—	1175,76	1270,93	4,83	7,87	,0785
56	,95254	—	146,97	240,58	6,39	41,57	,4145	6	,99063	—	1371,72	1467,04	4,68	6,82	,0680
100 + 55	,95318	—	149,64	243,22	6,42	41,11	,4100	100 + 5	,99194	—	1646,06	1741,54	4,52	5,74	,0572
54	,95381	—	152,38	245,95	6,43	40,66	,4054	4	,99331	—	2057,58	2153,22	4,36	4,64	,0463
53	,95444	—	155,25	248,80	6,45	40,19	,4008	3	,99475	—	2743,44	2839,24	4,20	3,52	,0351
52	,95507	—	158,24	251,77	6,47	39,72	,3961	2	,99602	—	4115,15	4211,11	4,04	2,37	,0237
51	,95571	—	161,36	254,85	6,51	39,24	,3913	1	,99784	—	8230,31	8326,40	3,91	1,20	,0120

TABLE I.   HEAT 66°.

I.	II.	III.	IV.	V.	VI.	VII.	VIII.	I.	II.	III.	IV.	V.	VI.	VII.	VIII.
Spirit and water by weight.	Specific gravity.	Spirit by measure.	Water by measure.	Bulk of mixture.	Diminution of bulk.	Quantity of Spirit per cent.	Decimal multipliers.	Spirit and Water by weight.	Specific gravity.	Spirit by measure.	Water by measure.	Bulk of mixture.	Diminution of bulk.	Quantity of Spirit per cent.	Decimal multipliers.
Sp. + W.								Sp. + W.							
100 + 0	,82214	100	—	100,00	—	100,00	,9965	100 + 50	,89433	100	41,13	137,89	3,24	72,52	,7227
1	,82446	—	0,82	100,72	0,10	99,29	,9895	51	,89525	—	41,95	138,67	3,28	72,12	,7187
2	,82672	—	1,65	101,44	0,21	98,58	,9825	52	,89613	—	42,78	139,45	3,33	71,71	,7147
3	,82892	—	2,47	102,16	0,31	97,88	,9755	53	,89702	—	43,60	140,22	3,38	71,31	,7107
4	,83107	—	3,29	102,89	0,40	97,19	,9686	54	,89789	—	44,42	141,00	3,42	70,92	,7068
100 + 5	,83315	—	4,11	103,62	0,49	96,51	,9618	100 + 55	,89875	—	45,24	141,78	3,46	70,53	,7028
6	,83519	—	4,94	104,35	0,59	95,84	,9551	56	,89959	—	46,06	142,56	3,50	70,14	,6990
7	,83717	—	5,76	105,08	0,68	95,17	,9484	57	,90042	—	46,88	143,35	3,53	69,76	,6952
8	,83912	—	6,58	105,82	0,76	94,50	,9418	58	,90124	—	47,71	144,13	3,58	69,38	,6914
9	,84102	—	7,41	106,56	0,85	93,85	,9353	59	,90204	—	48,53	144,91	3,62	69,00	,6877
100 + 10	,84286	—	8,23	107,30	0,93	93,20	,9288	100 + 60	,90284	—	49,35	145,70	3,65	68,64	,6840
11	,84465	—	9,05	108,04	1,01	92,56	,9224	61	,90362	—	50,17	146,48	3,69	68,27	,6803
12	,84641	—	9,87	108,79	1,08	91,92	,9160	62	,90439	—	51,00	147,26	3,74	67,90	,6767
13	,84813	—	10,69	109,54	1,15	91,29	,9098	63	,90514	—	51,82	148,05	3,77	67,54	,6731
14	,84981	—	11,51	110,29	1,22	90,67	,9036	64	,90589	—	52,64	148,83	3,81	67,19	,6695
100 + 15	,85145	—	12,34	111,04	1,30	90,05	,8975	100 + 65	,90663	—	53,47	149,62	3,85	66,84	,6660
16	,85307	—	13,16	111,79	1,37	89,45	,8914	66	,90737	—	54,29	150,40	3,89	66,49	,6625
17	,85467	—	13,98	112,55	1,43	88,85	,8854	67	,90809	—	55,11	151,19	3,92	66,14	,6591
18	,85624	—	14,80	113,30	1,50	88,26	,8795	68	,90881	—	55,94	151,97	3,97	65,79	,6557
19	,85778	—	15,63	114,06	1,57	87,67	,8737	69	,90952	—	56,76	152,76	4,00	65,46	,6523
100 + 20	,85928	—	16,45	114,82	1,63	87,09	,8679	100 + 70	,91023	—	57,58	153,55	4,03	65,13	,6490
21	,86076	—	17,27	115,57	1,70	86,52	,8622	71	,91091	—	58,40	154,33	4,07	64,80	,6457
22	,86221	—	18,09	116,33	1,76	85,96	,8566	72	,91159	—	59,23	155,12	4,11	64,46	,6424
23	,86363	—	18,92	117,09	1,83	85,40	,8511	73	,91225	—	60,05	155,91	4,14	64,13	,6392
24	,86502	—	19,74	117,85	1,89	84,85	,8456	74	,91291	—	60,87	156,70	4,17	63,82	,6360
100 + 25	,86639	—	20,57	118,62	1,95	84,30	,8401	100 + 75	,91356	—	61,70	157,48	4,22	63,50	,6328
26	,86774	—	21,39	119,38	2,01	83,76	,8348	76	,91421	—	62,52	158,27	4,25	63,18	,6296
27	,86907	—	22,21	120,14	2,07	83,23	,8295	77	,91484	—	63,34	159,06	4,28	62,87	,6265
28	,87037	—	23,04	120,91	2,13	82,70	,8242	78	,91547	—	64,16	159,85	4,31	62,56	,6234
29	,87165	—	23,86	121,68	2,18	82,18	,8190	79	,91610	—	64,99	160,64	4,35	62,25	,6203
100 + 30	,87291	—	24,68	122,44	2,24	81,67	,8139	100 + 80	,91670	—	65,81	161,43	4,38	61,95	,6173
31	,87415	—	25,50	123,20	2,30	81,17	,8088	81	,91732	—	66,63	162,22	4,41	61,64	,6143
32	,87537	—	26,33	123,97	2,36	80,66	,8038	82	,91791	—	67,45	163,01	4,44	61,34	,6113
33	,87658	—	27,15	124,74	2,41	80,16	,7989	83	,91851	—	68,27	163,80	4,47	61,05	,6083
34	,87776	—	27,97	125,52	2,45	79,67	,7940	84	,91910	—	69,00	164,59	4,50	60,76	,6054
100 + 35	,87892	—	28,80	126,28	2,52	79,18	,7891	100 + 85	,91967	—	69,92	165,39	4,53	60,46	,6026
36	,88006	—	29,61	127,05	2,56	78,70	,7844	86	,92023	—	70,74	166,18	4,56	60,18	,5997
37	,88117	—	30,44	127,82	2,62	78,23	,7796	87	,92078	—	71,57	166,97	4,60	59,89	,5969
38	,88227	—	31,26	128,59	2,67	77,76	,7749	88	,92133	—	72,39	167,76	4,63	59,61	,5941
39	,88336	—	32,08	129,36	2,72	77,30	,7703	89	,92187	—	73,21	168,56	4,65	59,33	,5913
100 + 40	,88443	—	32,90	130,14	2,76	76,84	,7658	100 + 90	,92241	—	74,03	169,35	4,68	59,05	,5885
41	,88549	—	33,72	130,91	2,81	76,38	,7612	91	,92294	—	74,86	170,14	4,72	58,77	,5858
42	,88654	—	34,55	131,68	2,87	75,93	,7568	92	,92347	—	75,68	170,93	4,75	58,50	,5831
43	,88758	—	35,37	132,45	2,92	75,50	,7524	93	,92400	—	76,50	171,72	4,78	58,23	,5804
44	,88860	—	36,19	133,23	2,96	75,06	,7480	94	,92452	—	77,32	172,52	4,80	57,96	,5777
100 + 45	,88959	—	37,02	134,00	3,02	74,62	,7436	100 + 95	,92504	—	78,14	173,31	4,83	57,70	,5750
46	,89058	—	37,84	134,78	3,06	74,19	,7394	96	,92554	—	78,96	174,09	4,87	57,44	,5725
47	,89154	—	38,66	135,56	3,10	73,77	,7352	97	,92604	—	79,78	174,88	4,90	57,18	,5699
48	,89249	—	39,48	136,33	3,15	73,35	,7310	98	,92654	—	80,60	175,67	4,93	56,92	,5673
49	,89342	—	40,31	137,11	3,20	72,93	,7268	99	,92703	—	81,43	176,47	4,96	56,66	,5647

TABLE II. HEAT 66°.

I. Water and Spirit by weight.	II. Specific gravity.	III. Spirit by measure.	IV. Water by measure.	V. Bulk of mixture.	VI. Diminution of bulk.	VII. Quantity of spirit per cent.	VIII. Decimal multipliers.	I. Water and spirit by weight.	II. Specific gravity.	III. Spirit by measure.	IV. Water by measure.	V. Bulk of mixture.	VI. Diminution of bulk.	VII. Quantity of spirit per cent.	VIII. Decimal multipliers.
W. + Sp.								W. + Sp.							
100+100	,92752	100	82,26	177,27	4,99	56,41	,5621	100+50	,95601	100	164,53	257,99	6,54	38,76	,3863
99	,92801	—	83,09	178,08	5,01	56,15	,5596	49	,95667	—	167,89	261,31	6,58	38,27	,3814
98	,92851	—	83,94	178,90	5,04	55,89	,5571	48	,95732	—	171,38	264,79	6,59	37,77	,3764
97	,92900	—	84,80	179,74	5,06	55,63	,5544	47	,95798	—	175,02	268,42	6,60	37,26	,3713
96	,92949	—	85,68	180,59	5,09	55,37	,5518	46	,95864	—	178,83	272,20	6,63	36,73	,3661
100+95	,92998	—	86,59	181,46	5,13	55,11	,5492	100+45	,95930	—	182,81	276,16	6,65	36,21	,3609
94	,93046	—	87,51	182,35	5,16	54,83	,5465	44	,95996	—	186,96	280,28	6,68	35,67	,3556
93	,93095	—	88,45	183,26	5,19	54,56	,5438	43	,96061	—	191,31	284,61	6,70	35,13	,3501
92	,93144	—	89,41	184,20	5,21	54,28	,5410	42	,96127	—	195,86	289,16	6,70	34,58	,3446
91	,93194	—	90,39	185,16	5,23	54,00	,5382	41	,96193	—	200,64	293,93	6,71	34,02	,3391
100+90	,93242	—	91,40	186,13	5,27	53,72	,5354	100+40	,96259	—	205,66	298,94	6,72	33,45	,3334
89	,93295	—	92,42	187,13	5,29	53,43	,5326	39	,96325	—	210,93	304,20	6,73	32,87	,3276
88	,93347	—	93,47	188,16	5,31	53,14	,5297	38	,96392	—	216,48	309,74	6,74	32,28	,3217
87	,93399	—	94,55	189,20	5,35	52,85	,5268	37	,96459	—	222,33	315,59	6,74	31,68	,3158
86	,93452	—	95,65	190,27	5,38	52,56	,5238	36	,96526	—	228,51	321,76	6,75	31,07	,3097
100+85	,93505	—	96,78	191,36	5,42	52,25	,5207	100+35	,96593	—	235,04	328,29	6,75	30,46	,3035
84	,93559	—	97,93	192,49	5,44	51,95	,5178	34	,96661	—	241,95	335,21	6,74	29,83	,2973
83	,93614	—	99,11	193,64	5,47	51,64	,5147	33	,96728	—	249,28	342,59	6,69	29,19	,2909
82	,93669	—	100,32	194,82	5,50	51,33	,5116	32	,96797	—	257,07	350,40	6,67	28,54	,2844
81	,93725	—	101,56	196,02	5,54	51,02	,5084	31	,96865	—	265,36	358,69	6,67	27,88	,2778
100+80	,93781	—	102,83	197,25	5,58	50,70	,5052	100+30	,96935	—	274,21	367,53	6,68	27,21	,2711
79	,93836	—	104,13	198,52	5,61	50,37	,5020	29	,97004	—	283,66	377,00	6,66	26,52	,2643
78	,93892	—	105,46	199,82	5,64	50,04	,4987	28	,97074	—	293,80	387,15	6,65	25,83	,2574
77	,93947	—	106,83	201,16	5,67	49,71	,4954	27	,97145	—	304,68	398,09	6,59	25,12	,2504
76	,94003	—	108,24	202,54	5,70	49,37	,4920	26	,97216	—	316,40	409,83	6,57	24,40	,2432
100+75	,94058	—	109,68	203,95	5,73	49,03	,4886	100+25	,97288	—	329,05	422,52	6,53	23,67	,2359
74	,94116	—	111,15	205,40	5,75	48,69	,4852	24	,97362	—	342,76	436,28	6,48	22,92	,2285
73	,94174	—	112,67	206,88	5,79	48,34	,4817	23	,97436	—	357,67	451,24	6,43	22,16	,2209
72	,94232	—	114,24	208,43	5,81	47,98	,4782	22	,97513	—	373,92	467,54	6,38	21,39	,2132
71	,94290	—	115,85	210,00	5,85	47,62	,4746	21	,97591	—	391,73	485,41	6,32	20,60	,2053
100+70	,94349	—	117,51	211,62	5,89	47,26	,4709	100+20	,97670	—	411,32	505,05	6,27	19,80	,1973
69	,94409	—	119,21	213,28	5,93	46,89	,4672	19	,97750	—	432,96	526,78	6,18	18,98	,1892
68	,94469	—	120,96	215,00	5,96	46,51	,4635	18	,97831	—	457,02	550,90	6,12	18,15	,1809
67	,94529	—	122,77	216,78	5,99	46,13	,4597	17	,97916	—	483,91	577,87	6,04	17,31	,1725
66	,94590	—	124,64	218,61	6,03	45,75	,4559	16	,98002	—	514,15	608,20	5,95	16,45	,1639
100+65	,94652	—	126,56	220,49	6,07	45,35	,4520	100+15	,98091	—	548,43	642,57	5,86	15,57	,1551
64	,94714	—	128,53	222,43	6,10	44,95	,4480	14	,98181	—	587,60	681,86	5,74	14,67	,1462
63	,94776	—	130,56	224,44	6,12	44,56	,4441	13	,98276	—	632,80	727,16	5,64	13,75	,1371
62	,94838	—	132,67	226,51	6,16	44,15	,4400	12	,98374	—	685,53	780,01	5,52	12,82	,1278
61	,94900	—	134,85	228,65	6,20	43,74	,4359	11	,98476	—	747,85	842,46	5,39	11,87	,1183
100+60	,94963	—	137,10	230,86	6,24	43,32	,4317	100+10	,98580	—	822,64	917,36	5,28	10,90	,1086
59	,95026	—	139,42	233,16	6,26	42,89	,4275	9	,98690	—	914,05	1008,91	5,14	9,91	,0988
58	,95090	—	141,82	235,53	6,29	42,46	,4232	8	,98806	—	1028,30	1123,31	4,99	8,90	,0887
57	,95154	—	144,31	237,98	6,33	42,01	,4188	7	,98927	—	1175,20	1270,35	4,85	7,87	,0784
56	,95218	—	146,90	240,53	6,37	41,57	,4144	6	,99051	—	1371,07	1466,36	4,71	6,82	,0680
100+55	,95283	—	149,57	243,17	6,40	41,12	,4098	100+5	,99182	—	1645,28	1740,73	4,55	5,74	,0572
54	,95346	—	152,30	245,89	6,41	40,67	,4053	4	,99320	—	2056,61	2152,21	4,40	4,65	,0463
53	,95410	—	155,17	248,74	6,43	40,20	,4007	3	,99464	—	2742,15	2837,91	4,24	3,52	,0351
52	,95473	—	158,16	251,72	6,44	39,73	,3960	2	,99615	—	4113,21	4209,13	4,08	2,38	,0237
51	,95538	—	161,28	254,79	6,49	39,24	,3912	1	,99773	—	8226,42	8322,48	3,94	1,20	,0120

TABLE I.   HEAT 67°.

I. Spirit and water by weight.	II. Specific gravity.	III. Spirit by measure.	IV. Water by measure.	V. Bulk of mixture.	VI. Diminution of bulk.	VII. Quantity of spirit per cent.	VIII. Decimal multipliers.
Sp. + W.							
100 + 0	,82167	100	—	100,00	—	100,00	,9959
1	,82399	—	0,82	100,72	0,10	99,29	,9889
2	,82625	—	1,65	101,44	0,21	98,58	,9819
3	,82845	—	2,47	102,16	0,31	97,88	,9750
4	,83059	—	3,29	102,89	0,40	97,20	,9681
100 + 5	,83268	—	4,11	103,62	0,49	96,51	,9612
6	,83471	—	4,94	104,35	0,59	95,84	,9546
7	,83669	—	5,76	105,08	0,68	95,17	,9479
8	,83864	—	6,58	105,82	0,76	94,50	,9413
9	,84054	—	7,40	106,56	0,84	93,85	,9347
100 + 10	,84237	—	8,22	107,30	0,92	93,20	,9282
11	,84417	—	9,04	108,04	1,00	92,56	,9219
12	,84593	—	9,87	108,79	1,08	91,92	,9155
13	,84765	—	10,69	109,54	1,15	91,29	,9093
14	,84932	—	11,50	110,29	1,21	90,67	,9031
100 + 15	,85096	—	12,33	111,04	1,29	90,06	,8970
16	,85259	—	13,16	111,79	1,37	89,45	,8909
17	,85419	—	13,98	112,55	1,43	88,85	,8849
18	,85576	—	14,80	113,30	1,50	88,26	,8790
19	,85730	—	15,62	114,05	1,57	87,67	,8732
100 + 20	,85880	—	16,44	114,81	1,63	87,10	,8674
21	,86028	—	17,26	115,57	1,69	86,53	,8617
22	,86174	—	18,09	116,32	1,77	85,96	,8562
23	,86316	—	18,91	117,08	1,83	85,40	,8506
24	,86455	—	19,73	117,85	1,88	84,85	,8451
100 + 25	,86592	—	20,56	118,61	1,95	84,30	,8397
26	,86727	—	21,38	119,37	2,01	83,77	,8343
27	,86860	—	22,20	120,14	2,06	83,24	,8290
28	,86991	—	23,03	120,90	2,13	82,71	,8238
29	,87119	—	23,85	121,67	2,18	82,19	,8186
100 + 30	,87245	—	24,67	122,44	2,23	81,68	,8135
31	,87369	—	25,49	123,20	2,29	81,17	,8083
32	,87490	—	26,31	123,97	2,34	80,67	,8033
33	,87612	—	27,14	124,74	2,40	80,17	,7985
34	,87729	—	27,96	125,51	2,45	79,68	,7936
100 + 35	,87845	—	28,78	126,27	2,51	79,19	,7887
36	,87959	—	29,59	127,04	2,55	78,71	,7840
37	,88070	—	30,42	127,82	2,60	78,24	,7792
38	,88180	—	31,24	128,59	2,65	77,77	,7745
39	,88289	—	32,07	129,36	2,71	77,31	,7699
100 + 40	,88395	—	32,89	130,13	2,76	76,84	,7654
41	,88502	—	33,71	130,90	2,81	76,39	,7608
42	,88607	—	34,54	131,67	2,87	75,94	,7564
43	,88711	—	35,36	132,45	2,91	75,50	,7520
44	,88813	—	36,18	133,22	2,96	75,06	,7476
100 + 45	,88912	—	37,00	133,99	3,01	74,63	,7433
46	,89012	—	37,82	134,77	3,05	74,20	,7390
47	,89118	—	38,64	135,55	3,09	73,77	,7348
48	,89203	—	39,46	136,32	3,14	73,35	,7306
49	,89297	—	40,29	137,10	3,19	72,94	,7265
100 + 50	,89387	100	41,11	137,88	3,23	72,53	,7223
51	,89480	—	41,93	138,66	3,27	72,12	,7183
52	,89568	—	42,76	139,44	3,32	71,71	,7143
53	,89657	—	43,58	140,21	3,37	71,31	,7103
54	,89744	—	44,40	140,99	3,41	70,92	,7064
100 + 55	,89830	—	45,22	141,77	3,45	70,53	,7025
56	,89914	—	46,04	142,55	3,49	70,14	,6987
57	,89997	—	46,86	143,34	3,52	69,76	,6949
58	,90079	—	47,69	144,12	3,57	69,38	,6911
59	,90160	—	48,51	144,90	3,61	69,01	,6874
100 + 60	,90239	—	49,33	145,68	3,65	68,64	,6836
61	,90317	—	50,15	146,47	3,68	68,27	,6800
62	,90394	—	50,98	147,25	3,73	67,91	,6763
63	,90470	—	51,80	148,04	3,76	67,55	,6727
64	,90545	—	52,62	148,82	3,80	67,19	,6692
100 + 65	,90616	—	53,45	149,61	3,84	66,84	,6657
66	,90693	—	54,27	150,39	3,88	66,49	,6622
67	,90765	—	55,09	151,18	3,91	66,14	,6588
68	,90837	—	55,91	151,96	3,95	65,80	,6554
69	,90909	—	56,73	152,75	3,98	65,46	,6520
100 + 70	,90978	—	57,55	153,53	4,02	65,13	,6487
71	,91048	—	58,37	154,31	4,06	64,80	,6454
72	,91116	—	59,20	155,11	4,09	64,47	,6421
73	,91181	—	60,02	155,90	4,12	64,14	,6389
74	,91248	—	60,84	156,69	4,15	63,82	,6357
100 + 75	,91312	—	61,67	157,47	4,20	63,50	,6325
76	,91377	—	62,49	158,26	4,23	63,19	,6293
77	,91440	—	63,31	159,05	4,26	62,87	,6262
78	,91503	—	64,13	159,83	4,30	62,56	,6231
79	,91566	—	64,96	160,63	4,33	62,24	,6200
100 + 80	,91625	—	65,78	161,42	4,36	61,95	,6170
81	,91688	—	66,60	162,21	4,39	61,65	,6140
82	,91747	—	67,42	162,99	4,43	61,35	,6110
83	,91807	—	68,24	163,78	4,46	61,06	,6080
84	,91866	—	69,06	164,58	4,48	60,77	,6051
100 + 85	,91924	—	69,89	165,37	4,52	60,47	,6023
86	,91980	—	70,71	166,16	4,55	60,18	,5994
87	,92035	—	71,53	166,95	4,58	59,90	,5966
88	,92090	—	72,36	167,74	4,62	59,62	,5938
89	,92144	—	73,18	168,54	4,64	59,34	,5910
100 + 90	,92198	—	74,00	169,33	4,67	59,06	,5982
91	,92252	—	74,83	170,12	4,71	58,78	,5855
92	,92306	—	75,65	170,91	4,74	58,51	,5828
93	,92358	—	76,47	171,70	4,77	58,24	,5801
94	,92410	—	77,29	172,50	4,79	57,97	,5774
100 + 95	,92462	—	78,10	173,29	4,81	57,71	,5747
96	,92512	—	78,92	174,07	4,85	57,45	,5721
97	,92562	—	79,74	174,86	4,80	57,19	,5696
98	,92611	—	80,56	175,65	4,91	56,93	,5670
99	,92660	—	81,39	176,45	4,94	56,67	,5644

TABLE II.   HEAT 67°.

I. Water and spirit by weight.	II. Specific gravity.	III. Spirit by measure.	IV. Water by measure.	V. Bulk of mixture.	VI. Diminution of bulk.	VII. Quantity of spirit per cent.	VIII. Decimal multipliers.
W. + Sp.							
100+100	,92709	100	82,22	177,25	4,97	56,42	,5619
99	,92758	—	83,05	178,06	4,99	56,16	,5594
98	,92808	—	83,90	178,88	5,02	55,90	,5568
97	,92857	—	84,77	179,72	5,05	55,64	,5542
96	,92906	—	85,64	180,57	5,07	55,38	,5516
100+95	,92956	—	86,55	181,44	5,11	55,12	,5489
94	,93004	—	87,47	182,33	5,14	54,84	,5462
93	,93053	—	88,41	183,24	5,17	54,57	,5435
92	,93102	—	89,37	184,18	5,19	54,29	,5407
91	,93152	—	90,35	185,14	5,21	54,01	,5379
100+90	,93200	—	91,36	186,10	5,26	53,73	,5351
89	,93253	—	92,38	187,11	5,27	53,44	,5323
88	,93305	—	93,43	188,13	5,30	53,15	,5294
87	,93357	—	94,51	189,18	5,33	52,86	,5265
86	,93410	—	95,61	190,25	5,36	52,56	,5235
100+85	,93463	—	96,74	191,34	5,40	52,26	,5205
84	,93517	—	97,89	192,46	5,43	51,96	,5175
83	,93572	—	99,06	193,61	5,45	51,65	,5144
82	,93627	—	100,27	194,79	5,48	51,34	,5113
81	,93683	—	101,51	195,99	5,52	51,02	,5082
100+80	,93740	—	102,78	197,22	5,56	50,70	,5050
79	,93795	—	104,08	198,49	5,59	50,38	,5018
78	,93851	—	105,41	199,79	5,62	50,05	,4985
77	,93906	—	106,78	201,13	5,65	49,72	,4952
76	,93962	—	108,19	202,51	5,68	49,38	,4918
100+75	,94018	—	109,63	203,92	5,71	49,04	,4884
74	,94076	—	111,10	205,37	5,73	48,70	,4850
73	,94134	—	112,62	206,85	5,77	48,34	,4815
72	,94192	—	114,19	208,39	5,80	47,99	,4780
71	,94251	—	115,80	209,97	5,83	47,63	,4744
100+70	,94310	—	117,46	211,59	5,87	47,27	,4707
69	,94370	—	119,15	213,25	5,90	46,89	,4670
68	,94430	—	120,90	214,97	5,93	46,51	,4633
67	,94491	—	122,71	216,75	5,96	46,14	,4595
66	,94552	—	124,58	218,57	6,01	45,76	,4557
100+65	,94614	—	126,50	220,45	6,05	45,37	,4518
64	,94676	—	128,47	222,39	6,08	44,97	,4478
63	,94738	—	130,50	224,40	6,10	44,56	,4439
62	,94800	—	132,61	226,47	6,14	44,15	,4398
61	,94862	—	134,78	228,61	6,17	43,75	,4357
100+60	,94928	—	137,03	230,82	6,21	43,33	,4315
59	,94989	—	139,36	233,11	6,25	42,90	,4273
58	,95053	—	141,75	235,48	6,27	42,46	,4230
57	,95117	—	144,24	237,93	6,31	42,02	,4186
56	,95182	—	146,83	240,48	6,35	41,58	,4142
100+55	,95247	—	149,50	243,12	6,38	41,13	,4096
54	,95311	—	152,23	245,84	6,39	40,67	,4051
53	,95375	—	155,09	248,68	6,41	40,21	,4005
52	,95439	—	158,09	251,66	6,43	39,73	,3958
51	,95504	—	161,21	254,73	6,48	39,25	,3910
W. + Sp.							
100+50	,95568	100	164,45	257,93	6,52	38,77	,3861
49	,95634	—	167,81	261,25	6,56	38,28	,3812
48	,95699	—	171,30	264,72	6,58	37,77	,3762
47	,95765	—	174,94	268,35	6,59	37,26	,3711
46	,95831	—	178,74	272,13	6,61	36,74	,3660
100+45	,95898	—	182,72	276,09	6,63	36,22	,3608
44	,95964	—	186,87	280,21	6,66	35,68	,3555
43	,96031	—	191,22	284,54	6,68	35,14	,3500
42	,96097	—	195,77	289,09	6,68	34,59	,3445
41	,96163	—	200,55	293,85	6,70	34,03	,3390
100+40	,96230	—	205,56	298,85	6,71	33,46	,3333
39	,96297	—	210,83	304,11	6,72	32,88	,3275
38	,96364	—	216,38	309,65	6,73	32,29	,3216
37	,96431	—	222,23	315,50	6,73	31,69	,3157
36	,96498	—	228,40	321,67	6,73	31,08	,3096
100+35	,96566	—	234,93	328,19	6,74	30,47	,3035
34	,96634	—	241,84	335,11	6,73	29,84	,2972
33	,96702	—	249,16	342,48	6,68	29,20	,2908
32	,96771	—	256,95	350,28	6,67	28,55	,2843
31	,96840	—	265,23	358,58	6,65	27,89	,2777
100+30	,96911	—	274,08	367,41	6,67	27,22	,2711
29	,96980	—	283,53	376,87	6,66	26,53	,2643
28	,97050	—	293,66	387,03	6,63	25,84	,2573
27	,97122	—	304,54	397,96	6,58	25,13	,2503
26	,97194	—	316,25	409,69	6,56	24,41	,2431
100+25	,97266	—	328,89	422,37	6,52	23,68	,2358
24	,97341	—	342,60	436,12	6,48	22,93	,2284
23	,97416	—	357,50	451,07	6,43	22,17	,2208
22	,97493	—	373,74	467,36	6,38	21,39	,2131
21	,97572	—	391,55	485,22	6,32	20,61	,2052
100+20	,97652	—	411,12	504,85	6,27	19,80	,1973
19	,97732	—	432,76	526,57	6,19	18,99	,1892
18	,97814	—	456,80	550,67	6,13	18,16	,1808
17	,97899	—	483,68	577,63	6,05	17,31	,1724
16	,97984	—	513,91	607,94	5,97	16,45	,1638
100+15	,98076	—	548,17	642,30	5,87	15,57	,1551
14	,98166	—	587,32	681,57	5,75	14,67	,1462
13	,98261	—	632,50	726,84	5,64	13,76	,1370
12	,98360	—	685,21	779,67	5,54	12,82	,1277
11	,98462	—	747,50	842,08	5,42	11,87	,1183
100+10	,98566	—	822,25	916,95	5,30	10,90	,1086
9	,98677	—	913,62	1008,46	5,16	9,91	,0987
8	,98787	—	1027,81	1122,80	5,01	8,91	,0887
7	,98914	—	1174,65	1269,77	4,88	7,88	,0784
6	,99039	—	1370,42	1465,69	4,73	6,82	,0679
100+5	,99170	—	1644,51	1739,93	4,57	5,74	,0572
4	,99308	—	2055,63	2151,21	4,42	4,65	,0463
3	,99435	—	2740,85	2836,58	4,27	3,53	,0351
2	,99604	—	4111,27	4207,16	4,11	2,38	,0237
1	,99762	—	8221,55	8318,57	3,98	1,20	,0120

I. Spirit and water by weight.	II. Specific gravity.	III. Spirit by measure.	IV. Water by measure.	V. Bulk of mixture.	VI. Diminution of bulk.	VII. Quantity of spirit per cent.	VIII. Decimal multipliers.
Sp. + W.							
100 + 0	,82119	100	—	100,00	—	100,00	,9954
1	,82351	—	0,92	100,72	0,10	99,29	,9883
2	,82577	—	1,64	101,44	0,20	98,58	,9813
3	,82797	—	2,47	102,16	0,31	97,88	,9744
4	,83011	—	3,29	102,89	0,40	97,20	,9676
100 + 5	,83221	—	4,11	103,62	0,49	96,51	,9607
6	,83423	—	4,94	104,35	0,59	95,84	,9540
7	,83621	—	5,76	105,08	0,68	95,17	,9474
8	,83816	—	6,58	105,82	0,76	94,50	,9408
9	,84006	—	7,40	106,56	0,84	93,85	,9343
100 + 10	,84189	—	8,22	107,30	0,92	93,20	,9277
11	,84369	—	9,04	108,04	1,00	92,56	,9214
12	,84544	—	9,86	108,79	1,07	91,92	,9150
13	,84716	—	10,68	109,54	1,14	91,29	,9088
14	,84884	—	11,50	110,29	1,21	90,67	,9026
100 + 15	,85048	—	12,33	111,04	1,29	90,06	,8965
16	,85211	—	13,15	111,79	1,36	89,45	,8904
17	,85371	—	13,97	112,55	1,42	88,86	,8844
18	,85528	—	14,79	113,30	1,49	88,26	,8785
19	,85682	—	15,61	114,05	1,56	87,68	,8727
100 + 20	,85832	—	16,43	114,81	1,62	87,10	,8670
21	,85980	—	17,25	115,57	1,68	86,53	,8613
22	,86126	—	18,08	116,32	1,76	85,97	,8557
23	,86269	—	18,90	117,08	1,82	85,41	,8502
24	,86408	—	19,72	117,84	1,88	84,85	,8446
100 + 25	,86545	—	20,55	118,61	1,94	84,31	,8392
26	,86630	—	21,37	119,37	2,00	83,77	,8338
27	,86813	—	22,19	120,13	2,06	83,24	,8285
28	,86944	—	23,02	120,90	2,12	82,71	,8234
29	,87073	—	23,84	121,66	2,18	82,19	,8182
100 + 30	,87198	—	24,66	122,43	2,23	81,68	,8130
31	,87323	—	25,47	123,19	2,28	81,17	,8079
32	,87444	—	26,30	123,96	2,34	80,67	,8029
33	,87565	—	27,12	124,73	2,39	80,17	,7981
34	,87683	—	27,94	125,50	2,44	79,68	,7932
100 + 35	,87799	—	28,77	126,27	2,50	79,19	,7883
36	,87912	—	29,58	127,04	2,54	78,71	,7836
37	,88023	—	30,41	127,81	2,60	78,24	,7788
38	,88133	—	31,23	128,58	2,65	77,77	,7741
39	,88242	—	32,05	129,35	2,70	77,31	,7695
100 + 40	,88348	—	32,87	130,13	2,74	76,85	,7650
41	,88455	—	33,69	130,89	2,80	76,39	,7604
42	,88560	—	34,52	131,67	2,85	75,94	,7560
43	,88664	—	35,34	132,44	2,90	75,50	,7516
44	,88766	—	36,16	133,21	2,95	75,06	,7472
100 + 45	,88866	—	36,99	133,99	3,00	74,63	,7429
46	,88966	—	37,80	134,76	3,04	74,20	,7386
47	,89062	—	38,62	135,54	3,08	73,78	,7344
48	,89157	—	39,44	136,31	3,13	73,36	,7303
49	,89251	—	40,27	137,09	3,18	72,94	,7261
100 + 50	,89342	100	41,09	137,87	3,22	72,53	,7220
51	,89434	—	41,91	138,65	3,26	72,13	,7180
52	,89523	—	42,74	139,43	3,31	71,72	,7140
53	,89612	—	43,56	140,20	3,36	71,32	,7099
54	,89699	—	44,38	140,98	3,40	70,93	,7060
100 + 55	,89785	—	45,20	141,76	3,44	70,54	,7021
56	,89869	—	46,02	142,54	3,48	70,15	,6983
57	,89952	—	46,84	143,33	3,51	69,77	,6945
58	,90034	—	47,66	144,11	3,55	69,39	,6908
59	,90115	—	48,48	144,89	3,59	69,02	,6871
100 + 60	,90194	—	49,31	145,67	3,64	68,65	,6833
61	,90272	—	50,13	146,46	3,67	68,28	,6796
62	,90349	—	50,95	147,24	3,71	67,91	,6760
63	,90425	—	51,77	148,02	3,75	67,55	,6724
64	,90500	—	52,59	148,81	3,78	67,20	,6689
100 + 65	,90574	—	53,42	149,59	3,83	66,85	,6654
66	,90648	—	54,24	150,37	3,87	66,50	,6619
67	,90721	—	55,06	151,16	3,90	66,15	,6585
68	,90793	—	55,88	151,94	3,94	65,81	,6551
69	,90865	—	56,70	152,73	3,97	65,47	,6517
100 + 70	,90934	—	57,53	153,52	4,01	65,14	,6484
71	,91004	—	58,35	154,30	4,05	64,81	,6451
72	,91072	—	59,17	155,09	4,08	64,48	,6418
73	,91138	—	59,99	155,88	4,11	64,15	,6386
74	,91204	—	60,81	156,67	4,14	63,83	,6354
100 + 75	,91268	—	61,64	157,45	4,19	63,51	,6322
76	,91332	—	62,46	158,24	4,22	63,19	,6290
77	,91396	—	63,28	159,03	4,25	62,88	,6259
78	,91459	—	64,10	159,82	4,28	62,57	,6228
79	,91521	—	64,93	160,61	4,32	62,26	,6197
100 + 80	,91581	—	65,75	161,40	4,35	61,96	,6167
81	,91643	—	66,57	162,19	4,38	61,66	,6137
82	,91703	—	67,39	162,97	4,42	61,36	,6107
83	,91763	—	68,21	163,77	4,44	61,06	,6077
84	,91822	—	69,03	164,56	4,47	60,77	,6048
100 + 85	,91881	—	69,85	165,35	4,50	60,48	,6020
86	,91937	—	70,67	166,14	4,53	60,19	,5992
87	,91992	—	71,50	166,93	4,57	59,91	,5963
88	,92047	—	72,32	167,72	4,60	59,62	,5935
89	,92101	—	73,14	168,52	4,62	59,34	,5907
100 + 90	,92155	—	73,96	169,31	4,65	59,06	,5879
91	,92209	—	74,79	170,10	4,69	58,79	,5851
92	,92262	—	75,61	170,89	4,72	58,51	,5825
93	,92315	—	76,43	171,68	4,75	58,25	,5797
94	,92367	—	77,25	172,48	4,77	57,98	,5793
100 + 95	,92419	—	78,07	173,27	4,80	57,71	,5743
96	,92469	—	78,88	174,05	4,83	57,45	,5716
97	,92519	—	79,70	174,84	4,86	57,19	,5692
98	,92568	—	80,52	175,63	4,89	56,93	,5663
99	,92617	—	81,35	176,43	4,92	56,67	,5643

I. Water and spirit by weight.	II. Specific gravity.	III. Spirit by measure.	IV. Water by measure.	V. Bulk of mixture.	VI. Diminution of bulk.	VII. Quantity of spirit per cent.	VIII. Decimal multipliers.	I. Water and spirit by weight.	II. Specific gravity.	III. Spirit by measure.	IV. Water by measure.	V. Bulk of mixture.	VI. Diminution of bulk.	VII. Quantity of spirit per cent.	VIII. Decimal multipliers.
W. + Sp.								**W. + Sp.**							
100+100	,92666	100	82,18	177,23	4,95	56,42	,5616	100 + 50	,95535	100	164,37	257,87	6,50	38,78	,3860
99	,92715	—	83,01	178,04	4,97	56,17	,5591	49	,95601	—	167,73	261,19	6,54	38,29	,3811
98	,92765	—	83,86	178,86	5,00	55,91	,5566	48	,95667	—	171,22	264,66	6,56	37,78	,3761
97	,92814	—	84,73	179,69	5,04	55,64	,5539	47	,95733	—	174,86	268,28	6,58	37,27	,3710
9?	,92864	—	85,60	180,54	5,06	55,38	,5513	46	,95799	—	178,66	272,06	6,60	36,75	,3659
100 + 95	,92913	—	86,51	181,41	5,10	55,12	,5487	100 + 45	,95866	—	182,64	276,02	6,62	36,23	,3606
94	,92962	—	87,43	182,30	5,13	54,85	,5460	44	,95933	—	186,78	280,14	6,64	35,69	,3553
93	,93011	—	88,37	183,22	5,15	54,58	,5433	43	,96000	—	191,13	284,47	6,66	35,15	,3499
92	,93060	—	89,33	184,15	5,18	54,30	,5405	42	,96067	—	195,68	289,01	6,67	34,60	,3444
91	,93110	—	90,31	185,11	5,20	54,02	,5377	41	,96133	—	200,46	293,77	6,69	34,04	,3389
100 + 90	,93158	—	91,31	186,08	5,23	53,74	,5349	100 + 40	,96201	—	205,46	298,77	6,69	33,47	,3332
89	,93211	—	92,33	187,09	5,24	53,45	,5321	39	,96268	—	210,73	304,03	6,70	32,89	,3274
88	,93263	—	93,39	188,11	5,28	53,16	,5292	38	,96336	—	216,28	309,56	6,72	32,30	,3216
87	,93315	—	94,46	189,15	5,31	52,87	,5263	37	,96403	—	222,12	315,41	6,71	31,70	,3156
86	,93368	—	95,56	190,22	5,34	52,57	,5233	36	,96470	—	228,29	321,58	6,71	31,09	,3095
100 + 85	,93421	—	96,69	191,31	5,38	52,27	,5203	100 + 35	,96539	—	234,82	328,09	6,73	30,48	,3034
84	,93475	—	97,84	192,43	5,41	51,97	,5173	34	,96607	—	241,72	335,01	6,71	29,85	,2971
83	,93530	—	99,01	193,58	5,43	51,66	,5142	33	,96676	—	249,04	342,37	6,67	29,21	,2908
82	,93586	—	100,22	194,76	5,46	51,34	,5111	32	,96745	—	256,83	350,17	6,68	28,56	,2843
81	,93642	—	101,46	195,96	5,50	51,03	,5080	31	,96815	—	265,11	358,46	6,65	27,90	,2777
100 + 80	,93698	—	102,73	197,19	5,54	50,71	,5048	100 + 30	,96886	—	273,95	367,29	6,66	27,23	,2710
79	,93754	—	104,03	198,46	5,57	50,39	,5016	29	,96956	—	283,40	376,75	6,65	26,54	,2642
78	,93810	—	105,36	199,76	5,60	50,06	,4983	28	,97027	—	293,52	386,91	6,61	25,85	,2573
77	,93865	—	106,73	201,10	5,63	49,72	,4950	27	,97099	—	304,39	397,83	6,56	25,14	,2502
76	,93922	—	108,14	202,48	5,66	49,38	,4916	26	,97172	—	316,10	409,56	6,54	24,42	,2430
100 + 75	,93978	—	109,58	203,89	5,69	49,04	,4882	100 + 25	,97245	—	328,74	422,22	6,52	23,69	,2357
74	,94036	—	111,05	205,34	5,71	48,70	,4848	24	,97320	—	342,44	435,96	6,48	22,94	,2283
73	,94094	—	112,57	206,82	5,75	48,35	,4813	23	,97396	—	357,33	450,90	6,43	22,18	,2208
72	,94153	—	114,14	208,36	5,78	47,99	,4778	22	,97474	—	373,57	467,18	6,39	21,40	,2131
71	,94212	—	115,75	209,94	5,81	47,64	,4742	21	,97553	—	391,36	485,03	6,33	20,61	,2052
100 + 70	,94271	—	117,40	211,55	5,85	47,27	,4705	100 + 20	,97634	—	410,93	504,65	6,28	19,81	,1972
69	,94331	—	119,10	213,21	5,89	46,90	,4668	19	,97714	—	432,56	526,36	6,20	19,00	,1891
68	,94392	—	120,84	214,93	5,91	46,52	,4631	18	,97797	—	456,59	550,45	6,14	18,17	,1808
67	,94453	—	122,65	216,71	5,94	46,15	,4593	17	,97882	—	483,45	577,39	6,06	17,32	,1724
66	,94514	—	124,52	218,53	5,99	45,76	,4555	16	,97970	—	513,67	607,69	5,98	16,46	,1638
100 + 65	,94576	—	126,44	220,41	6,03	45,37	,4516	100 + 15	,98060	—	547,91	642,03	5,88	15,58	,1550
64	,94638	—	128,41	222,35	6,06	44,97	,4477	14	,98151	—	587,04	681,28	5,76	14,68	,1461
63	,94700	—	130,44	224,35	6,09	44,57	,4437	13	,98246	—	632,20	726,53	5,67	13,76	,1370
62	,94762	—	132,54	226,43	6,11	44,16	,4396	12	,98346	—	684,89	779,33	5,56	12,83	,1277
61	,94825	—	134,72	228,56	6,16	43,75	,4355	11	,98448	—	747,15	841,71	5,44	11,88	,1183
100 + 60	,94888	—	136,97	230,77	6,20	43,34	,4313	100 + 10	,98553	—	821,86	916,54	5,32	10 91	,1086
59	,94952	—	139,29	233,07	6,22	42,91	,4271	9	,98664	—	913,19	1008,01	5,18	9,92	,0987
58	,95016	—	141,68	235,44	6,24	42,47	,4228	8	,98780	—	1027,33	1122,29	5,04	8,91	,0887
57	,95080	—	144,17	237,88	6,29	42,03	,4184	7	,98901	—	1174,10	1269,19	4,91	7,88	,0784
56	,95146	—	146,76	240,43	6,33	41,59	,4140	6	,99027	—	1369,78	1465,02	4,76	6,83	,0679
100 + 55	,95211	—	149,43	243,07	6,36	41,14	,4095	100 + 5	,99158	—	1643,74	1739,13	4,61	5,75	,0572
54	,95275	—	152,16	245,79	6,37	40,68	,4049	4	,99295	—	2054,67	2150,23	4,44	4,65	,0463
53	,95340	—	155,02	248,63	6,39	40,21	,4003	3	,99441	—	2739,56	2835,26	4,30	3,53	,0351
52	,95405	—	158,01	251,60	6,41	39,74	,3956	2	,99593	—	4109,34	4205,19	4,15	2,38	,0237
51	,95470	—	161,13	254,67	6,46	39,26	,3908	1	,99751	—	8218,68	8314,68	4,00	1,20	,0120

TABLE I.   HEAT 69°.

I. Spirit and water by weight.	II. Specific gravity.	III. Spirit by measure.	IV. Water by measure.	V. Bulk of mixture.	VI. Diminution of bulk.	VII. Quantity of spirit per cent.	VIII. Decimal multipliers.
Sp. + W.							
100 + 0	,82071	100	—	100,00	—	100,00	,9948
1	,82303	—	0,82	100,72	0,10	99,29	,9877
2	,82529	—	1,64	101,44	0,20	98,58	,9807
3	,82749	—	2,46	102,16	0,30	97,89	,9738
4	,82963	—	3,28	102,88	0,40	97,20	,9670
100 + 5	,83173	—	4,11	103,62	0,49	96,51	,9601
6	,83375	—	4,93	104,34	0,59	95,84	,9534
7	,83573	—	5,75	105,08	0,67	95,17	,9468
8	,83768	—	6,58	105,82	0,76	94,51	,9402
9	,83957	—	7,40	106,55	0,85	93,85	,9337
100 + 10	,84141	—	8,22	107,29	0,93	93,20	,9272
11	,84320	—	9,03	108,04	0,99	92,56	,9208
12	,84495	—	9,86	108,79	1,07	91,93	,9145
13	,84666	—	10,68	109,54	1,14	91,30	,9083
14	,84835	—	11,49	110,28	1,21	90,68	,9021
100 + 15	,84999	—	12,32	111,04	1,28	90,06	,8960
16	,85163	—	13,15	111,79	1,36	89,46	,8899
17	,85323	—	13,96	112,54	1,42	88,86	,8839
18	,85480	—	14,78	113,29	1,49	88,27	,8780
19	,85634	—	15,61	114,05	1,56	87,68	,8722
100 + 20	,85784	—	16,42	114,81	1,61	87,10	,8665
21	,85932	—	17,24	115,56	1,68	86,53	,8608
22	,86078	—	18,07	116,31	1,76	85,97	,8552
23	,86221	—	18,89	117,07	1,82	85,41	,8497
24	,86361	—	19,71	117,84	1,87	84,86	,8442
100 + 25	,86498	—	20,54	118,60	1,94	84,31	,8388
26	,86633	—	21,36	119,36	2,00	83,78	,8334
27	,86766	—	22,18	120,13	2,05	83,25	,8281
28	,86897	—	23,00	120,89	2,11	82,72	,8229
29	,87026	—	23,83	121,66	2,17	82,20	,8177
100 + 30	,87151	—	24,64	122,42	2,22	81,69	,8126
31	,87276	—	25,46	123,19	2,27	81,18	,8075
32	,87398	—	26,28	123,96	2,32	80,68	,8025
33	,87518	—	27,11	124,73	2,38	80,18	,7976
34	,87636	—	27,93	125,50	2,43	79,68	,7927
100 + 35	,87752	—	28,75	126,26	2,49	79,19	,7879
36	,87865	—	29,57	127,03	2,54	78,71	,7832
37	,87976	—	30,39	127,80	2,59	78,24	,7784
38	,88086	—	31,21	128,58	2,63	77,77	,7737
39	,88195	—	32,04	129,35	2,69	77,31	,7691
100 + 40	,88301	—	32,86	130,12	2,74	76,85	,7646
41	,88408	—	33,68	130,89	2,79	76,39	,7600
42	,88513	—	34,51	131,66	2,85	75,95	,7556
43	,88617	—	35,33	132,43	2,90	75,51	,7512
44	,88719	—	36,15	133,21	2,94	75,07	,7468
100 + 45	,88818	—	36,97	133,98	2,99	74,64	,7425
46	,88919	—	37,78	134,76	3,02	74,21	,7382
47	,89016	—	38,60	135,53	3,07	73,78	,7340
48	,89111	—	39,42	136,30	3,12	73,36	,7299
49	,89205	—	40,25	137,08	3,17	72,95	,7257
Sp. + W.							
100 + 50	,89297	100	41,07	137,86	3,21	72,54	,7216
51	,89389	—	41,89	138,64	3,25	72,13	,7176
52	,89478	—	42,72	139,42	3,30	71,72	,7136
53	,89567	—	43,54	140,19	3,35	71,32	,7096
54	,89654	—	44,36	140,97	3,39	70,93	,7057
100 + 55	,89740	—	45,18	141,75	3,43	70,54	,7018
56	,89824	—	46,00	142,53	3,47	70,15	,6980
57	,89907	—	46,82	143,32	3,50	69,77	,6942
58	,89989	—	47,64	144,10	3,54	69,39	,6904
59	,90070	—	48,46	144,88	3,58	69,02	,6867
100 + 60	,90149	—	49,29	145,66	3,63	68,66	,6829
61	,90227	—	50,11	146,45	3,66	68,29	,6793
62	,90304	—	50,93	147,23	3,70	67,92	,6756
63	,90380	—	51,75	148,01	3,74	67,56	,6720
64	,90455	—	52,57	148,80	3,77	67,20	,6685
100 + 65	,90529	—	53,39	149,58	3,81	66,85	,6650
66	,90603	—	54,21	150,36	3,85	66,50	,6616
67	,90677	—	55,03	151,15	3,88	66,15	,6582
68	,90749	—	55,85	151,93	3,92	65,81	,6548
69	,90821	—	56,67	152,71	3,96	65,48	,6514
100 + 70	,90890	—	57,50	153,51	3,99	65,14	,6481
71	,90960	—	58,32	154,28	4,04	64,81	,6448
72	,91028	—	59,14	155,07	4,07	64,48	,6415
73	,91094	—	59,96	155,86	4,10	64,15	,6383
74	,91160	—	60,78	156,66	4,12	63,83	,6351
100 + 75	,91224	—	61,61	157,44	4,17	63,51	,6319
76	,91288	—	62,43	158,23	4,20	63,20	,6287
77	,91352	—	63,25	159,01	4,24	62,88	,6256
78	,91415	—	64,07	159,80	4,27	62,57	,6225
79	,91477	—	64,90	160,59	4,31	62,26	,6194
100 + 80	,91537	—	65,72	161,38	4,34	61,96	,6164
81	,91599	—	66,54	162,17	4,37	61,66	,6134
82	,91659	—	67,36	162,95	4,41	61,36	,6104
83	,91719	—	68,18	163,75	4,43	61,07	,6074
84	,91778	—	69,00	164,54	4,46	60,77	,6045
100 + 85	,91837	—	69,82	165,33	4,49	60,48	,6017
86	,91894	—	70,64	166,12	4,52	60,20	,5988
87	,91949	—	71,46	166,91	4,55	59,91	,5960
88	,92004	—	72,29	167,70	4,59	59,63	,5932
89	,92058	—	73,11	168,50	4,61	59,35	,5904
100 + 90	,92112	—	73,93	169,29	4,64	59,07	,5876
91	,92166	—	74,75	170,08	4,67	58,79	,5849
92	,92219	—	75,57	170,87	4,70	58,52	,5822
93	,92272	—	76,39	171,66	4,73	58,25	,5795
94	,92324	—	77,22	172,40	4,76	57,98	,5769
100 + 95	,92376	—	78,03	173,25	4,78	57,72	,5742
96	,92426	—	78,84	174,03	4,81	57,46	,5716
97	,92476	—	79,66	174,82	4,84	57,20	,5691
98	,92525	—	80,48	175,61	4,87	56,94	,5665
99	,92574	—	81,31	176,41	4,90	56,68	,5639

TABLE 11. HEAT 69°.

I. Water and spirit by weight.	II. Specific gravity.	III. Spirit by measure.	IV. Water by measure.	V. Bulk of mixture.	VI. Diminution of bulk.	VII. Quantity of spirit per cent.	VIII. Decimal multipliers.
W. + Sp.							
100+100	,92623	100	82,14	177,21	4.93	56,43	,5613
99	,92672	—	82,97	178,02	4,95	56,17	,5588
98	,92722	—	83,82	178,84	4,98	55,91	,5563
97	,92772	—	84,69	179,67	5,02	55,65	,5537
96	,92822	—	85,57	180,52	5,05	55,39	,5511
100 + 95	,92870	—	86,47	181,39	5,08	55,13	,5484
94	,92920	—	87,39	182,28	5,11	54,86	,5457
93	,92969	—	88,33	183,20	5,13	54,58	,5430
92	,93018	—	89,29	184,13	5,16	54,31	,5402
91	,93068	—	90,27	185,08	5,19	54,03	,5374
100 + 90	,93116	—	91,27	186,06	5,21	53,74	,5346
89	,93169	—	92,29	187,06	5,23	53,46	,5318
88	,93221	—	93,34	188,09	5,25	53,17	,5290
87	,93273	—	94,42	189,12	5,30	52,87	,5260
86	,93326	—	95,51	190,19	5,32	52,58	,5210
100 + 85	,93379	—	96,64	191,28	5,36	52,28	,5200
84	,93434	—	97,79	192,40	5,39	51,98	,5170
83	,93489	—	98,96	193,55	5,41	51,67	,5140
82	,93545	—	100,17	194,73	5,44	51,35	,5109
81	,93601	—	101,41	195,93	5,48	51,04	,5078
100 + 80	,93657	—	102,68	197,16	5,52	50,72	,5046
79	,93713	—	103,98	198,43	5,55	50,39	,5013
78	,93769	—	105,31	199,73	5,58	50,06	,4981
77	,93825	—	106,68	201,07	5,61	49,73	,4948
76	,93882	—	108,09	202,45	5,64	49,39	,4914
100 + 75	,93938	—	109,53	203,86	5,67	49,05	,4880
74	,93996	—	111,00	205,31	5,69	48,71	,4846
73	,94055	—	112,52	206,79	5,73	48,36	,4811
72	,94114	—	114,09	208,32	5,77	48,00	,4776
71	,94173	—	115,70	209,90	5,80	47,64	,4740
100 + 70	,94232	—	117,35	211,51	5,84	47,28	,4703
69	,94293	—	119,04	213,17	5,87	46,91	,4666
68	,94354	—	120,79	214,90	5,89	46,53	,4629
67	,94415	—	122,60	216,67	5,93	46,16	,4591
66	,94476	—	124,46	218.49	5,97	45,77	,4553
100 + 65	,94538	—	126,38	220,37	6,01	45,38	,4514
64	,94600	—	128,35	222,31	6,04	44,98	,4475
63	,94662	—	130,38	224,31	6,07	44,58	,4435
62	,94725	—	132,48	226,38	6,10	44,17	,4394
61	,94788	—	134,66	228,52	6,14	43,76	,4353
100 + 60	,94851	—	136,91	230,74	6,17	43,34	,4311
59	,94915	—	139,22	233,02	6,20	42,91	,4269
58	,94979	—	141,62	235,39	6,23	42,48	,4226
57	,95043	—	144,11	237,84	6,27	42,04	,4183
56	,95110	—	146,69	240,38	6,31	41,60	,4138
100 + 55	,95175	—	149,36	243,02	6,34	41,15	,4093
54	,95241	—	152,09	245,74	6,35	40,69	,4048
53	,95305	—	154,95	248,58	6,37	40,22	,4002
52	,95371	—	157,94	251,54	6,40	39,75	,3955
51	,95437	—	161,05	254,61	6,44	39,27	,3907

I. Water and spirit by weight.	II. Specific gravity.	III. Spirit by measure.	IV. Water by measure.	V. Bulk of mixture.	VI. Diminution of bulk.	VII. Quantity of spirit per cent.	VIII. Decimal multipliers.
W. + Sp.							
100 + 50	,95502	100	164,29	257,81	6,48	38,79	,3859
49	,95568	—	167,65	261,13	6,52	38,30	,3810
48	,95634	—	171,14	264,60	6,54	37,79	,3760
47	,95701	—	174,78	268,21	6,57	37,28	,3709
46	,95767	—	178,58	271,99	6,59	36,76	,3658
100 + 45	,95834	—	182,56	275,95	6,61	36,24	,3605
44	,95902	—	186,69	280,07	6,62	35,70	,3552
43	,95969	—	191,04	284,40	6,64	35,16	,3498
42	,96037	—	195,59	288,94	6,65	34,61	,3443
41	,96104	—	200,36	293,69	6,67	34,05	,3387
100 + 40	,96172	—	205,36	298,09	6,67	33,48	,3330
39	,96240	—	210,63	303,94	6,69	32,90	,3273
38	,96308	—	216,18	309,48	6,70	32,31	,3215
37	,96375	—	222,02	315,32	6,70	31,71	,3155
36	,96443	—	228,18	321,49	6,69	31,10	,3094
100 + 35	,96511	—	234,71	327,99	6,72	30,49	,3033
34	,96580	—	241,61	334,91	6,70	29,86	,2970
33	,96650	—	248,93	342,26	6,67	29,22	,2907
32	,96720	—	256,71	350,05	6,66	28,57	,2842
31	,96790	—	264,99	358,35	6,64	27,91	,2776
100 + 30	,96861	—	273,82	367,17	6,65	27,24	,2709
29	,96932	—	283,27	376,63	6,64	26,55	,2641
28	,97004	—	293,38	386,79	6,59	25,86	,2572
27	,97077	—	304,25	397,70	6,55	25,15	,2502
26	,97150	—	315,95	409,43	6,52	24,43	,2430
100 + 25	,97224	—	328,59	422,07	6.52	23,70	,2357
24	,97300	—	342,28	435,80	6,48	22,95	,2283
23	,97376	—	357,16	450,73	6,43	22,19	,2207
22	,97455	—	373,40	467,00	6,40	21,41	,2130
21	,97534	—	391,18	484.84	6,34	20,62	,2052
100 + 20	,97615	—	410,74	504,45	6,29	19,82	,1972
19	,97696	—	432,36	526,15	6,21	19,01	,1890
18	,97780	—	456,38	550,23	6,15	18,17	,1807
17	,97865	—	483,22	577,15	6,07	17,33	,1723
16	,97954	—	513,43	607,44	5,99	16,46	,1637
100 + 15	,98044	—	547,65	641,76	5,89	15,58	,1550
14	,98136	—	586,77	680,99	5,78	14,68	,1461
13	,98232	—	631,90	726,22	5,68	13,77	,1370
12	,98332	—	684,57	778,99	5,58	12,84	,1277
11	,98434	—	746,80	841,34	5,46	11,88	,1182
100 + 10	,98540	—	821,48	916,13	5,35	10,91	,1086
9	,98651	—	912,76	1007,56	5,20	9,92	,0987
8	,98767	—	1026,85	1121,78	5,07	8,91	,0887
7	,98889	—	1173,55	1268,61	4,94	7,88	,0784
6	,99015	—	1369,14	1464,35	4,79	6,83	,0679
100 + 5	,99146	—	1042,97	1738,33	4,64	5,75	,0572
4	,99284	—	2053,70	2149,23	4,47	4,05	,0463
3	,99429	—	2738,28	2833,95	4,33	3,53	,0351
2	,99581	—	4107,42	4203,24	4,18	2,38	,0237
1	,99740	—	8214,83	8310,80	4,03	1,20	,0120

TABLE I.   HEAT 70°.

I. Spirit and water by weight.	II. Specific gravity.	III. Spirit by measure	IV. Water by measure.	V. Bulk of mixture.	VI. Diminution of bulk.	VII. Quantity of spirit per cent.	VIII. Decimal multipliers.	I. Spirit and water by weight.	II. Specific gravity.	III. Spirit by measure	IV. Water by measure.	V. Bulk of mixture.	VI. Diminution of bulk.	VII. Quantity of spirit per cent.	VIII. Decimal multipliers.
Sp. + W.								Sp. + W.							
100 + 0	,82023	100	—	100,00	—	100,00	,9942	100 + 50	,89252	100	41,05	137,85	3,20	72,54	,7212
1	,82255	—	0,82	100,72	0,10	99.29	,9871	51	,89343	—	41,88	138,63	3,25	72,14	,7172
2	,82481	—	1,64	101,44	0,20	98,58	,9801	52	,89433	—	42,70	139,41	3,29	71,73	,7132
3	,82701	—	2,46	102,16	0,30	97,89	,9732	53	,89522	—	43,52	140,18	3,34	71.33	,7092
4	,82915	—	3,28	102.88	0,40	97,20	,9664	54	,89609	—	44,34	140 96	3,38	70,94	,7053
100 + 5	,83124	—	4,10	103,61	0,49	96,52	,9596	100 + 55	,89695	—	45,16	141 74	3,42	70,55	,7014
6	,83327	—	4,93	104,34	0,59	95,84	,9529	56	,89779	—	45,98	142,52	3.46	70,16	,6976
7	,83525	—	5,75	105,08	0,67	95,17	,9462	57	,89862	—	46.80	143,31	3,49	69,78	,6938
8	,83719	—	6,57	105,81	0,76	94,51	,9396	58	,89944	—	47,62	144,09	3,53	69.40	,6900
9	,83908	—	7,39	106,55	0,84	93,85	,9331	59	,90025	—	48,44	144 87	3,57	69.03	,6863
100 + 10	,84092	—	8,21	107,29	0,92	93,20	,9266	100 + 60	,90104	—	49.27	145,65	3,62	68,66	,6826
11	,84271	—	9,03	108,04	0,99	92,56	,9202	61	,90182	—	50,09	146,44	3,65	68,29	,6789
12	,84446	—	9,85	108,78	1,07	91,93	,9139	62	,90259	—	50,91	147,22	3,69	67,92	,0753
13	,84618	—	10,67	109,54	1,13	91,30	,9077	63	,90335	—	51,73	148,00	3,73	67,56	,6717
14	,84786	—	11,49	110,28	1,21	90,68	,9015	64	,90410	—	52,55	148,79	3,76	67 21	,6682
100 + 15	,84951	—	12,32	111,04	1,28	90,06	,8954	100 + 65	,90484	—	53,37	149,57	3,80	66,86	,6647
16	,85114	—	13,14	111,79	1,35	89,46	,8894	66	,90558	—	54,19	150,35	3,84	66,51	,6612
17	,85274	—	13,96	112,54	1,42	88,86	,8834	67	,90632	—	55,01	151,14	3,87	66,16	,6578
18	,85431	—	14,78	113,29	1,49	88,27	,8775	68	,90705	—	55,83	151,92	3,91	65,83	,6544
19	,85585	—	15,60	114,04	1,56	87,68	,8717	69	,90777	—	56,65	152,70	3,95	65,49	,6510
100 + 20	,85736	—	16,42	114,80	1,62	87,11	,8660	100 + 70	,90847	—	57.48	153,49	3,99	65,15	,6477
21	,85884	—	17,24	115,56	1,68	86,54	,8603	71	,90916	—	58,30	154,27	4,03	64,82	,6444
22	,86030	—	18,06	116,31	1,75	85,97	,8547	72	,90984	—	59,12	155,06	4,06	64,49	,6411
23	,86173	—	18,88	117,07	1,81	85,41	,8492	73	,91050	—	59,94	155,85	4,09	64,16	,6379
24	,86313	—	19,71	117,83	1,88	84,86	,8437	74	,91116	—	60,76	156,64	4,12	63,84	,6347
100 + 25	,86451	—	20,53	118,60	1,93	84,32	,8383	100 + 75	,91181	—	61 58	157,42	4 16	63,52	,6316
26	,86586	—	21,35	119,36	1,99	83,78	,8330	76	,91245	—	62,40	158,21	4 19	63,21	,6284
27	,86719	—	22,17	120,12	2,05	83,25	,8277	77	,91308	—	63.22	159,00	4,22	62,89	,6253
28	,86850	—	22,99	120,89	2,10	82,72	,8225	78	,91370	—	64.04	159,79	4,25	62,58	,6222
29	,86979	—	23,81	121,65	2,16	82,20	,8173	79	,91432	—	64,87	160,58	4,29	62,27	,6191
100 + 30	,87105	—	24,63	122,41	2,22	81,69	,8122	100 + 80	,91493	—	65,69	161.37	4,32	61,97	,6161
31	,87229	—	25,45	123,18	2,27	81,18	,8071	81	,91554	—	66,51	162,16	4,35	61,67	,6131
32	,87351	—	26,27	123,95	2,32	80,68	,8021	82	,91615	—	67,33	162,94	4,39	61,37	,6101
33	,87471	—	27,10	124,72	2,38	80,18	,7972	83	,91675	—	68,15	163,73	4,42	61,08	,6072
34	,87589	—	27,92	125,49	2,43	79,69	,7923	84	,91734	—	68.97	164,52	4,45	60,78	,6043
100 + 35	,87705	—	28,74	126,25	2,49	79,20	,7875	100 + 85	,91793	—	69,79	165.31	4,48	60,49	,6014
36	,87818	—	29,56	127,02	2,54	78,72	,7827	86	,91850	—	70,61	166,10	4,51	60,20	,5985
37	,87929	—	30,38	127,80	2,58	78,25	,7780	87	,91906	—	71 43	166,89	4,54	59,92	,5957
38	,88039	—	31,20	128,57	2,63	77,78	,7733	88	,91961	—	72,26	167,68	4,58	59,64	,5929
39	,88147	—	32,02	129,34	2,68	77,32	,7687	89	,92015	—	73,08	168,48	4,60	59,36	,5901
100 + 40	,88254	—	32,84	130,11	2,73	76,86	,7641	100 + 90	,92069	—	73,90	169,27	4,63	59,08	,5874
41	,88361	—	33,66	130,88	2,78	76,40	,7596	91	,92123	—	74,72	170,06	4,66	58,80	,5846
42	,88466	—	34,49	131,65	2,84	75,95	,7551	92	,92176	—	75,54	170,85	4,69	58,53	,5819
43	,88570	—	35,31	132,43	2,88	75,51	,7507	93	,92229	—	76,36	171,64	4 72	58,26	,5792
44	,88672	—	36,13	133,20	2,93	75,07	,7464	94	,92281	—	77.18	172,44	4,74	57,99	,5766
100 + 45	,88773	—	36,95	133,97	2,98	74,64	,7421	100 + 95	,92333	—	78,00	173,23	4 77	57,73	,5739
46	,88872	—	37,77	134,75	3,02	74,21	,7378	96	,92383	—	78,81	174,01	4 80	57,47	,5713
47	,88969	—	38,59	135,52	3,07	73,79	,7336	97	,92413	—	79,63	174,80	4,83	57,21	,5688
48	,89065	—	39,41	136,30	3,11	73,37	,7295	98	,92482	—	80,45	175,59	4,86	56,95	,5662
49	,89159	—	40,23	137,07	3,16	72,95	,7253	99	,92531	··	81,28	176,39	4,89	56,69	,5636

TABLE II. HEAT 70°.

I. Water and spirit by weight.	II. Specific gravity.	III. Spirit by measure.	IV. Water by measure.	V. Bulk of mixture.	VI. Diminution of bulk.	VII. Quantity of spirit per cent.	VIII. Decimal multipliers.	I. Water and spirit by weight.	II. Specific gravity.	III. Spirit by measure.	IV. Water by measure.	V. Bulk of mixture.	VI. Diminution of bulk.	VII. Quantity of spirit per cent.	VIII. Decimal multipliers.
W. + Sp.								W. + Sp.							
100+100	,92580	100	82,11	177,19	4,92	56,44	,5611	100 + 50	,95469	100	164,22	257,75	6,47	38,80	,3857
99	,92629	—	82,94	178,00	4,94	56,18	,5585	49	,95535	—	167,57	261,07	6,50	38,30	,3808
98	,92679	—	83,79	178,82	4,97	55,92	,5560	48	,95601	—	171,06	264,53	6,53	37,80	,3758
97	,92729	—	84,66	179,65	5,01	55,66	,5534	47	,95668	—	174,70	268,15	6,55	37,29	,3707
96	,92779	—	85,54	180,50	5,04	55,40	,5508	46	,95735	—	178,50	271,93	6,57	36,77	,3656
100 + 95	,92828	—	86,43	181,37	5,06	55,14	,5482	100 + 45	,95802	—	182,47	275,88	6,59	36,25	,3604
94	,92877	—	87,35	182,26	5,09	54,86	,5455	44	,95870	—	186,61	280,00	6,61	35,71	,3551
93	,92926	—	88,29	183,18	5,11	54,59	,5428	43	,95938	—	190,95	284,33	6,62	35,17	,3497
92	,92976	—	89,25	184,11	5,14	54,31	,5400	42	,96006	—	195,50	288,87	6,63	34,62	,3442
91	,93026	—	90,23	185,06	5,17	54,03	,5372	41	,96074	—	200,27	293,62	6,65	34,06	,3386
100 + 90	,93076	—	91,23	186,04	5,19	53,75	,5344	100 + 40	,96143	—	205,27	298,60	6,67	33,49	,3330
89	,93127	—	92,25	187,04	5,21	53,46	,5316	39	,96211	—	210,54	303,86	6,68	32,91	,3272
88	,93179	—	93,30	188,06	5,24	53,17	,5287	38	,96279	—	216,08	309,40	6,68	32,32	,3214
87	,93231	—	94,37	189,10	5,27	52,88	,5258	37	,96347	—	221,92	315,24	6,68	31,72	,3154
86	,93284	—	95,47	190,17	5,30	52,58	,5228	36	,96415	—	228,08	321,41	6,67	31,11	,3094
100 + 85	,93337	—	96,60	191,26	5,30	52,28	,5198	100 + 35	,96484	—	234,60	327,90	6,70	30,50	,3032
84	,93392	—	97,75	192,37	5,38	51,98	,5168	34	,96553	—	241,50	334,81	6,69	29,87	,2969
83	,93447	—	98,92	193,52	5,40	51,67	,5137	33	,96622	—	248,82	342,15	6,67	29,23	,2906
82	,93503	—	100,13	194,70	5,43	51,36	,5106	32	,96694	—	256,59	349,94	6,65	28,58	,2841
81	,93559	—	101,37	195,90	5,47	51,04	,5075	31	,96765	—	264,87	358,23	6,64	27,91	,2775
100 + 80	,93616	—	102,64	197,13	5,51	50,73	,5043	100 + 30	,96836	—	273,70	367,05	6,65	27,24	,2709
79	,93672	—	103,94	198,40	5,54	50,40	,5011	29	,96908	—	283,14	376,51	6,63	26,56	,2641
78	,93728	—	105,27	199,70	5,57	50,07	,4978	28	,96980	—	293,25	386,67	6,58	25,86	,2571
77	,93784	—	106,64	201,04	5,60	49,74	,4945	27	,97054	—	304,11	397,57	6,54	25,15	,2501
76	,93841	—	108,04	202,42	5,62	49,40	,4912	26	,97128	—	315,81	409,30	6,51	24,43	,2429
100 + 75	,93898	—	109,48	203,83	5,65	49,06	,4878	100 + 25	,97203	—	328,44	421,92	6,52	23,70	,2356
74	,93956	—	110,95	205,28	5,67	48,72	,4844	4. 24	,97279	—	342,12	435,64	6,48	22,95	,2282
73	,94015	—	112,47	206,76	5,71	48,37	,4809	23	,97356	—	357,00	450,56	6,44	22,19	,2207
72	,94074	—	114,04	208,29	5,75	48,01	,4774	22	,97435	—	373,23	466,83	6,40	21,42	,2130
71	,94133	—	115,65	209,86	5,79	47,65	,4738	21	,97515	—	390,00	484,65	6,35	20,63	,2051
100 + 70	,94193	—	117,30	211,48	5,82	47,29	,4701	100 + 20	,97596	—	410,55	504,26	6,29	19,83	,1971
69	,94254	—	118,99	213,14	5,85	46,92	,4664	19	,97678	—	432,16	525,93	6,23	19,01	,1890
68	,94315	—	120,74	214,86	5,88	46,54	,4627	18	,97762	—	456,17	550,01	6,16	18,18	,1807
67	,94376	—	122,55	216,63	5,92	46,16	,4589	17	,97848	—	483,00	576,92	6,08	17,33	,1723
66	,94438	—	124,40	218,45	5,95	45,78	,4551	16	,97937	—	513,19	607,19	6,00	16,47	,1637
100 + 65	,94500	—	126,32	220,33	5,99	45,39	,4512	100 + 15	,98028	—	547,40	641,49	5,91	15,59	,1550
64	,94562	—	128,29	222,27	6,02	44,99	,4473	14	,98121	—	586,50	680,70	5,80	14,69	,1461
63	,94624	—	130,32	224,27	6,05	44,59	,4433	13	,98217	—	631,61	725,91	5,70	13,77	,1370
62	,94687	—	132,42	226,34	6,08	44,18	,4393	12	,98317	—	684,25	778,65	5,60	12,84	,1277
61	,94750	—	134,60	228,48	6,12	43,77	,4352	11	,98420	—	746,45	840,97	5,48	11,89	,1182
100 + 60	,94813	—	136,85	230,69	6,16	43,35	,4310	100 + 10	,98527	—	821,10	915,72	5,38	10,92	,1086
59	,94877	—	139,16	232,97	6,19	42,92	,4268	9	,98638	—	912,33	1007,11	5,22	9,93	,0987
58	,94942	—	141,56	235,34	6,22	42,49	,4225	8	,98754	—	1026,37	1121,28	5,09	8,92	,0887
57	,95007	—	144,05	237,80	6,25	42,05	,4181	7	,98876	—	1173,00	1268,04	4,96	7,89	,0784
56	,95073	—	146,63	240,34	6,29	41,61	,4137	6	,99002	—	1368,50	1463,68	4,82	6,83	,0679
100 + 55	,95139	—	149,29	242,97	6,32	41,16	,4092	100 + 5	,99134	—	1642,20	1737,53	4,67	5,75	,0572
54	,95205	—	152,02	245,69	6,33	40,70	,4046	4	,99272	—	2052,74	2148,24	4,50	4,65	,0463
53	,95270	—	154,88	248,53	6,35	40,23	,4000	3	,99417	—	2737,00	2832,64	4,36	3,53	,0351
52	,95336	—	157,87	251,48	6,39	39,76	,3953	2	,99569	—	4105,50	4201,29	4,21	2,38	,0237
51	,95403	—	160,98	254,56	6,42	39,28	,3905	1	,99728	—	8210,99	8306,93	4,06	1,20	,0120

TABLE I.   HEAT 71°.

I. Spirit and water by weight.	II. Specific gravity.	III. Spirit by measure.	IV. Water by measure.	V. Bulk of mixture.	VI. Diminution of bulk.	VII. Quantity of Spirit per cent.	VIII. Decimal multipliers.
Sp. + W.							
100 + 0	,81975	100	—	100,00	—	100,00	,9936
1	,82206	—	0,82	100,72	0,10	99,29	,9866
2	,82432	—	1,64	101,44	0,20	98,58	,9796
3	,82652	—	2,46	102,16	0,30	97,89	,9727
4	,82866	—	3,28	102,88	0,40	97,20	,9659
100 + 5	,83075	—	4,10	103,61	0,49	96,52	,9590
6	,83278	—	4,93	104,34	0,59	95,84	,9523
7	,83477	—	5,75	105,08	0,67	95,17	,9457
8	,83671	—	6,57	105,81	0,76	94,51	,9391
9	,83860	—	7,39	106,55	0,84	93,85	,9326
100 + 10	,84044	—	8,21	107,29	0,92	93,20	,9261
11	,84223	—	9,03	108,04	0,99	92,56	,9197
12	,84398	—	9,85	108,78	1,07	91,93	,9134
13	,84570	—	10,67	109,53	1,14	91,30	,9072
14	,84738	—	11,48	110,28	1,20	90,68	,9010
100 + 15	,84903	—	12,31	111,03	1,28	90,06	,8949
16	,85066	—	13,14	111,78	1,36	89,46	,8889
17	,85226	—	13,95	112,53	1,42	88,86	,8830
18	,85383	—	14,77	113,29	1,48	88,27	,8771
19	,85537	—	15,59	114,04	1,55	87,68	,8713
100 + 20	,85688	—	16,41	114,80	1,61	87,11	,8655
21	,85836	—	17,23	115,56	1,67	86,54	,8598
22	,85982	—	18,05	116,31	1,74	85,97	,8542
23	,86125	—	18,87	117,07	1,80	85,42	,8488
24	,86265	—	19,70	117,83	1,87	84,87	,8433
100 + 25	,86403	—	20,52	118,59	1,93	84,32	,8379
26	,86539	—	21,34	119,35	1,99	83,78	,8325
27	,86672	—	22,16	120,12	2,04	83,25	,8272
28	,86802	—	22,98	120,88	2,10	82,72	,8220
29	,86931	—	23,80	121,65	2,15	82,20	,8168
100 + 30	,87057	—	24,62	122,41	2,21	81,69	,8117
31	,87181	—	25,44	123,18	2,26	81,18	,8067
32	,87303	—	26,26	123,94	2,32	80,68	,8017
33	,87423	—	27,09	124,71	2,38	80,18	,7968
34	,87541	—	27,91	125,49	2,42	79,69	,7919
100 + 35	,87658	—	28,73	126,25	2,48	79,20	,7870
36	,87771	—	29,55	127,02	2,53	78,72	,7823
37	,87882	—	30,37	127,79	2,58	78,25	,7776
38	,87992	—	31,19	128,56	2,63	77,78	,7719
39	,88100	—	32,01	129,33	2,68	77,32	,7683
100 + 40	,88206	—	32,83	130,10	2,73	76,86	,7637
41	,88314	—	33,65	130,87	2,78	76,40	,7592
42	,88419	—	34,47	131,64	2,83	75,96	,7547
43	,88523	—	35,29	132,42	2,87	75,52	,7503
44	,88625	—	36,11	133,19	2,92	75,08	,7460
100 + 45	,88726	—	36,93	133,96	2,97	74,65	,7417
46	,88825	—	37,75	134,74	3,01	74,22	,7375
47	,88922	—	38,57	135,51	3,06	73,79	,7332
48	,89018	—	39,39	136,29	3,10	73,37	,7291
49	,89113	—	40,21	137,06	3,15	72,96	,7249

I. Spirit and Water by weight.	II. Specific gravity.	III. Spirit by measure.	IV. Water by measure.	V. Bulk of mixture.	VI. Diminution of bulk.	VII. Quantity of Spirit per cent.	VIII. Decimal multipliers.
Sp. + W							
100 + 50	,89206	100	41,03	137,84	3,19	72,55	,7208
51	,89297	—	41,86	138,62	3,24	72,14	,7169
52	,89387	—	42,68	139,40	3,28	71,74	,7129
53	,89476	—	43,50	140,17	3,33	71,34	,7089
54	,89563	—	44,31	140,95	3,36	70,95	,7050
100 + 55	,89649	—	45,13	141,73	3,40	70,56	,7011
56	,89733	—	45,95	142,51	3,44	70,17	,6973
57	,89816	—	46,77	143,29	3,48	69,79	,6935
58	,89898	—	47,59	144,07	3,52	69,40	,6897
59	,89979	—	48,41	144,85	3,56	69,03	,6859
100 + 60	,90058	—	49,24	145,63	3,61	68,66	,6822
61	,90136	—	50,06	146,42	3,64	68,30	,6786
62	,90213	—	50,88	147,20	3,68	67,93	,6750
63	,90289	—	51,70	147,98	3,72	67,57	,0714
64	,90364	—	52,52	148,77	3,75	67,21	,6679
100 + 65	,90438	—	53,34	149,55	3,79	66,86	,6644
66	,90512	—	54,16	150,33	3,83	66,52	,6609
67	,90586	—	54,98	151,12	3,86	66,17	,6575
68	,90659	—	55,80	151,90	3,90	65,83	,6541
69	,90731	—	56,62	152,68	3,94	65,49	,6507
100 + 70	,90801	—	57,45	153,47	3,98	65,15	,6474
71	,90870	—	58,27	154,26	4,01	64,82	,6441
72	,90938	—	59,09	155,05	4,04	64,49	,6408
73	,91005	—	59,91	155,83	4,08	64,17	,6376
74	,91071	—	60,73	156,62	4,11	63,85	,6344
100 + 75	,91135	—	61,55	157,41	4,14	63,53	,6312
76	,91200	—	62,37	158,19	4,18	63,21	,6281
77	,91263	—	63,19	158,98	4,21	62,90	,6250
78	,91325	—	64,01	159,77	4,24	62,59	,6219
79	,91388	—	64,83	160,56	4,27	62,28	,6188
100 + 80	,91449	—	65,65	161,35	4,30	61,98	,6158
81	,91510	—	66,47	162,14	4,33	61,67	,6128
82	,91571	—	67,29	162,92	4,37	61,38	,6099
83	,91631	—	68,11	163,71	4,40	61,08	,6070
84	,91690	—	68,93	164,50	4,43	60,79	,6040
100 + 85	,91749	—	69,75	165,30	4,45	60,50	,6011
86	,91806	—	70,57	166,09	4,48	60,21	,5983
87	,91862	—	71,39	166,88	4,51	59,92	,5954
88	,91917	—	72,22	167,67	4,55	59,64	,5927
89	,91971	—	73,04	168,46	4,58	59,36	,5899
100 + 90	,92025	—	73,86	169,25	4,61	59,08	,5871
91	,92079	—	74,68	170,04	4,64	58,81	,5844
92	,92132	—	75,50	170,83	4,67	58,54	,5817
93	,92185	—	76,32	171,63	4,69	58,27	,5790
94	,92227	—	77,14	172,42	4,72	58,00	,5764
100 + 95	,92289	—	77,96	173,21	4,75	57,73	,5737
96	,92339	—	78,77	173,99	4,78	57,47	,5711
97	,92389	—	79,59	174,78	4,81	57,21	,5686
98	,92439	—	80,41	175,57	4,84	56,95	,5660
99	,92488	—	81,24	176,37	4,87	56,70	,5634

TABLE II.   HEAT 71°.

I. Water and spirit by weight.	II. Specific gravity.	III. Spirit by measure.	IV. Water by measure.	V. Bulk of mixture.	VI. Diminution of bulk.	VII. Quantity of spirit per cent.	VIII. Decimal multipliers.	I. Water and spirit by weight.	II. Specific gravity.	III. Spirit by measure.	IV. Water by measure.	V. Bulk of mixture.	VI. Diminution of bulk.	VII. Quantity of spirit per cent.	VIII. Decimal multipliers.
W. + Sp.								W. + Sp.							
100+100	,92536	100	82,07	177,17	4,90	56,44	,5608	100 + 50	,95433	100	164,13	257,70	6,43	38,81	,3856
99	,92586	—	82,90	177,97	4,93	56,18	,5583	49	,95501	—	167,49	261,01	6,48	38,31	,3807
98	,92636	—	83,75	178,79	4,96	55,92	,5557	48	,95567	—	170,98	264,47	6,51	37,81	,3757
97	,92686	—	84,62	179,62	5,00	55,66	,5531	. 47	,95635	—	174,61	268,09	6,52	37,30	,3706
96	,92736	—	85,50	180,47	5,03	55,40	,5506	46	,95702	—	178,41	271,86	6,55	36,78	,3655
100 + 95	,92785	—	86,39	181,34	5,05	55,14	,5479	100 + 45	,95770	—	182,38	275,81	6,57	36,25	,3603
94	,92835	—	87,30	182,23	5,07	54,87	,5453	44	,95838	—	186,52	279,93	6,59	35,72	,3550
93	,92884	—	88,24	183,15	5,09	54,60	,5426	43	,95906	—	190,86	284,25	6,61	35,17	,3496
92	,92934	—	89,20	184,08	5,12	54,32	,5398	42	,95975	—	195,41	288,79	6,62	34,62	,3441
91	,92984	—	90,18	185,03	5,15	54,04	,5370	41	,96043	—	200,17	293,54	6,63	34,06	,3385
100 + 90	,93033	—	91,18	186,01	5,17	53,76	,5342	100 + 40	,96111	—	205,17	298,52	6,65	33,49	,3329
89	,93085	—	92,20	187,01	5,19	53,47	,5314	39	,96181	—	210,44	303,77	6,67	32,92	,3271
88	,93138	—	93,25	188,03	5,22	53,18	,5285	38	,96249	—	215,97	309,31	6,66	32,33	,3213
87	,93190	—	94,32	189,08	5,24	52,89	,5256	37	,96318	—	221,81	315,15	6,66	31,73	,3153
86	,93243	—	95,42	190,14	5,28	52,59	,5226	36	,96387	—	227,97	321,31	6,66	31,12	,3093
100 + 85	,93296	—	96,55	191,23	5,32	52,29	,5196	100 + 35	,96456	—	234,49	327,80	6,69	30,51	,3031
84	,93351	—	97,70	192,35	5,35	51,99	,5166	34	,96526	—	241,38	334,70	6,68	29,87	,2969
83	,93406	—	98,87	193,50	5,37	51,68	,5135	33	,96596	—	248,70	342,04	6,66	29,23	,2905
82	,93462	—	100,08	194,67	5,41	51,37	,5104	32	,96668	—	256,47	349,83	6,64	28,58	,2840
81	,93518	—	101,32	195,87	5,45	51,05	,5073	31	,96739	—	264,74	358,11	6,63	27,92	,2774
100 + 80	,93576	—	102,59	197,10	5,49	50,73	,5041	100 + 30	,96811	—	273,57	366,92	6,65	27,25	,2708
79	,93632	—	103,89	198,36	5,53	50,41	,5009	29	,96883	—	283,00	376,38	6,62	26,57	,2640
78	,93688	—	105,22	199,67	5,55	50,08	,4976	28	,96956	—	293,11	386,54	6,57	25,87	,2571
77	,93744	—	106,59	201,01	5,58	49,75	,4943	27	,97030	—	303,96	397,43	6,53	25,16	,2501
76	,93801	—	107,99	202,39	5,60	49,41	,4910	26	,97105	—	315,66	409,15	6,51	24,44	,2429
100 + 75	,93857	—	109,43	203,80	5,63	49,07	,4876	100 + 25	,97180	—	328,28	421,77	6,52	23,71	,2356
74	,93916	—	110,90	205,24	5,66	48,72	,4842	24	,97257	—	341,96	435,48	6,48	22,96	,2282
73	,93975	—	112,41	206,72	5,69	48,37	,4807	23	,97334	—	356,83	450,39	6,44	22,20	,2207
72	,94034	—	113,98	208,25	5,73	48,01	,4772	22	,97414	—	373,05	466,65	6,40	21,43	,2130
71	,94093	—	115,59	209,82	5,77	47,66	,4736	21	,97495	—	390,81	484,46	6,35	20,64	,2051
100 + 70	,94153	—	117,24	211,44	5,80	47,29	,4699	100 + 20	,97575	—	410,36	504,06	6,30	19,84	,1971
69	,94214	—	118,93	213,10	5,83	46,92	,4662	19	,97659	—	431,95	525,72	6,23	19,02	,1890
68	,94275	—	120,68	214,82	5,86	46,55	,4625	18	,97744	—	455,95	549,79	6,16	18,19	,1807
67	,94336	—	122,49	216,60	5,89	46,17	,4588	17	,97831	—	482,77	576,68	6,09	17,34	,1723
66	,94398	—	124,34	218,41	5,93	45,79	,4550	16	,97920	—	512,95	606,93	6,02	16,48	,1637
100 + 65	,94461	—	126,26	220,29	5,97	45,39	,4511	100 + 15	,98011	—	547,14	641,22	5,92	15,60	,1550
64	,94523	—	128,23	222,23	6,00	44,99	,4472	14	,98105	—	586,22	680,41	5,81	14,70	,1461
63	,94585	—	130,26	224,23	6,03	44,59	,4432	13	,98201	—	631,31	725,59	5,72	13,78	,1370
62	,94649	—	132,36	226,30	6,06	44,19	,4391	12	,98302	—	683,93	778,31	5,62	12,85	,1277
61	,94712	—	134,53	228,43	6,10	43,78	,4350	11	,98405	—	746,10	840,60	5,50	11,89	,1182
100 + 60	,94775	—	136,78	230,64	6,14	43,36	,4308	100 + 10	,98513	—	820,71	915,31	5,40	10,92	,1086
59	,94841	—	139,09	232,93	6,16	42,93	,4266	9	,98624	—	911,90	1006,65	5,25	9,93	,0987
58	,94905	—	141,49	235,30	6,19	42,50	,4224	8	,98740	—	1025,89	1120,77	5,12	8,92	,0887
57	,94971	—	143,98	237,75	6,23	42,06	,4180	7	,98861	—	1173,46	1267,46	5,00	7,89	,0784
56	,95037	—	146,56	240,29	6,27	41,61	,4136	6	,99089	—	1367,86	1463,01	4,85	6,84	,0679
100 + 55	,95102	—	149,21	242,91	6,30	41,17	,4091	100 + 5	,99121	—	1641,43	1736,73	4,70	5,76	,0572
54	,95169	—	151,95	245,63	6,32	40,71	,4045	4	,99259	—	2051,79	2147,26	4,53	4,66	,0463
53	,95234	—	154,80	248,47	6,33	40,24	,3999	3	,99404	—	2735,73	2834,34	4,39	3,53	,0351
52	,95301	—	157,80	251,42	6,38	39,77	,3952	2	,99557	—	4103,59	4199,35	4,24	2,38	,0237
51	,95368	—	160,90	254,50	6,40	39,29	,3904	1	,99716	—	8207,17	8303,07	4,10	1,21	,0120

I. Spirit and water by weight.	II. Specific gravity.	III. Spirit by measure.	IV. Water by measure.	V. Bulk of mixture.	VI. Diminution of bulk.	VII. Quantity of spirit per cent.	VIII. Decimal multipliers.
Sp. + W.							
100 + 0	,81927	100	—	100,00	—	100,00	,9930
1	,82157	—	0,82	100,72	0,10	99,29	,9860
2	,82383	—	1,64	101,44	0,20	98,58	,9790
3	,82603	—	2,46	102,16	0,30	97,89	,9721
4	,82817	—	3,28	102,88	0,40	97,20	,9653
100 + 5	,83026	—	4,10	103,61	0,49	96,52	,9585
6	,83229	—	4,92	104,34	0,58	95,84	,9518
7	,83428	—	5,74	105,08	0,66	95,17	,9452
8	,83623	—	6,56	105,81	0,75	94,51	,9386
9	,83812	—	7,39	106,55	0,84	93,85	,9321
100 + 10	,83995	—	8,20	107,29	0,91	93,20	,9256
11	,84175	—	9,02	108,04	0,98	92,56	,9192
12	,84350	—	9,84	108,78	1,06	91,93	,9129
13	,84522	—	10,66	109,53	1,13	91,30	,9067
14	,84690	—	11,48	110,28	1,20	90,68	,9005
100 + 15	,84854	—	12,31	111,03	1,28	90,06	,8944
16	,85018	—	13,13	111,78	1,35	89,46	,8885
17	,85177	—	13,95	112,53	1,42	88,86	,8825
18	,85334	—	14,76	113,29	1,47	88,27	,8766
19	,85489	—	15,58	114,04	1,54	87,68	,8708
100 + 20	,85640	—	16,40	114,80	1,60	87,11	,8650
21	,85788	—	17,22	115,55	1,67	86,54	,8594
22	,85933	—	18,04	116,30	1,74	85,98	,8537
23	,86077	—	18,86	117,06	1,80	85,42	,8483
24	,86217	—	19,69	117,82	1,87	84,87	,8429
100 + 25	,86356	—	20,51	118,59	1,92	84,33	,8374
26	,86491	—	21,33	119,35	1,98	83,79	,8321
27	,86624	—	22,15	120,11	2,04	83,26	,8268
28	,86754	—	22,97	120,88	2,09	82,73	,8216
29	,86883	—	23,79	121,64	2,15	82,21	,8164
100 + 30	,87008	—	24,61	122,40	2,21	81,70	,8113
31	,87133	—	25,43	123,17	2,26	81,19	,8062
32	,87255	—	26,25	123,94	2,31	80,69	,8013
33	,87375	—	27,07	124,71	2,36	80,19	,7963
34	,87494	—	27,89	125,48	2,41	79,70	,7914
100 + 35	,87610	—	28,71	126,24	2,47	79,21	,7866
36	,87724	—	29,53	127,01	2,52	78,73	,7819
37	,87835	—	30,35	127,78	2,57	78,26	,7772
38	,87944	—	31,17	128,55	2,62	77,79	,7715
39	,88053	—	31,99	129,33	2,66	77,33	,7679
100 + 40	,88159	—	32,81	130,10	2,71	76,87	,7633
41	,88267	—	33,63	130,87	2,76	76,41	,7588
42	,88372	—	34,46	131,64	2,82	75,96	,7543
43	,88476	—	35,27	132,41	2,86	75,52	,7499
44	,88578	—	36,09	133,18	2,91	75,08	,7456
100 + 45	,88679	—	36,91	133,95	2,96	74,65	,7413
46	,88778	—	37,73	134,73	3,00	74,22	,7371
47	,88875	—	38,55	135,50	3,05	73,80	,7329
48	,88971	—	39,37	136,28	3,09	73,38	,7287
49	,89066	—	40,19	137,05	3,14	72,96	,7246

I. Spirit and water by weight.	II. Specific gravity.	III. Spirit by measure.	IV. Water by measure.	V. Bulk of mixture.	VI. Diminution of bulk.	VII. Quantity of spirit per cent.	VIII. Decimal multipliers.
Sp. + W.							
100 + 50	,89159	100	41,01	137,83	3,18	72,55	,7205
51	,89250	—	41,84	138,61	3,23	72,15	,7165
52	,89341	—	42,66	139,39	3,27	71,74	,7125
53	,89430	—	43,48	140,16	3,32	71,34	,7085
54	,89517	—	44,29	140,94	3,35	70,95	,7046
100 + 55	,89602	—	45,11	141,72	3,39	70,56	,7007
56	,89687	—	45,93	142,50	3,43	70,17	,6969
57	,89770	—	46,75	143,28	3,47	69,79	,6931
58	,89852	—	47,57	144,06	3,51	69,41	,6893
59	,89933	—	48,39	144,84	3,55	69,04	,6855
100 + 60	,90011	—	49,22	145,62	3,60	68,67	,6819
61	,90090	—	50,04	146,41	3,63	68,30	,6782
62	,90166	—	50,86	147,19	3,67	67,93	,6746
63	,90242	—	51,68	147,97	3,71	67,57	,6710
64	,90317	—	52,50	148,76	3,74	67,22	,6675
100 + 65	,90391	—	53,32	149,54	3,78	66,87	,6640
66	,90466	—	54,14	150,32	3,82	66,52	,6606
67	,90540	—	54,96	151,11	3,85	66,17	,6572
68	,90613	—	55,78	151,89	3,89	65,83	,6538
69	,90685	—	56,60	152,67	3,93	65,50	,6504
100 + 70	,90755	—	57,42	153,46	3,96	65,16	,6471
71	,90824	—	58,24	154,25	3,99	64,83	,6438
72	,90892	—	59,06	155,03	4,03	64,50	,6405
73	,90959	—	59,88	155,82	4,06	64,17	,6373
74	,91025	—	60,70	156,61	4,09	63,85	,6341
100 + 75	,91090	—	61,52	157,39	4,13	63,53	,6309
76	,91155	—	62,34	158,18	4,16	63,22	,6278
77	,91218	—	63,16	158,96	4,20	62,90	,6247
78	,91280	—	63,98	159,76	4,22	62,59	,6216
79	,91343	—	64,80	160,55	4,25	62,29	,6185
100 + 80	,91405	—	65,62	161,34	4,28	61,98	,6155
81	,91460	—	66,44	162,13	4,31	61,68	,6125
82	,91527	—	67,26	162,91	4,35	61,38	,6095
83	,91587	—	68,08	163,69	4,39	61,09	,6067
84	,91646	—	68,90	164,49	4,41	60,79	,6038
100 + 85	,91704	—	69,72	165,28	4,44	60,50	,6009
86	,91761	—	70,54	166,07	4,47	60,22	,5980
87	,91818	—	71,36	166,86	4,50	59,93	,5952
88	,91873	—	72,19	167,65	4,54	59,65	,5924
89	,91927	—	73,01	168,44	4,57	59,37	,5896
100 + 90	,91981	—	73,83	169,23	4,60	59,09	,5868
91	,92035	—	74,65	170,02	4,63	58,82	,5841
92	,92088	—	75,47	170,81	4,66	58,54	,5814
93	,92141	—	76,29	171,61	4,68	58,27	,5787
94	,92193	—	77,11	172,40	4,71	58,01	,5760
100 + 95	,92244	—	77,93	173,19	4,74	57,74	,5734
96	,92295	—	78,73	173,97	4,76	57,48	,5709
97	,92345	—	79,55	174,76	4,79	57,22	,5683
98	,92395	—	80,37	175,55	4,82	56,96	,5657
99	,92445	—	81,20	176,35	4,85	56,71	,5631

TABLE II.   HEAT 72°.

	I. Water and spirit by weight	II. Specific gravity	III. Spirit by measure	IV. Water by measure	V. Bulk of mixture	VI. Diminution of bulk	VII. Quantity of spirit per cent	VIII. Decimal multipliers
W. + S.								
100+100	,92493	100	82,03	177,15	4,88	56,45	,5605	
	9?	,92543	—	82,86	177,95	4,91	56,19	,5580
	9-	,92593	—	83,71	178,77	4,94	55,93	,5555
	97	,9? 43	—	84,58	179,60	4,98	55,67	,5529
	9?	,92693	—	85,46	180,45	5,01	55,41	,5503
100+95	,92742	—	86,35	181,32	5,03	55,15	,5476	
	94	,92792	—	87,26	182,20	5,06	54,88	,5450
	93	,92841	—	88,20	183,13	5,07	54,61	,5423
	92	,92891	—	89,16	184,06	5,10	54,33	,5395
	91	,92941	—	90,14	185,01	5,13	54,05	,5367
100+90	,92991	—	91,14	185,99	5,15	53,77	,5339	
	89	,93043	—	92,16	186,98	5,18	53,48	,5311
	88	,93096	—	93,21	188,01	5,20	53,19	,5282
	87	,93145	—	94,28	189,05	5,23	52,90	,5253
	8?	,03201	—	95,38	190,11	5,27	52,60	,5224
100+85	,93255	—	96,51	191,20	5,31	52,30	,5194	
	84	,93311	—	97,66	192,32	5,34	52,00	,5164
	83	,93366	—	98,83	193,47	5,36	51,69	,5133
	82	,93422	—	100,04	194,64	5,40	51,37	,5102
	81	,93478	—	101,28	195,84	5,44	51,06	,5071
100+80	,93535	—	102,54	197,07	5,47	50,74	,5039	
	79	,93591	—	103,84	198,33	5,51	50,41	,5007
	78	,93647	—	105,17	199,64	5,53	50,08	,4974
	77	,93703	—	106,54	200,98	5,56	49,75	,4941
	76	,93760	—	107,94	202,35	5,59	49,42	,4908
100+75	,93810	—	109,38	203,70	5,62	49,08	,4874	
	74	,93875	—	110,85	205,21	5,64	48,73	,4840
	73	,93934	—	112,36	206,69	5,67	48,38	,4805
	72	,93993	—	113,92	208,22	5,70	48,02	,4770
	71	,94052	—	115,53	209,79	5,74	47,67	,4734
100+70	,94112	—	117,19	211,41	5,78	47,30	,4697	
	69	,94172	—	118,87	213,07	5,80	46,93	,4660
	68	,94234	—	120,62	214,79	5,83	46,55	,4623
	67	,94205	—	122,43	216,56	5,87	46,18	,4586
	66	,94358	—	124,28	218,38	5,90	45,79	,4548
100+65	,94421	—	126,20	220,25	5,93	45,40	,4509	
	64	,94483	—	128,17	222,19	5,98	45,00	,4470
	63	,94546	—	130,20	224,19	6,01	44,60	,4430
	62	,94610	—	132,50	226,26	6,04	44,20	,4389
	61	,94674	—	134,47	228,39	6,08	43,79	,4348
100+60	,94737	—	136,71	230,60	6,11	43,37	,4306	
	59	,94802	—	139,03	232,88	6,15	42,94	,4264
	58	,94868	—	141,42	235,25	6,17	42,50	,4221
	57	,94935	—	143,91	237,70	6,21	42,07	,4178
	56	,95000	—	146,49	240,24	6,25	41,62	,4134
100+55	,95065	—	149,14	242,86	6,28	41,17	,4089	
	54	,95132	—	151,86	245,58	6,30	40,71	,4043
	53	,95199	—	154,73	248,42	6,31	40,25	,3997
	52	,95265	—	157,72	251,37	6,35	39,78	,3950
	51	,95332	—	160,83	254,45	6,38	39,30	,3902

	I. Water and spirit by weight	II. Specific gravity	III. Spirit by measure	IV. Water by measure	V. Bulk of mixture	VI. Diminution of bulk	VII. Quantity of spirit per cent	VIII. Decimal multipliers
W + Sp.								
100+50	,95397	100	164,05	257,64	6,41	38,81	,3854	
	49	,95466	—	167,41	260,95	6,46	38,32	,3805
	48	,95533	—	170,90	264,41	6,49	37,81	,3755
	47	,95601	—	174,53	268,03	6,50	37,30	,3705
	46	,95669	—	178,12	271,79	6,53	36,79	,3654
100+45	,95737	—	182,29	275,74	6,55	36,26	,3601	
	44	,95805	—	186,44	279,86	6,58	35,73	,3549
	43	,95874	—	190,77	284,18	6,59	35,18	,3495
	42	,95943	—	195,32	288,71	6,61	34,63	,3440
	41	,96011	—	200,08	293,46	6,62	34,07	,3384
100+40	,96080	—	205,07	298,44	6,63	33,50	,3327	
	39	,96150	—	210,34	303,69	6,65	32,93	,3270
	38	,96219	—	215,87	309,22	6,65	32,34	,3212
	37	,96288	—	221,71	315,05	6,66	31,74	,3152
	36	,96358	—	227,86	321,20	6,66	31,13	,3092
100+35	,96428	—	234,38	327,70	6,68	30,51	,3030	
	34	,96498	—	241,27	334,59	6,68	29,88	,2968
	33	,96569	—	248,58	341,93	6,65	29,24	,2904
	32	,96641	—	256,35	349,72	6,63	28,59	,2839
	31	,96713	—	264,62	357,99	6,63	27,93	,2774
100+30	,96786	—	273,44	306,80	6,64	27,26	,2707	
	29	,96858	—	282,87	376,25	6,62	26,58	,2639
	28	,96931	—	292,97	386,40	6,57	25,88	,2570
	27	,97006	—	303,82	397,29	6,53	25,17	,2500
	26	,97081	—	315,51	409,00	6,51	24,45	,2428
100+25	,97157	—	328,13	421,62	6,51	23,72	,2355	
	24	,97234	—	341,80	435,32	6,48	22,97	,2281
	23	,97312	—	356,66	450,22	6,44	22,21	,2206
	22	,97392	—	372,87	466,47	6,40	21,44	,2129
	21	,97474	—	390,63	484,27	6,36	20,65	,2050
100+20	,97557	—	410,17	503,86	6,31	19,85	,1971	
	19	,97639	—	431,75	525,51	6,24	19,03	,1890
	18	,97725	—	455,74	549,57	6,17	18,20	,1807
	17	,97813	—	482,54	576,44	6,10	17,35	,1723
	16	,97902	—	512,71	606,68	6,03	16,48	,1637
100+15	,97994	—	546,88	640,95	5,93	15,60	,1549	
	14	,98088	—	585,95	680,12	5,83	14,70	,1460
	13	,98187	—	631,02	725,28	5,74	13,79	,1369
	12	,98286	—	683,61	777,97	5,64	12,85	,1276
	11	,98390	—	745,76	840,23	5,53	11,90	,1182
100+10	,98499	—	820,33	914,90	5,43	10,93	,1085	
	9	,98610	—	911,47	1006,20	5,27	9,54	,0987
	8	,98726	—	1025,41	1120,26	5,15	8,93	,0887
	7	,98848	—	1172,92	1266,88	5,04	7,84	,0784
	6	,99075	—	1367,22	1462,34	4,88	6,84	,0679
100+5	,99108	—	1640,67	1735,93	4,74	5,76	,0572	
	4	,99226	—	2050,83	2146,27	4,52	4,66	,0463
	3	,99391	—	2734,46	2830,04	4,42	3,53	,0351
	2	,99544	—	4101,68	4197,41	4,27	2,38	,0237
	1	,99704	—	8203,36	8299,21	4,15	1,21	,0120

TABLE I.   HEAT 73°.

I.	II.	III.	IV.	V.	VI.	VII.	VIII.	I.	II.	III.	IV.	V.	VI.	VII.	VIII.
Spirit and water by weight.	Specific gravity.	Spirit by measure.	Water by measure.	Bulk of mixture.	Diminution of bulk.	Quantity of spirit per cent.	Decimal multipliers.	Spirit and water by weight.	Specific gravity.	Spirit by measure.	Water by measure.	Bulk of mixture.	Diminution of bulk.	Quantity of spirit per cent.	Decimal multipliers.
Sp. + W.								Sp. + W.							
100 + 0	,81878	100	—	100,00	—	100,00	,9924	100 + 50	,89112	100	40,99	137,82	3,17	72,56	,7201
1	,82108	—	0,82	100,72	0,10	99,29	,9854	51	,89204	—	41,82	138,60	3,22	72,15	,7161
2	,82334	—	1,64	101,44	0,20	98,58	,9784	52	,89294	—	42,64	139,38	3,26	71,75	,7121
3	,82553	—	2,46	102,16	0,30	97,89	,9715	53	,89383	—	43,46	140,15	3,31	71,35	,7082
4	,82768	—	3,28	102,88	0,40	97,20	,9647	54	,89471	—	44,27	140,93	3,34	70,96	,7042
100 + 5	,82977	—	4,10	103,61	0,49	96,52	,9579	100 + 55	,89556	—	45,09	141,71	3,38	70,57	,7003
6	,83180	—	4,92	104,34	0,58	95,84	,9512	56	,89641	—	45,91	142,49	3,42	70,18	,6965
7	,83379	—	5,74	105,08	0,66	95,17	,9446	57	,89724	—	46,73	143,27	3,46	69,80	,6927
8	,83574	—	6,56	105,81	0,75	94,51	,9380	58	,89806	—	47,55	144,05	3,50	69,42	,6889
9	,83764	—	7,38	106,55	0,83	93,85	,9315	59	,89887	—	48,37	144,83	3,54	69,04	,6852
100 + 10	,83947	—	8,20	107,29	0,91	93,21	,9250	100 + 60	,89964	—	49,19	145,61	3,58	68,67	,6815
11	,84127	—	9,02	108,03	0,99	92,57	,9186	61	,90043	—	50,01	146,40	3,61	68,31	,6779
12	,84302	—	9,84	108,78	1,06	91,93	,9123	62	,90120	—	50,83	147,18	3,65	67,94	,6743
13	,84474	—	10,66	109,53	1,13	91,30	,9061	63	,90196	—	51,65	147,96	3,69	67,58	,6707
14	,84642	—	11,48	110,28	1,20	90,68	,9000	64	,90271	—	52,47	148,75	3,72	67,23	,6672
100 + 15	,84806	—	12,30	111,02	1,28	90,07	,8939	100 + 65	,90345	—	53,29	149,53	3,76	66,88	,6637
16	,84970	—	13,12	111,78	1,34	89,46	,8879	66	,90420	—	54,11	150,31	3,80	66,53	,6602
17	,85129	—	13,94	112,53	1,41	88,87	,8820	67	,90494	—	54,93	151,09	3,84	66,18	,6568
18	,85286	—	14,76	113,28	1,48	88,27	,8761	68	,90567	—	55,75	151,87	3,88	65,84	,6534
19	,85440	—	15,58	114,04	1,54	87,69	,8703	69	,90639	—	56,57	152,65	3,92	65,51	,6501
100 + 20	,85592	—	16,39	114,80	1,59	87,11	,8646	100 + 70	,90709	—	57,39	153,44	3,95	65,17	,6468
21	,85740	—	17,21	115,55	1,66	86,54	,8589	71	,90778	—	58,22	154,23	3,99	64,84	,6435
22	,85885	—	18,04	116,30	1,74	85,98	,8533	72	,90846	—	59,03	155,01	4,02	64,51	,6402
23	,86029	—	18,86	117,06	1,80	85,42	,8478	73	,90913	—	59,85	155,80	4,05	64,18	,6370
24	,86169	—	19,68	117,82	1,86	84,87	,8423	74	,90979	—	60,67	156,59	4,08	63,86	,6338
100 + 25	,86308	—	20,50	118,58	1,92	84,33	,8369	100 + 75	,91044	—	61,49	157,38	4,11	63,54	,6306
26	,86443	—	21,32	119,34	1,98	83,79	,8316	76	,91109	—	62,31	158,16	4,15	63,23	,6275
27	,86576	—	22,14	120,11	2,03	83,26	,8263	77	,91173	—	63,13	158,95	4,18	62,91	,6244
28	,86706	—	22,96	120,87	2,09	82,73	,8211	78	,91235	—	63,95	159,74	4,21	62,60	,6213
29	,86835	—	23,78	121,63	2,15	82,21	,8159	79	,91298	—	64,77	160,53	4,24	62,29	,6182
100 + 30	,86960	—	24,60	122,40	2,20	81,70	,8108	100 + 80	,91360	—	65,59	161,32	4,27	61,99	,6152
31	,87085	—	25,41	123,17	2,24	81,19	,8058	81	,91421	—	66,41	162,11	4,30	61,69	,6122
32	,87207	—	26,23	123,93	2,30	80,69	,8009	82	,91482	—	67,23	162,89	4,34	61,39	,6092
33	,87327	—	27,06	124,70	2,36	80,19	,7959	83	,91542	—	68,05	163,68	4,37	61,10	,6064
34	,87446	—	27,88	125,47	2,41	79,70	,7910	84	,91601	—	68,87	164,47	4,40	60,80	,6035
100 + 35	,87562	—	28,70	126,23	2,47	79,21	,7862	100 + 85	,91659	—	69,69	165,26	4,43	60,51	,6006
36	,87676	—	29,52	127,00	2,52	78,73	,7815	86	,91716	—	70,51	166,05	4,46	60,22	,5977
37	,87788	—	30,34	127,78	2,56	78,26	,7768	87	,91773	—	71,33	166,84	4,49	59,94	,5949
38	,87897	—	31,16	128,55	2,61	77,79	,7711	88	,91829	—	72,15	167,63	4,52	59,66	,5921
39	,88006	—	31,98	129,32	2,66	77,33	,7675	89	,91883	—	72,97	168,42	4,55	59,38	,5893
100 + 40	,88112	—	32,80	130,09	2,71	76,87	,7629	100 + 90	,91937	—	73,79	169,21	4,58	59,10	,5865
41	,88220	—	33,61	130,86	2,75	76,41	,7584	91	,91991	—	74,61	170,00	4,61	58,82	,5838
42	,88325	—	34,44	131,63	2,81	75,97	,7539	92	,92044	—	75,43	170,79	4,64	58,55	,5811
43	,88429	—	35,26	132,41	2,85	75,53	,7495	93	,92097	—	76,25	171,59	4,66	58,28	,5784
44	,88531	—	36,08	133,17	2,91	75,09	,7452	94	,92149	—	77,07	172,38	4,69	58,01	,5758
100 + 45	,88632	—	36,90	133,94	2,96	74,66	,7409	100 + 95	,92199	—	77,89	173,17	4,72	57,75	,5732
46	,88731	—	37,71	134,72	2,99	74,23	,7367	96	,92251	—	78,70	173,95	4,75	57,49	,5706
47	,88828	—	38,53	135,49	3,04	73,80	,7325	97	,92301	—	79,52	174,74	4,78	57,23	,5680
48	,88924	—	39,35	136,27	3,08	73,38	,7283	98	,92351	—	80,33	175,53	4,80	56,97	,5654
49	,89019	—	40,17	137,04	3,13	72,97	,7242	99	,92402	—	81,16	176,33	4,83	56,71	,5629

TABLE II.   HEAT 73°.

I. Water and spirit by weight.	II. Specific gravity.	III. Spirit by measure.	IV. Water by measure.	V. Bulk of mixture.	VI. Diminution of bulk.	VII. Quantity of spirit per cent.	VIII. Decimal multipliers.
W. + Sp.							
100+100	,92450	100	81,99	177,13	4,86	56,45	,5603
99	,92500	—	82,82	177,93	4,89	56,20	,5578
98	,92550	—	83,67	178,75	4,92	55,94	,5552
97	,92600	—	84,54	179,58	4,96	55,68	,5526
96	,92650	—	85,42	180,43	4,99	55,41	,5500
100+95	,92699	—	86,31	181,29	5,02	55,15	,5474
94	,92749	—	87,22	182,18	5,04	54,88	,5447
93	,92798	—	68,16	183,10	5,06	54,61	,5420
92	,92848	—	89,12	184,03	5,09	54,33	,5393
91	,92899	—	90,10	184,98	5,12	54,05	,5365
100+90	,92949	—	91,10	185,96	5,14	53,77	,5337
89	,93001	—	92,12	186,95	5,17	53,48	,5309
88	,93054	—	93,17	187,98	5,19	53,19	,5280
87	,93106	—	94,24	189,02	5,22	52,90	,5251
86	,93160	—	95,33	190,08	5,25	52,61	,5222
100+85	,93214	—	96,46	191,17	5,29	52,31	,5191
84	,93270	—	97,61	192,29	5,32	52,01	,5161
83	,93325	—	98,78	193,44	5,34	51,70	,5131
82	,93381	—	99,99	194,61	5,38	51,38	,5100
81	,93437	—	101,23	195,81	5,42	51,07	,5069
100+80	,93495	—	102,49	197,04	5,45	50,75	,5037
79	,93550	—	103,79	198,30	5,49	50,42	,5005
78	,93606	—	105,12	199,61	5,51	50,09	,4972
77	,93662	—	106,49	200,94	5,55	49,76	,4939
76	,93719	—	107,89	202,32	5,57	49,43	,4906
100+75	,93775	—	109,33	203,73	5,60	49,09	,4872
74	,93834	—	110,80	205,18	5,62	48,74	,4838
73	,93893	—	112,31	206,65	5,66	48,39	,4803
72	,93952	—	113,87	208,18	5,69	48,03	,4768
71	,94011	—	115,48	209,76	5,72	47,67	,4732
100+70	,94071	—	117,13	211,37	5,76	47,31	,4695
69	,94132	—	118,82	213,03	5,79	46,94	,4658
68	,94193	—	120,57	214,76	5,81	46,56	,4621
67	,94255	—	122,37	216,52	5,85	46,18	,4584
66	,94318	—	124,22	218,34	5,88	45,80	,4546
100+65	,94381	—	126,14	220,21	5,93	45,41	,4507
64	,94443	—	128,11	222,15	5,96	45,01	,4468
63	,94507	—	130,14	224,15	5,99	44,61	,4428
62	,94571	—	132,24	226,22	6,02	44,20	,4387
61	,94635	—	134,41	228,35	6,06	43,79	,4346
100+60	,94702	—	136,65	230,55	6,10	43,37	,4304
59	,94764	—	138,97	232,84	6,13	42,94	,4262
58	,94831	—	141,36	235,20	6,16	42,51	,4220
57	,94897	—	143,84	237,65	6,19	42,07	,4176
56	,94963	—	146,42	240,19	6,23	41,63	,4132
100+55	,95029	—	149,07	242,81	6,26	41,18	,4087
54	,95096	—	151,81	245,53	6,28	40,72	,4042
53	,95162	—	154,66	248,37	6,29	40,26	,3996
52	,95229	—	157,64	251,31	6,34	39,79	,3949
51	,95296	—	160,76	254,40	6,36	39,31	,3901

I. Water and spirit by weight.	II. Specific gravity.	III. Spirit by measure.	IV. Water by measure.	V. Bulk of mixture.	VI. Diminution of bulk.	VII. Quantity of spirit per cent.	VIII. Decimal multipliers.
W. + Sp.							
100+50	,95362	100	163,98	257,58	6,40	38,82	,3853
49	,95431	—	167,33	260,89	6,44	38,33	,3803
48	,95499	—	170,82	264,35	6,47	37,82	,3753
47	,95567	—	174,45	267,97	6,48	37,31	,3703
46	,95636	—	178,24	271,73	6,51	36,80	,3652
100+45	,95704	—	182,21	275,67	6,54	36,27	,3600
44	,95772	—	186,35	279,79	6,56	35 74	,3548
43	,95842	—	190,68	284,11	6,57	35,19	,3494
42	,95911	—	195,23	288,64	6,59	34,64	,3439
41	,95980	—	199,99	293,38	6,61	34,08	,3383
100+40	,96049	—	204,98	298,36	6,62	33,51	,3326
39	,96119	—	210,24	303,60	6,64	32,94	,3269
38	,96189	—	215,77	309,13	6,64	32,35	,3211
37	,96259	—	221,60	314,96	6,64	31,75	,3151
36	,96329	—	227,75	321,10	6,65	31,14	,3091
100+35	,96400	—	234,27	327,60	6,67	30,52	,3029
34	,96471	—	241,16	334,49	6,67	29,89	,2967
33	,96542	—	248,46	341,82	6,64	29,25	,2904
32	,96614	—	256,23	349,60	6,63	28,60	,2839
31	,96687	—	264,50	357,87	6,63	27,94	,2773
100+30	,96760	—	273,31	366,68	6,63	27,27	,2706
29	,96833	—	282,74	376,13	6,61	26,59	,2639
28	,96907	—	292,83	386,26	6,57	25,89	,2569
27	,96982	—	303,68	397,14	6,54	25,18	,2499
26	,97057	—	315,36	408,85	6,51	24,46	,2428
100+25	,97134	—	327,98	421,47	6,51	23,73	,2355
24	,97211	—	341,64	435,16	6,48	22,98	,2281
23	,97290	—	356,49	450,05	6,44	22,23	,2205
22	,97371	—	372,70	466,30	6,40	21,44	,2128
21	,97453	—	390,45	484,09	6,36	20,65	,2050
100+20	,97535	—	409,98	503,67	6,33	19,84	,1970
19	,97620	—	431,55	525,30	6,25	19,04	,1889
18	,97706	—	455,53	549,35	6,18	18,20	,1806
17	,97795	—	482,32	576,20	6,12	17,35	,1722
16	,97885	—	512,47	606,43	6,04	16,49	,1636
100+15	,97977	—	546,63	640,68	5,95	15,61	,1549
14	,98072	—	585,68	679,83	5,85	14,71	,1460
13	,98172	—	630,73	724,97	5,76	13,79	,1369
12	,98271	—	683,29	777,63	5,66	12,86	,1276
11	,98375	—	745,41	839,86	5,55	11,90	,1182
100+10	,98481	—	819,95	914,49	5,46	10,93	,1085
9	,98596	—	911,05	1005,75	5,30	9,94	,0987
8	,98712	—	1024,93	1119,76	5,18	8,93	,0888
7	,98814	—	1172,39	1266,31	5,08	7,90	,0784
6	,99061	—	1366,58	1461,68	4,90	6,84	,0679
100+5	,99204	—	1639,91	1735,13	4,78	5,76	,0572
4	,99231	—	2049,88	2145,29	4,59	4,66	,0463
3	,99378	—	2733,18	2828,73	4,45	3,53	,0351
2	,99531	—	4099,78	4195,43	4,31	2,38	,0237
1	,99691	—	8199,55	8295,35	4,20	1 21	,0120

Sp. + W.

I. Spirit and water by weight.	II. Specific gravity.	III. Spirit by measure.	IV. Water by measure.	V. Bulk of mixture.	VI. Diminution of bulk.	VII. Quantity of spirit per cent.	VIII. Decimal multipliers.
100 + 0	,81829	100	—	100,00	—	100,00	,9919
1	,82059	—	0,82	100,72	0,10	99,29	,9848
2	,82284	—	1,64	101,44	0,20	98,58	,9778
3	,82504	—	2,46	102,16	0,30	97,89	,9709
4	,82719	—	3,28	102,88	0,40	97,20	,9641
100 + 5	,82927	—	4,10	103,61	0,49	96,52	,9573
6	,83131	—	4,92	104,34	0,58	95,84	,9506
7	,83330	—	5,74	105,07	0,66	95,17	,9440
8	,83525	—	6,56	105,81	0,75	94,51	,9374
9	,83715	—	7,38	106,54	0,84	93,86	,9309
100 + 10	,83899	—	8,20	107,28	0,92	93,21	,9245
11	,84079	—	9,01	108,03	0,98	92,57	,9181
12	,84254	—	9,83	108,78	1,05	91,93	,9118
13	,84426	—	10,65	109,52	1,13	91,30	,9056
14	,84594	—	11,47	110,27	1,20	90,68	,8995
100 + 15	,84758	—	12,30	111,02	1,28	90,07	,8934
16	,84921	—	13,12	111,77	1,35	89,47	,8874
17	,85080	—	13,93	112,52	1,41	88,87	,8815
18	,85237	—	14,75	113,28	1,47	88,28	,8756
19	,85391	—	15,57	114,03	1,54	87,69	,8698
100 + 20	,85544	—	16,39	114,79	1,60	87,11	,8641
21	,85691	—	17,20	115,54	1,66	86,54	,8584
22	,85837	—	18,03	116,30	1,73	85,98	,8528
23	,85981	—	18,85	117,05	1,80	85,43	,8473
24	,86121	—	19,67	117,81	1,86	84,88	,8419
100 + 25	,86260	—	20,49	118,58	1,91	84,33	,8365
26	,86395	—	21,31	119,34	1,97	83,79	,8311
27	,86528	—	22,13	120,10	2,03	83,26	,8258
28	,86658	—	22,95	120,87	2,08	82,74	,8206
29	,86787	—	23,77	121,63	2,14	82,22	,8155
100 + 30	,86912	—	24,59	122,39	2,20	81,70	,8104
31	,87037	—	25,40	123,16	2,24	81,19	,8053
32	,87159	—	26,22	123,93	2,29	80,69	,8004
33	,87279	—	27,04	124,70	2,34	80,20	,7954
34	,87398	—	27,86	125,47	2,39	79,71	,7906
100 + 35	,87514	—	28,68	126,23	2,45	79,22	,7858
36	,87628	—	29,50	127,00	2,50	78,74	,7811
37	,87740	—	30,32	127,77	2,55	78,27	,7764
38	,87850	—	31,15	128,54	2,61	77,80	,7707
39	,87959	—	31,96	129,31	2,65	77,33	,7671
100 + 40	,88065	—	32,78	130,09	2,69	76,87	,7625
41	,88173	—	33,60	130,86	2,74	76,42	,7580
42	,88278	—	34,43	131,63	2,80	75,97	,7535
43	,88382	—	35,24	132,40	2,84	75,53	,7491
44	,88484	—	36,06	133,17	2,89	75,09	,7448
100 + 45	,88585	—	36,88	133,94	2,94	74,66	,7405
46	,88684	—	37,69	134,71	2,98	74,23	,7363
47	,88781	—	38,51	135,49	3,02	73,81	,7321
48	,88877	—	39,33	136,26	3,07	73,39	,7279
49	,88972	—	40,15	137,03	3,12	72,97	,7238

Sp. + W.

I. Spirit and water by weight.	II. Specific gravity.	III. Spirit by measure.	IV. Water by measure.	V. Bulk of mixture.	VI. Diminution of bulk.	VII. Quantity of spirit per cent.	VIII. Decimal multipliers.
100 + 50	,89065	100	40,97	137,81	3,16	72,56	,7197
51	,89157	—	41,80	138,59	3,21	72,16	,7157
52	,89248	—	42,62	139,37	3,25	71,75	,7117
53	,89337	—	43,44	140,14	3,30	71,35	,7078
54	,89425	—	44,25	140,92	3,33	70,96	,7039
100 + 55	,89510	—	45,07	141,70	3,37	70,57	,7000
56	,89595	—	45,89	142,48	3,41	70,18	,6962
57	,89678	—	46,71	143,26	3,45	69,80	,6924
58	,89760	—	47,53	144,04	3,49	69,42	,6886
59	,89840	—	48,35	144,82	3,53	69,05	,6849
100 + 60	,89918	—	49,17	145,60	3,57	68,68	,6812
61	,89997	—	49,99	146,39	3,60	68,31	,6776
62	,90073	—	50,81	147,17	3,64	67,94	,6739
63	,90149	—	51,63	147,95	3,68	67,58	,6703
64	,90224	—	52,45	148,74	3,71	67,23	,6668
100 + 65	,90299	—	53,27	149,52	3,75	66,88	,6633
66	,90374	—	54,09	150,30	3,79	66,53	,6599
67	,90448	—	54,91	151,08	3,83	66,18	,6565
68	,90521	—	55,72	151,86	3,86	65,84	,6531
69	,90593	—	56,54	152,64	3,90	65,51	,6498
100 + 70	,90663	—	57,36	153,42	3,94	65,17	,6464
71	,90732	—	58,19	154,22	3,97	64,84	,6431
72	,90800	—	59,00	155,00	4,00	64,51	,6399
73	,90867	—	59,82	155,79	4,03	64,19	,6367
74	,90933	—	60,64	156,58	4,06	63,86	,6335
100 + 75	,90999	—	61,46	157,36	4,10	63,54	,6303
76	,91063	—	62,28	158,15	4,13	63,23	,6272
77	,91127	—	63,10	158,93	4,17	62,91	,6241
78	,91190	—	63,92	159,72	4,20	62,60	,6210
79	,91253	—	64,74	160,51	4,23	62,30	,6179
100 + 80	,91316	—	65,56	161,30	4,26	61,99	,6149
81	,91376	—	66,38	162,09	4,29	61,69	,6119
82	,91437	—	67,20	162,87	4,33	61,40	,6089
83	,91497	—	68,02	163,66	4,36	61,10	,6061
84	,91556	—	68,84	164,46	4,38	60,81	,6032
100 + 85	,91614	—	69,66	165,24	4,42	60,52	,6003
86	,91671	—	70,48	166,03	4,45	60,23	,5974
87	,91728	—	71,30	166,82	4,48	59,94	,5946
88	,91784	—	72,12	167,61	4,51	59,64	,5918
89	,91839	—	72,94	168,40	4,54	59,38	,5890
100 + 90	,91893	—	73,76	169,19	4,57	59,10	,5862
91	,91947	—	74,58	169,98	4,60	58,83	,5835
92	,92000	—	75,39	170,77	4,62	58,56	,5808
93	,92053	—	76,21	171,57	4,64	58,29	,5781
94	,92105	—	77,03	172,36	4,67	58,02	,5755
100 + 95	,92155	—	77,85	173,15	4,70	57,75	,5728
96	,92207	—	78,66	173,93	4,73	57,49	,5703
97	,92257	—	79,48	174,72	4,76	57,23	,5677
98	,92307	—	80,30	175,51	4,79	56,97	,5651
99	,92358	—	81,12	176,31	4,81	56,72	,5626

TABLE II.   HEAT 74°.

I. Water and Spirit by weight.	II. Specific gravity.	III. Spirit by measure.	IV. Water by measure.	V. Bulk of mixture.	VI. Diminution of bulk.	VII. Quantity of spirit per cent.	VIII. Decimal multipliers.
W. + Sp.							
100+100	,92407	100	81,95	177,11	4,84	56,46	,5600
99	,92457	—	82,78	177,91	4,87	56,20	,5575
98	,92507	—	83,63	178,73	4,90	55,94	,5549
97	,92557	—	84,50	179,56	4,94	55,68	,5524
96	,92607	—	85,38	180,41	4,97	55,42	,5598
100+95	,92656	—	86,27	181,27	5,00	55,16	,5471
94	,92706	—	87,18	182,16	5,02	54,89	,5445
93	,92756	—	88,12	183,08	5,04	54,62	,5418
92	,92806	—	89,08	184,00	5,08	54,34	,5390
91	,92856	—	90,06	184,96	5,10	54,06	,5362
100+90	,92907	—	91,06	185,94	5,12	53,78	,5334
89	,92959	—	92,08	186,92	5,16	53,49	,5306
88	,93012	—	93,13	187,95	5,18	53,20	,5277
87	,93065	—	94,20	188,99	5,21	52,91	,5248
86	,93119	—	95,29	190,05	5,24	52,62	,5219
100+85	,93173	—	96,41	191,14	5,27	52,32	,5189
84	,93229	—	97,57	192,26	5,31	52,02	,5159
83	,93285	—	98,74	193,41	5,33	51,71	,5129
82	,93341	—	99,94	194,58	5,36	51,39	,5098
81	,93397	—	101,18	195,78	5,40	51,08	,5067
100+80	,93455	—	102,44	197,01	5,43	50,76	,5035
79	,93509	—	103,74	198,27	5,47	50,43	,5003
78	,93566	—	105,07	199,58	5,49	50,10	,4970
77	,93622	—	106,44	200,91	5,53	49,77	,4937
76	,93679	—	107,84	202,28	5,56	49,44	,4903
100+75	,93735	—	109,28	203,70	5,58	49,09	,4869
74	,93794	—	110,75	205,14	5,61	48,75	,4835
73	,93852	—	112,26	206,62	5,64	48,40	,4801
72	,93911	—	113,82	208,15	5,67	48,04	,4766
71	,93970	—	115,43	209,72	5,71	47,68	,4730
100+70	,94031	—	117,08	211,34	5,74	47,32	,4693
69	,94091	—	118,77	213,00	5,77	46,95	,4656
68	,94153	—	120,51	214,72	5,79	46,57	,4619
67	,94215	—	122,31	216,49	5,82	46,19	,4582
66	,94278	—	124,17	218,30	5,87	45,81	,4544
100+65	,94341	—	126,08	220,17	5,91	45,42	,4505
64	,94404	—	128,05	222,11	5,94	45,02	,4466
63	,94468	—	130,08	224,11	5,97	44,62	,4426
62	,94532	—	132,18	226,18	6,00	44,21	,4385
61	,94597	—	134,35	228,31	6,04	43,80	,4344
100+60	,94661	—	136,59	230,51	6,08	43,38	,4303
59	,94727	—	138,90	232,79	6,11	42,95	,4261
58	,94794	—	141,30	235,15	6,15	42,52	,4218
57	,94860	—	143,77	237,60	6,17	42,08	,4175
56	,94927	—	146,35	240,14	6,21	41,64	,4131
100+55	,94993	—	149,00	242,76	6,24	41,19	,4086
54	,95060	—	151,74	245,48	6,26	40,73	,4041
53	,95126	—	154,59	248,32	6,27	40,27	,3995
52	,95193	—	157,57	251,26	6,31	39,80	,3948
51	,95260	—	160,69	254,34	6,35	39,32	,3900
100+50	,95327	100	163,91	257,52	6,39	38,83	,3852
49	,95396	—	167,25	260,83	6,42	38,34	,3802
48	,95465	—	170,74	264,29	6,45	37,83	,3753
47	,95533	—	174,37	267,91	6,46	37,32	,3703
46	,95602	—	178,16	271,67	6,49	36,81	,3651
100+45	,95671	—	182,12	275,60	6,52	36,28	,3599
44	,95740	—	186,26	279,72	6,54	35,75	,3546
43	,95810	—	190,59	284,04	6,55	35,20	,3492
42	,95879	—	195,14	288,56	6,58	34,65	,3438
41	,95949	—	199,90	293,30	6,60	34,09	,3382
100+40	,96018	—	204,89	298,28	6,61	33,52	,3325
39	,96089	—	210,14	303,51	6,63	32,95	,3268
38	,96159	—	215,67	309,04	6,63	32,36	,3210
37	,96230	—	221,50	314,86	6,64	31,76	,3150
36	,96301	—	227,65	321,00	6,65	31,15	,3090
100+35	,96372	—	234,16	327,50	6,66	30,53	,3029
34	,96444	—	241,05	334,39	6,66	29,90	,2966
33	,96515	—	248,35	341,71	6,64	29,26	,2903
32	,96588	—	256,11	349,49	6,62	28,61	,2838
31	,96661	—	264,38	357,75	6,63	27,95	,2773
100+30	,96734	—	273,18	366,56	6,62	27,28	,2706
29	,96808	—	282,61	376,00	6,61	26,60	,2638
28	,96883	—	292,70	386,13	6,57	25,90	,2569
27	,96958	—	303,54	397,00	6,54	25,19	,2499
26	,97033	—	315,21	408,70	6,51	24,47	,2427
100+25	,97110	—	327,83	421,32	6,51	23,74	,2354
24	,97188	—	341,48	435,01	6,47	22,99	,2280
23	,97268	—	356,33	449,89	6,44	22,23	,2205
22	,97350	—	372,53	466,07	6,41	21,45	,2128
21	,97432	—	390,27	483,91	6,36	20,66	,2049
100+20	,97515	—	409,79	503,48	6,31	19,86	,1970
19	,97601	—	431,35	525,10	6,25	19,04	,1889
18	,97688	—	455,32	549,13	6,19	18,21	,1806
17	,97777	—	482,10	575,97	6,13	17,36	,1722
16	,97868	—	512,23	605,18	6,05	16,50	,1636
100+15	,97960	—	546,38	640,41	5,97	15,62	,1549
14	,98056	—	585,41	679,54	5,87	14,72	,1460
13	,98157	—	630,44	724,66	5,78	13,80	,1369
12	,98256	—	682,97	777,29	5,68	12,86	,1276
11	,98360	—	745,07	839,49	5,58	11,91	,1181
100+10	,98469	—	819,57	914,08	5,49	10,94	,1085
9	,98582	—	910,63	1005,30	5,33	9,95	,0987
8	,98698	—	1024,46	1119,26	5,20	8,93	,0886
7	,98820	—	1171,86	1265,74	5,12	7,90	,0784
6	,99047	—	1365,95	1461,02	4,93	6,85	,0679
100+5	,99080	—	1639,15	1734,34	4,81	5,77	,0572
4	,99218	—	2048,03	2144,30	4,63	4,66	,0463
3	,99365	—	2731,91	2827,43	4,48	3,53	,0351
2	,99518	—	4097,87	4193,52	4,35	2,38	,0237
1	,99678	—	8195,74	8291,50	4,24	1,21	,0120

Mr. GILPIN's *Tables*

TABLE I.   HEAT 75°.

I. Spirit and water by weight.	II. Specific gravity.	III. Spirit by measure.	IV. Water by measure.	V. Bulk of mixture.	VI. Diminution of bulk.	VII. Quantity of spirit per cent.	VIII. Decimal multipliers.
Sp. + W.							
100 + 0	,81780	100	—	100,00	—	100,00	,9913
1	,82010	—	0,82	100,72	0,10	99,29	,9842
2	,82235	—	1,64	101,44	0,20	98,58	,9772
3	,82455	—	2,46	102,16	0,30	97,89	,9703
4	,82669	—	3,28	102,88	0,40	97,21	,9635
100 + 5	,82878	—	4,10	103,61	0,49	96,52	,9567
6	,83082	—	4,91	104,34	0,57	95,84	,9500
7	,83281	—	5,73	105,07	0,66	95,17	,9434
8	,83476	—	6,55	105,81	0,74	94,51	,9368
9	,83666	—	7,37	106,54	0,83	93,86	,9303
100 + 10	,83851	—	8,19	107,28	0,91	93,21	,9239
11	,84031	—	9,01	108,03	0,98	92,57	,9176
12	,84206	—	9,83	108,77	1,06	91,94	,9113
13	,84378	—	10,65	109,52	1,13	91,31	,9051
14	,84546	—	11,47	110,27	1,20	90,69	,8990
100 + 15	,84710	—	12,29	111,02	1,27	90,07	,8930
16	,84872	—	13,11	111,77	1,34	89,47	,8870
17	,85031	—	13,93	112,52	1,41	88,87	,8810
18	,85188	—	14,75	113,28	1,47	88,28	,8751
19	,85342	—	15,56	114,03	1,53	87,69	,8693
100 + 20	,85496	—	16,38	114,79	1,59	87,12	,8636
21	,85642	—	17,20	115,54	1,66	86,55	,8579
22	,85788	—	18,02	116,30	1,72	85,99	,8523
23	,85932	—	18,84	117,05	1,79	85,43	,8468
24	,86073	—	19,66	117,81	1,85	84,88	,8414
100 + 25	,86212	—	20,48	118,57	1,91	84,34	,8360
26	,86347	—	21,30	119,33	1,97	83,80	,8306
27	,86480	—	22,12	120,10	2,02	83,27	,8253
28	,86610	—	22,94	120,86	2,08	82,74	,8201
29	,86738	—	23,76	121,62	2,14	82,22	,8150
100 + 30	,86864	—	24,58	122,39	2,19	81,71	,8099
31	,86988	—	25,39	123,16	2,23	81,20	,8049
32	,87110	—	26,21	123,92	2,29	80,70	,7999
33	,87231	—	27,03	124,69	2,34	80,20	,7950
34	,87350	—	27,85	125,46	2,39	79,71	,7901
100 + 35	,87466	—	28,67	126,22	2,45	79,22	,7853
36	,87580	—	29,49	126,99	2,50	78,74	,7806
37	,87692	—	30,31	127,76	2,55	78,27	,7759
38	,87802	—	31,13	128,53	2,60	77,80	,7712
39	,87911	—	31,95	129,31	2,64	77,34	,7666
100 + 40	,88018	—	32,77	130,08	2,69	76,88	,7621
41	,88125	—	33,59	130,85	2,74	76,42	,7576
42	,88230	—	34,41	131,62	2,79	75,98	,7531
43	,88334	—	35,22	132,39	2,83	75,54	,7487
44	,88437	—	36,04	133,16	2,88	75,10	,7444
100 + 45	,88538	—	36,86	133,93	2,93	74,67	,7401
46	,88637	—	37,68	134,70	2,98	74,24	,7359
47	,88734	—	38,50	135,48	3,02	73,81	,7317
48	,88830	—	39,32	136,25	3,07	73,39	,7275
49	,88925	—	40,14	137,03	3,11	72,98	,7234
100 + 50	,89018	100	40,96	137,80	3,16	72,57	,7193
51	,89110	—	41,78	138,58	3,20	72,16	,7153
52	,89201	—	42,60	139,36	3,24	71,76	,7113
53	,89290	—	43,42	140,13	3,29	71,36	,7074
54	,89378	—	44,23	140,91	3,32	70,97	,7035
100 + 55	,89464	—	45,05	141,69	3,36	70,58	,6996
56	,89548	—	45,87	142,47	3,40	70,19	,6958
57	,89631	—	46,69	143,25	3,44	69,81	,6920
58	,89713	—	47,51	144,03	3,48	69,43	,6882
59	,89793	—	48,33	144,81	3,52	69,05	,6845
100 + 60	,89872	—	49,15	145,59	3,56	68,68	,6808
61	,89950	—	49,97	146,38	3,59	68,32	,6772
62	,90026	—	50,79	147,16	3,63	67,95	,6736
63	,90102	—	51,61	147,94	3,67	67,59	,6700
64	,90177	—	52,43	148,73	3,70	67,24	,6665
100 + 65	,90252	—	53,25	149,51	3,74	66,89	,6630
66	,90327	—	54,07	150,29	3,78	66,54	,6595
67	,90401	—	54,89	151,07	3,82	66,19	,6561
68	,90474	—	55,70	151,85	3,85	65,85	,6527
69	,90546	—	56,52	152,63	3,89	65,52	,6494
100 + 70	,90617	—	57,34	153,41	3,93	65,19	,6461
71	,90686	—	58,16	154,20	3,96	64,85	,6428
72	,90754	—	58,98	154,99	3,99	64,52	,6395
73	,90821	—	59,80	155,78	4,02	64,20	,6363
74	,90887	—	60,62	156,56	4,06	63,87	,6331
100 + 75	,90952	—	61,44	157,35	4,09	63,55	,6300
76	,91017	—	62,26	158,14	4,12	63,24	,6268
77	,91081	—	63,08	158,92	4,16	62,92	,6237
78	,91145	—	63,90	159,71	4,19	62,61	,6206
79	,91208	—	64,71	160,50	4,21	62,31	,6176
100 + 80	,91270	—	65,53	161,28	4,25	62,00	,6146
81	,91331	—	66,35	162,07	4,28	61,70	,6116
82	,91392	—	67,17	162,86	4,31	61,40	,6087
83	,91452	—	67,99	163,65	4,34	61,11	,6058
84	,91511	—	68,81	164,44	4,37	60,81	,6029
100 + 85	,91569	—	69,63	165,22	4,41	60,52	,6000
86	,91626	—	70,45	166,01	4,44	60,24	,5971
87	,91683	—	71,27	166,80	4,47	59,95	,5943
88	,91739	—	72,09	167,59	4,50	59,67	,5915
89	,91794	—	72,91	168,38	4,53	59,39	,5887
100 + 90	,91849	—	73,73	169,17	4,56	59,11	,5860
91	,91902	—	74,55	169,96	4,59	58,84	,5832
92	,91955	—	75,36	170,75	4,61	58,57	,5805
93	,92006	—	76,18	171,55	4,63	58,30	,5779
94	,92060	—	77,00	172,34	4,66	58,03	,5752
100 + 95	,92111	—	77,82	173,13	4,69	57,76	,5726
96	,92162	—	78,63	173,91	4,72	57,50	,5700
97	,92213	—	79,45	174,70	4,75	57,24	,5674
98	,92263	—	80,27	175,49	4,78	56,98	,5648
99	,92314	—	81,09	176,29	4,80	56,73	,5623

TABLE II.   HEAT 75°.

I. Water and spirit by weight.	II. Specific gravity.	III. Spirit by measure.	IV. Water by measure.	V. Bulk of mixture.	VI. Diminution of bulk.	VII. Quantity of spirit per cent.	VIII. Decimal multipliers.
W. + Sp.							
100+100	,92364	100	81,92	177,09	4,83	56,47	,5598
99	,92414	—	82,75	177,89	4,86	56,21	,5572
98	,92464	—	83,60	178,71	4 89	55,95	,5547
97	,92513	—	84,46	179,54	4,92	55,69	,5521
96	,92563	—	85,34	180,39	4,95	55,43	,5495
100+95	,92613	—	86,23	181,25	4,98	55,17	,5469
94	,92663	—	87,14	182,14	5,00	54,90	,5442
93	,92713	—	88,08	183,05	5,03	54,63	,5415
92	,92763	—	89,04	183,98	5,06	54,35	,5388
91	,92814	—	90,02	184,94	5,08	54,07	,5360
100+90	,92865	—	91,02	185,91	5,11	53,79	,5332
89	,92917	—	92,04	186,90	5,14	53,50	,5304
88	,92979	—	93,09	187,92	5,17	53,21	,5275
87	,93023	—	94,16	188,96	5,20	52,92	,5246
86	,93077	—	95,25	190,02	5,23	52,62	,5217
100+85	,93132	—	96,37	191,11	5,26	52,32	,5187
84	,93188	—	97,52	192,23	5,29	52,02	,5157
83	,93244	—	98,69	193,38	5,31	51,71	,5126
82	,93300	—	99,90	194,55	5,35	51,40	,5095
81	,93356	—	101,14	195,75	5,39	51,08	,5064
100+80	,93413	—	102,40	196,98	5,42	50,77	,5032
79	,93469	—	103,69	198,24	5,45	50,44	,5000
78	,93525	—	105,02	199,54	5,48	50,11	,4967
77	,93581	—	106,39	200,85	5,51	49,77	,4934
76	,93638	—	107,79	202,25	5,54	49,44	,4901
100+75	,93695	—	109,23	203,66	5,57	49,10	,4867
74	,93753	—	110,70	205,10	5,60	48,75	,4833
73	,93811	—	112,21	206,59	5,62	48,40	,4798
72	,93870	—	113,77	208,12	5,65	48,05	,4763
71	,93929	—	115,38	209,69	5,69	47,69	,4727
100+70	,93989	—	117,03	211,31	5,72	47,32	,4691
69	,94050	—	118,72	212,97	5,75	46,95	,4654
68	,94112	—	120,46	214,68	5,78	46,58	,4617
67	,94174	—	122,26	216,45	5,81	46,20	,4580
66	,94237	—	124,12	218,27	5,85	45,82	,4542
100+65	,94301	—	126,03	220,14	5,89	45,43	,4503
64	,94364	—	127,99	222,07	5,92	45,03	,4464
63	,94428	—	130,02	224,07	5,95	44,63	,4424
62	,94493	—	132,12	226,14	5,98	44,23	,4383
61	,94558	—	134,29	228,27	6,02	43,81	,4342
100+60	,94623	—	136,53	230,47	6,06	43,39	,4301
59	,94689	—	138,84	232,74	6,10	42,96	,4259
58	,94756	—	141,23	235,10	6,13	42,53	,4216
57	,94823	—	143,71	237,55	6,16	42,09	,4173
56	,94890	—	146,28	240,09	6,19	41,65	,4129
100+55	,94957	—	148,93	242,71	6,22	41,20	,4084
54	,95023	—	151,67	245,43	6,24	40,74	,4039
53	,95090	—	154,52	248,26	6,26	40,28	,3993
52	,95157	—	157,50	251,21	6,29	39,81	,3946
51	,95224	—	160,61	254,28	6,33	39,33	,3898

I. Water and spirit by weight.	II. Specific gravity.	III. Spirit by measure.	IV. Water by measure.	V. Bulk of mixture.	VI. Diminution of bulk.	VII. Quantity of spirit per cent.	VIII. Decimal multipliers.
W. + Sp.							
100+50	,95292	100	163,84	257,46	6,38	38,84	,3850
49	,95361	—	167,18	260,78	6,40	38,35	,3801
48	,95430	—	170,66	264,24	6,42	37,84	,3751
47	,95499	—	174,29	267,85	6,44	37,33	,3701
46	,95568	—	178,08	271,61	6,47	36,82	,3650
100+45	,95638	—	182,04	275,53	6,51	36,29	,3598
44	,95707	—	186,18	279,65	6,53	35,76	,3545
43	,95777	—	190,51	283,97	6,54	35,21	,3491
42	,95847	—	195,05	288,49	6,56	34,66	,3436
41	,95917	—	199,81	293,23	6,58	34,10	,3380
100+40	,95987	—	204,80	298,20	6,60	33,53	,3324
39	,96058	—	210,05	303,43	6,62	32,96	,3267
38	,96129	—	215,57	308,95	6,62	32,37	,3209
87	,96200	—	221,40	314,76	6,64	31,77	,3149
36	,96272	—	227,55	320,90	6,65	31,16	,3089
100+35	,96344	—	234,05	327,41	6,64	30,54	,3028
34	,96416	—	240,94	334,29	6,65	29,91	,2965
33	,96488	—	248,24	341,61	6,63	29,27	,2902
32	,96561	—	256,00	349,38	6,62	28,62	,2837
31	,96634	—	264,26	357,64	6,62	27,96	,2772
100+30	,96708	—	273,06	366,44	6,62	27,29	,2705
29	,96783	—	282,48	375,88	6,60	26,60	,2637
28	,96858	—	292,57	386,00	6,57	25,91	,2568
26	,96935	—	303,40	396,86	6,54	25,20	,2498
100+25	,97086	—	327,68	421,17	6,51	23,74	,2354
24	,97165	—	341,33	434,86	6,47	22,99	,2280
23	,97246	—	356,17	449,73	6,44	22,23	,2204
22	,97328	—	372,36	466,96	6,40	21,46	,2127
21	,97411	—	389,00	483,73	6,36	20,67	,2049
100+20	,97495	—	409,60	503,29	6,31	19,87	,1970
19	,97581	—	431,15	524,90	6,25	19,05	,1889
18	,97669	—	455,11	548,91	6,20	18,21	,1806
17	,97750	—	481,88	575,74	6,14	17,36	,1722
16	,97850	—	511,99	605,93	6,06	16,50	,1636
100+15	,97943	—	546,13	640,15	5,98	15,62	,1549
14	,98039	—	585,14	679,25	5,89	14,72	,1460
13	,98138	—	630,15	724,35	5,80	13,80	,1369
12	,98240	—	682,66	776,96	5,70	12,87	,1276
11	,98345	—	744,72	839,12	5,60	11,92	,1181
100+10	,98454	—	819,19	913,68	5,51	10,94	,1085
9	,98567	—	910,21	1004,85	5,36	9,95	,0987
8	,98684	—	1023,99	1118,76	5,23	8,94	,0886
7	,98807	—	1170,33	1265,46	5,16	7,90	,0784
6	,98933	—	1365,32	1460,36	4,96	6,85	,0679
100+5	,99066	—	1638,39	1733,55	4,84	5,77	,0572
4	,99205	—	2047,98	2143,32	4,66	4,66	,0462
3	,99351	—	2730,64	2826,13	4,51	3,53	,0351
2	,99504	—	4095,96	4191,58	4,38	2,38	,0237
1	,99664	—	8191,92	8287,65	4,27	1,21	,0120

I. Spirit and water by weight.	II. Specific gravity.	III. Spirit by measure.	IV. Water by measure.	V. Bulk of mixture.	VI. Diminution of bulk.	VII. Quantity of spirit *per cent.*	VIII. Decimal multipliers.
Sp. + W.							
100 + 0	,81730	100	—	100,00	—	100,00	,9907
1	,81961	—	0,82	100,72	0,10	99,29	,9836
2	,82185	—	1,64	101,44	0,20	98,58	,9767
3	,82406	—	2,46	102,16	0,30	97,89	,9698
4	,82620	—	3,28	102,88	0,40	97,21	,9630
100 + 5	,82829	—	4,10	103,61	0,49	96,52	,9562
6	,83033	—	4,91	104,34	0,57	95,84	,9495
7	,83232	—	5,73	105,07	0,66	95,17	,9429
8	,83427	—	6,55	105,81	0,74	94,51	,9363
9	,83617	—	7,37	106,54	0,83	93,86	,9298
100 + 10	,83801	—	8,19	107,28	0,91	93,21	,9234
11	,83982	—	9,01	108,03	0,98	92,57	,9171
12	,84157	—	9,83	108,77	1,06	91,94	,9108
13	,84329	—	10,64	109,52	1,12	91,31	,9046
14	,84498	—	11,46	110,27	1,19	90,69	,8985
100 + 15	,84662	—	12,28	111,02	1,26	90,07	,8925
16	,84824	—	13,11	111,77	1,34	89,47	,8865
17	,84983	—	13,92	112,52	1,40	88,87	,8805
18	,85140	—	14,74	113,27	1,47	88,28	,8746
19	,85293	—	15,56	114,03	1,53	87,70	,8688
100 + 20	,85447	—	16,37	114,79	1,58	87,12	,8631
21	,85593	—	17,19	115,53	1,66	86,55	,8575
22	,85739	—	18,01	116,29	1,72	85,99	,8519
23	,85883	—	18,83	117,04	1,79	85,43	,8464
24	,86024	—	19,65	117,81	1,84	84,88	,8409
100 + 25	,86163	—	20,47	118,57	1,90	84,34	,8355
26	,86298	—	21,29	119,33	1,96	83,80	,8302
27	,86431	—	22,11	120,09	2,02	83,27	,8249
28	,86562	—	22,93	120,86	2,07	82,74	,8197
29	,86690	—	23,74	121,61	2.13	82,22	,8146
100 + 30	,86816	—	24,57	122,38	2,19	81,71	,8095
31	,86940	—	25,38	123,15	2,23	81,20	,8045
32	,87062	—	26,20	123,92	2,28	80,70	,7995
33	,87184	—	27,02	124,68	2,34	80,21	,7945
34	,87303	—	27,84	125,45	2,39	79.72	,7897
100 + 35	,87419	—	28,66	126,21	2,45	79,23	,7849
36	,87533	—	29,48	126,98	2,50	78,75	,7802
37	,87644	—	30,30	127,75	2,55	78,27	,7755
38	,87754	—	31,12	128,52	2,60	77,80	,7708
39	,87863	—	31,93	129,30	2,63	77,34	,7662
100 + 40	,87970	—	32,75	130,07	2,68	76,89	,7617
41	,88077	—	33,57	130,84	2,73	76,43	,7572
42	,88182	—	34,39	131,61	2,78	75,98	,7527
43	,88287	—	35,20	132,38	2,82	75,54	,7483
44	,88390	—	36,02	133,15	2,87	75,10	,7440
100 + 45	,88490	—	36,84	133,92	2,92	74,67	,7397
46	,88590	—	37,66	134,69	2,97	74,24	,7355
47	,88687	—	38,48	135,47	3,01	73,82	,7313
48	,88783	—	39,30	136,24	3,06	73,40	,7272
49	,88878	—	40,12	137,02	3,10	72,98	,7231

I. Spirit and water by weight.	II. Specific gravity.	III. Spirit by measure.	IV. Water by measure.	V. Bulk of mixture.	VI. Diminution of bulk.	VII. Quantity of spirit *per cent.*	VIII. Decimal multipliers.
Sp. + W.							
100 + 50	,88970	100	40,94	137,79	3,15	72,57	,7190
51	,89063	—	41,76	138,57	3,19	72,17	,7150
52	,89154	—	42,58	139,34	3,24	71,77	,7110
53	,89243	—	43,39	140,12	3,27	71,37	,7071
54	,89331	—	44,21	140,90	3,31	70,98	,7032
100 + 55	,89417	—	45,03	141,67	3,36	70,59	,6993
56	,89501	—	45,85	142,45	3,40	70,20	,6955
57	,89584	—	46,66	143,23	3.43	69,82	,6917
58	,89666	—	47,48	144,01	3,47	69,44	,6879
59	,89747	—	48,30	144,79	3,51	69,06	,6842
100 + 60	,89826	—	49,12	145,57	3,55	68,69	,6805
61	,89904	—	49,94	146,36	3,58	68,33	,6769
62	,89980	—	50,76	147,14	3,62	67,96	,6733
63	,90056	—	51,58	147,92	3,66	67,60	,6697
64	,90131	—	52,40	148,71	3,69	67,24	,6662
100 + 65	,90205	—	53,22	149,49	3,73	66,89	,6627
66	,90281	—	54,04	150,27	3,77	66,54	,6592
67	,90355	—	54,86	151,05	3,81	66,20	,6558
68	,90428	—	55,67	151,83	3,84	65,86	,6524
69	,90500	—	56,49	152,61	3,88	65,52	,6491
100 + 70	,90570	—	57,31	153,39	3,92	65,19	,6458
71	,90640	—	58,13	154,19	3,94	64,86	,6425
72	,90708	—	58,95	154,97	3,98	64,53	,6392
73	,90775	—	59,77	155,76	4,01	64,20	,6360
74	,90841	—	60,59	156,54	4,05	63,88	,6328
100 + 75	,90907	—	61,41	157,33	4,08	63,56	,6297
76	,90972	—	62,23	158,12	4,11	63,25	,6265
77	,91036	—	63,05	158,90	4,15	62,93	,6234
78	,91100	—	63,87	159,69	4,18	62,62	,6203
79	,91163	—	64,68	160,48	4,20	62,31	,6173
100 + 80	,91226	—	65,50	161,26	4,24	62,01	,6143
81	,91286	—	66,32	162,05	4,27	61,71	,6113
82	,91347	—	67,14	162,84	4,30	61,41	,6084
83	,91407	—	67,96	163,63	4,33	61,12	,6055
84	,91466	—	68,77	164,42	4,35	60,82	,6026
100 + 85	,91524	—	69,59	165,21	4,38	60,53	,5997
86	,91581	—	70,41	165,99	4,42	60,24	,5968
87	,91638	—	71,24	166,78	4,46	59,96	,5940
88	,91694	—	72,05	167,57	4,48	59,68	,5912
89	,91749	—	72,87	168,36	4,51	59,40	,5884
100 + 90	,91804	—	73,69	169,15	4,54	59,12	,5857
91	,91857	—	74,51	169,94	4,57	58,84	,5830
92	,91911	—	75,32	170,73	4,59	58,57	,5803
93	,91964	—	76,14	171,52	4,62	58,30	,5776
94	,92016	—	76,97	172,32	4,65	58,03	,5749
100 + 95	,92067	—	77,78	173,11	4,67	57,77	,5723
96	,92118	—	78,59	173,88	4,71	57,51	,5698
97	,92169	—	79,41	174,67	4,74	57,25	,5672
98	,92219	—	80,23	175,46	4,77	56,99	,5646
99	,92270	—	81,05	176,26	4,79	56,73	,5621

TABLE II.   HEAT 76°

I. Water and spirit by weight.	II. Specific gravity.	III. Spirit by measure.	IV. Water by measure.	V. Bulk of mixture.	VI. Diminution of bulk.	VII. Quantity of spirit per cent.	VIII. Decimal multipliers.
W. + Sp.							
100+100	,92320	100	81,88	177,06	4,82	56,48	,5595
99	,92370	—	82,71	177,86	4,85	56,22	,5569
98	,92420	—	83,56	178,68	4,88	55,96	,5544
97	,92469	—	84,42	179,51	4,91	55,70	,5518
96	,92519	—	85,30	180,36	4,94	55,44	,5492
100+95	,92569	—	86,19	181,22	4,97	55,18	,5466
94	,92619	—	87,10	182,11	4,99	54,91	,5440
93	,92669	—	88,03	183,02	5,01	54,64	,5413
92	,92720	—	88,99	183,95	5,04	54,36	,5386
91	,92771	—	89,97	184,91	5,06	54,08	,5358
100+90	,92822	—	90,97	185,89	5,08	53,80	,5330
89	,92874	—	91,99	186,87	5,12	53,51	,5302
88	,92927	—	93,04	187,90	5,14	53,22	,5273
87	,92980	—	94,11	188,94	5,17	52,93	,5244
86	,93034	—	95,20	190,00	5,20	52,63	,5214
100+85	,93089	—	96,32	191,08	5,24	52,33	,5184
84	,93145	—	97,47	192,20	5,27	52,03	,5154
83	,93201	—	98,64	193,35	5,29	51,72	,5124
82	,93258	—	99,85	194,52	5,33	51,41	,5093
81	,93314	—	101,09	195,72	5,37	51,09	,5062
100+80	,93370	—	102,35	196,94	5,41	50,77	,5030
79	,93427	—	103,64	198,20	5,44	50,44	,4998
78	,93483	—	104,97	199,51	5,46	50,12	,4965
77	,93540	—	106,34	200,85	5,49	49,79	,4932
76	,93597	—	107,74	222,21	5,53	49,45	,4899
100+75	,93653	—	109,17	203,62	5,55	49,10	,4865
74	,93712	—	110,64	205,06	5,58	48,76	,4831
73	,93770	—	112,15	206,56	5,59	48,41	,4796
72	,93825	—	113,71	208,09	5,62	48,06	,4761
71	,93888	—	115,32	209,65	5,67	47,70	,4725
100+70	,93949	—	116,97	211,27	5,70	47,33	,4689
69	,94010	—	118,66	212,93	5,73	46,96	,4653
68	,94072	—	120,40	214,64	5,76	46,59	,4616
67	,94134	—	122,20	216,41	5,79	46,21	,4578
66	,94197	—	124,06	218,22	5,84	45,82	,4540
100+65	,94262	—	125,97	220,10	5,87	45,43	,4501
64	,94325	—	127,93	222,03	5,90	45,04	,4462
63	,94389	—	129,96	224,03	5,93	44,64	,4422
62	,94454	—	132,05	226,10	5,95	44,23	,4382
61	,94520	—	134,22	228,22	6,00	43,82	,4341
100+60	,94585	—	136,46	230,43	6,03	43,40	,4299
59	,94651	—	138,77	232,70	6,07	42,97	,4258
58	,94718	—	141,16	235,06	6,10	42,54	,4215
57	,94785	—	143,64	237,50	6,14	42,10	,4172
56	,94852	—	146,21	240,03	6,18	41,66	,4128
100+55	,94920	—	148,86	242,65	6,21	41,21	,4083
54	,94986	—	151,59	245,37	6,22	40,75	,4038
53	,95053	—	154,44	248,20	6,24	40,29	,3992
52	,95121	—	157,43	251,15	6,28	39,82	,3945
51	,95188	—	160,53	254,23	6,30	39,33	,3897
100+50	,95255	100	163,76	257,40	6,36	38,85	,3849
49	,95325	—	167,10	260,71	6,39	38,36	,3800
48	,95395	—	170,58	264,17	6,41	37,85	,3750
47	,95464	—	174,21	267,78	6,43	37,34	,3700
46	,95534	—	177,99	271,54	6,45	36,82	,3649
100+45	,95603	—	181,95	275,46	6,49	36,30	,3596
44	,95674	—	186,09	279,57	6,52	35,77	,3544
43	,95744	—	190,42	283,89	6,53	35,22	,3490
42	,95814	—	194,95	288,41	6,54	34,67	,3435
41	,95885	—	199,71	293,15	6,56	34,11	,3379
100+40	,95955	—	204,70	298,11	6,59	33,54	,3323
39	,96026	—	209,95	303,34	6,61	32,96	,3266
38	,96098	—	215,47	308,86	6,61	32,37	,3208
37	,96169	—	221,29	314,67	6,62	31,77	,3148
36	,96242	—	227,44	320,81	6,63	31,17	,3088
100+35	,96314	—	233,94	327,31	6,63	30,57	,3027
34	,96386	—	240,82	334,19	6,63	29,92	,2965
33	,96459	—	248,12	341,50	6,62	29,28	,2901
32	,96532	—	255,88	349,27	6,61	28,63	,2836
31	,96606	—	264,13	357,53	6,60	27,97	,2771
100+30	,96680	—	272,93	366,32	6,61	27,30	,2704
29	,96756	—	282,34	375,75	6,59	26,61	,2637
28	,96831	—	292,43	385,86	6,57	25,91	,2568
27	,96907	—	303,25	396,72	6,53	25,21	,2498
26	,96984	—	314,92	408,40	6,52	24,49	,2426
100+25	,97062	—	327,52	421,00	6,50	23,75	,2353
24	,97141	—	341,17	434,70	6,47	23,00	,2279
23	,97223	—	356,00	449,56	6,44	22,24	,2204
22	,97305	—	372,18	466,78	6,40	21,47	,2127
21	,97389	—	389,90	483,54	6,36	20,68	,2049
100+20	,97473	—	409,40	503,09	6,31	19,88	,1969
19	,97560	—	430,94	524,69	6,25	19,06	,1888
18	,97648	—	454,89	548,69	6,20	18,22	,1806
17	,97739	—	481,65	575,51	6,14	17,37	,1722
16	,97830	—	511,75	605,68	6,07	16,51	,1636
100+15	,97924	—	545,87	639,88	5,99	15,63	,1548
14	,98020	—	584,86	678,96	5,90	14,73	,1459
13	,98120	—	629,85	724,04	5,81	13,81	,1368
12	,98222	—	682,34	776,62	5,72	12,87	,1276
11	,98328	--	744,32	838,76	5,61	11,92	,1181
100+10	,98436	—	818,80	913,28	5,52	10,95	,1085
9	,98550	—	909,78	1004,40	5,38	9,96	,0986
8	,98668	—	1023,51	1118,26	5,25	8,94	,0886
7	,98791	—	1169,71	1264,60	5,17	7,90	,0783
6	,98920	—	1364,68	1459,69	4,99	6,85	,0679
100+ 5	,99051	—	1637,62	1732,75	4,87	5,77	,0572
4	,99191	—	2047,03	2142,33	4,70	4,67	,0462
3	,99301	—	2729,36	2824,82	4,54	3,50	,0351
2	,99490	—	4094,04	4189,62	4,42	2,39	,0236
1	,99650	--	8188,08	8283,78	4,30	1,21	,0120

TABLE I. HEAT 77°.

I. Spirit and water by weight.	II. Specific gravity.	III. Spirit by measure.	IV. Water by measure.	V. Bulk of mixture.	VI. Diminution of bulk.	VII. Quantity of spirit per cent.	VIII. Decimal multipliers.
Sp. + W.							
100 + 0	,81680	100	—	100,00	—	100,00	,9901
1	,81911	—	0 82	100,72	0,10	99,29	,9830
2	,82136	—	1,64	101,44	0,20	98,59	,9761
3	,82357	—	2,46	102,16	0,30	97,89	,9692
4	,82571	—	3,28	102,88	0,40	97,21	,9624
100 + 5	,82780	—	4,10	103,61	0,49	96,52	,9556
6	,82984	—	4,91	104,34	0,57	95,84	,9490
7	,83183	—	5,73	105,07	0,66	95,17	,9423
8	,83378	—	6,55	105,81	0,74	94,51	,9357
9	,83568	—	7,36	106,54	0,82	93,86	,9292
100 + 10	,83752	—	8,18	107,28	0,91	93,21	,9229
11	,83933	—	9,00	108,02	0,98	92,57	,9166
12	,84108	—	9,82	108,77	1,05	91,94	,9103
13	,84280	—	10,64	109,51	1,13	91,31	,9041
14	,84449	—	11,46	110,26	1,20	90,69	,8980
100 + 15	,84613	—	12,28	111,01	1,27	90,08	,8919
16	,84775	—	13,10	111,76	1,34	89,47	,8859
17	,84934	—	13,91	112,51	1,40	88,87	,8800
18	,85091	—	14,73	113,27	1,46	88,28	,8741
19	,85244	—	15,55	114,02	1,53	87,70	,8683
100 + 20	,85397	—	16,36	114,78	1,58	87,12	,8626
21	,85544	—	17,18	115,53	1,65	86,55	,8570
22	,85690	—	18,00	116,29	1,71	85,99	,8514
23	,85834	—	18,82	117,04	1,78	85,44	,8459
24	,85975	—	19,64	117,80	1,84	84,89	,8404
100 + 25	,86113	—	20,46	118,56	1,90	84,34	,8350
26	,86249	—	21,28	119,32	1,96	83,80	,8297
27	,86382	—	22,10	120,09	2,01	83,27	,8244
28	,86514	—	22,92	120,85	2,07	82,75	,8192
29	,86642	—	23,73	121,61	2,12	82,23	,8141
100 + 30	,86768	—	24,56	122,38	2,18	81,72	,8090
31	,86892	—	25,36	123,14	2,22	81,21	,8040
32	,87015	—	26,19	123,91	2,28	80,71	,7990
33	,87136	—	27,01	124,67	2,34	80,21	,7941
34	,87256	—	27,83	125,44	2,39	79,72	,7893
100 + 35	,87372	—	28,65	126,20	2,45	79,23	,7845
36	,87485	—	29,46	126,97	2,49	78,75	,7797
37	,87596	—	30,28	127,74	2,54	78,27	,7750
38	,87706	—	31,10	128,51	2,59	77,81	,7704
39	,87815	—	31,92	129,29	2,63	77,35	,7658
100 + 40	,87921	—	32,74	130,06	2,68	76,89	,7613
41	,88029	—	33,56	130,83	2,73	76,43	,7568
42	,88134	—	34,37	131,60	2,77	75,99	,7523
43	,88239	—	35,19	132,37	2,82	75,55	,7479
44	,88343	—	36,01	133,14	2,87	75,11	,7436
100 + 45	,88442	—	36,83	133,91	2,92	74,68	,7393
46	,88543	—	37,64	134,68	2,96	74,25	,7351
47	,88640	—	38,46	135,46	3,00	73,82	,7309
48	,88736	—	39,28	136,23	3,05	73,40	,7268
49	,88831	—	40,10	137,01	3,09	72,99	,7227
100 + 50	,88922	100	40,92	137,78	3,14	72,58	,7186
51	,89017	—	41,74	138,56	3,18	72,17	,7146
52	,89106	—	42,56	139,33	3,23	71,77	,7106
53	,89195	—	43,37	140,11	3,26	71,37	,7067
54	,89283	—	44,19	140,89	3,30	70,98	,7028
100 + 55	,89369	—	45,01	141,66	3,35	70,59	,6989
56	,89453	—	45,83	142,44	3,39	70,21	,6951
57	,89536	—	46,64	143,22	3,42	69,82	,6913
58	,89619	—	47,45	144,00	3,46	69,44	,6875
59	,89700	—	48,28	144,78	3,50	69,06	,6838
100 + 60	,89780	—	49,10	145,56	3,54	68,69	,6801
61	,89858	—	49,92	146,35	3,57	68,33	,6766
62	,89934	—	50,74	147,13	3,61	67,96	,6730
63	,90010	—	51,56	147,91	3,65	67,60	,6694
64	,90085	—	52,38	148,69	3,69	67,25	,6659
100 + 65	,90159	—	53,19	149,48	3,71	66,90	,6623
66	,90235	—	54,02	150,26	3,76	66,55	,6589
67	,90309	—	54,84	151,04	3,80	66,20	,6555
68	,90382	—	55,65	151,81	3,84	65,86	,6521
69	,90454	—	56,47	152,60	3,87	65,53	,6488
100 + 70	,90523	—	57,28	153,38	3,90	65,19	,6454
71	,90594	—	58,11	154,17	3,94	64,86	,6422
72	,90662	—	58,92	154,96	3,96	64,53	,6389
73	,90729	—	59,74	155,75	3,99	64,20	,6357
74	,90795	—	60,56	156,53	4,03	63,88	,6325
100 + 75	,90862	—	61,38	157,31	4,07	63,57	,6293
76	,90926	—	62,20	158,11	4,09	63,25	,6262
77	,90990	—	63,02	158,89	4,13	62,93	,6231
78	,91054	—	63,84	159,67	4,17	62,62	,6200
79	,91117	—	64,65	160,46	4,19	62,32	,6170
100 + 80	,91181	—	65,47	161,24	4,23	62,02	,6140
81	,91241	—	66,29	162,04	4,25	61,71	,6110
82	,91301	—	67,11	162,82	4,29	61,41	,6081
83	,91361	—	67,92	163,61	4,31	61,13	,6052
84	,91420	—	68,74	164,40	4,34	60,83	,6025
100 + 85	,91478	—	69,56	165,19	4,37	60,54	,5994
86	,91536	—	70,38	165,98	4,40	60,25	,5965
87	,91593	—	71,20	166,76	4,44	59,97	,5937
88	,91649	—	72,02	167,55	4,47	59,68	,5909
89	,91704	—	72,84	168,34	4,50	59,40	,5881
100 + 90	,91758	—	73,66	169,13	4,53	59,13	,5854
91	,91812	—	74,48	169,92	4,56	58,85	,5827
92	,91866	—	75,29	170,71	4,58	58,58	,5800
93	,91920	—	76,11	171,50	4,61	58,31	,5773
94	,91972	—	76,93	172,29	4,64	58,05	,5747
100 + 95	,92023	—	77,75	173,08	4,67	57,77	,5720
96	,92074	—	78,56	173,86	4,70	57,51	,5695
97	,92125	—	79,37	174,65	4,72	57,26	,5669
98	,92175	—	80,20	175,44	4,76	57,00	,5643
99	,92226	—	81,01	176,23	4,78	56,74	,5618

TABLE II.   HEAT 77°.

I. Water and spirit by weight.	II. Specific gravity.	III. Spirit by measure.	IV. Water by measure.	V. Bulk of mixture.	VI. Diminution of bulk.	VII. Quantity of spirit per cent.	VIII. Decimal multipliers.
W. + Sp.							
100 + 100	,92275	100	81,84	177.04	4,80	56,48	,5592
99	,92326	—	82,67	177,84	4,83	56,23	,5567
98	,92375	—	83,52	178,66	4 86	55,97	,5541
97	,92425	—	84.38	179,49	4,89	55,71	,5516
96	,92475	—	85.26	180.34	4,92	55,44	,5490
100 + 95	,92525	—	86,15	181,20	4,95	55,18	,5464
94	,92575	—	87,06	182,09	4,97	54,92	,5437
93	,92625	—	87,99	183,00	4,99	54,65	,5410
92	,92676	—	88,95	183,93	5,02	54,37	,5383
91	,92727	—	89,93	184,89	5,04	54,08	,5355
100 + 90	,92778	—	90,93	185,80	5,07	53,80	,5327
89	,92830	—	91,95	186,84	5,11	53,51	,5299
88	,92883	—	93,00	187,87	5,13	53,23	,5270
87	,92937	—	94,07	188.91	5,16	52,94	,5241
86	,92991	—	95,16	189,97	5.19	52,64	,5212
100 + 85	,93046	—	96,28	191,05	5,23	52,34	,5182
84	,93102	—	97,43	192,17	5,26	52,04	,5152
83	,93158	—	98,60	193,32	5,28	51,73	,5121
82	93215	—	99.80	194,48	5,32	51,42	,5090
81	,93272	—	101.04	195.68	5 36	51,10	,5059
100 + 80	,93327	—	102,30	196.91	5,39	50,78	,5028
79	,33385	—	103.59	198,17	5 42	50,45	,4996
78	,93441	—	104.92	199,47	5,45	50,12	,4963
77	,93498	—	106 29	200.81	5,48	49,80	,4930
76	,93555	—	107,69	202,18	5,51	49,46	,4897
100 + 75	,93611	—	109,12	203,59	5,53	49,11	,4863
74	,93670	—	110,59	205 03	5,56	48,77	,4829
73	,93728	—	112,10	206,52	5,58	48,42	,4794
72	,93784	—	113,66	208.06	5,60	48,07	,4759
71	,93847	—	115,27	209 62	5,65	47,67	,4723
100 + 70	,93908	—	116.92	211,24	5,68	47,34	,4687
69	,93969	—	118,60	212,90	5.70	46,97	,4651
68	,94031	—	120,35	214,60	5,75	46.59	,4614
67	,94094	—	122,14	216,37	5,77	46,22	,4576
66	,94157	—	124,00	218,18	5 82	45,83	,4538
100 + 65	,94221	—	125,91	220 05	5,86	45.44	,4499
64	,94285	—	127,87	221,99	5,88	45,04	,4460
63	,94350	—	129 90	223.98	5,92	44,64	,4421
62	,94415	—	131,99	226,05	5,94	44,23	,4380
61	,94481	—	134,16	228,18	5,98	43,82	,4339
100 + 60	,94540	—	136,39	230,38	6,01	43,41	,4297
59	,94613	—	138,70	232,65	6,05	42,98	,4256
58	,94680	—	141,09	235.01	6,08	42,55	,4213
57	,94747	—	143,57	237,45	6,12	42,11	,4170
56	,94814	—	146,14	239,98	6,16	41,67	,4126
100 + 55	,94882	—	148,79	242,60	6,19	41,21	,4081
54	,94949	—	151,52	245,31	6,21	40,76	,4036
53	,95016	—	154,37	248,15	6,22	40,30	,3990
52	,95084	—	157,36	251,09	6,27	39,83	,3943
51	,95151	—	160,45	254,18	6,27	39,34	,3895
W. + Sp.							
100 + 50	,95219	100	163,68	257,34	6,34	38,86	,3847
49	,95289	—	167,02	260.64	6,38	38,36	,3798
48	,95359	—	170,50	264,11	6,39	37,86	,3748
47	,95429	—	174.13	267,71	6,42	37,35	,3698
46	,95500	—	177,91	271,46	6,45	36,83	,3647
100 + 45	,95569	—	181,86	275,39	6,47	36,31	,3595
44	,95640	—	186,00	279 50	6,50	35,77	,3542
43	,95710	—	190,33	283,81	6,52	35,23	,3489
42	,95781	—	194,86	288,34	6,52	34,68	,3434
41	,95852	—	199,61	293,07	6,54	34,12	,3378
100 + 40	,95922	—	204,00	298,03	6,57	33,55	,3322
39	,95994	—	209,85	303,26	6,59	32,97	,3265
38	,96066	—	215,37	308,77	6,60	32,38	,3207
37	,96138	—	221,19	314,59	6,60	31,78	,3147
36	,96211	—	227,34	320.72	6,62	31,18	,3087
100 + 35	,96280	—	233,83	327,21	6,62	30,56	,3026
34	,96356	—	240,71	334,09	6,62	29,93	,2964
33	,96429	—	248,00	341,40	6,60	29,29	,2900
32	,96503	—	255,76	349,16	6,60	28,64	,2835
31	,96577	—	264,01	357,41	6,60	27,98	,2770
100 + 30	,96652	—	272,80	366,20	6,60	27,31	,2704
29	,96728	—	282,21	375,63	6,58	26,62	,2636
28	,96804	—	292,29	385,73	6,56	25,92	,2567
27	,96881	—	303,11	396,58	6,53	25,22	,2497
26	,96958	—	314,77	408,25	6,52	24,50	,2426
100 + 25	,97038	—	327,37	420,87	6,50	23,76	,2352
24	,97117	—	341,01	434,54	6.47	23,01	,2279
23	,97199	—	355,83	449,39	6 44	22,25	,2203
22	,97282	—	372 00	466,60	6,40	21,48	,2126
21	,97310	—	389,72	483.35	6,37	20,69	,2048
100 + 20	,97451	—	409,21	502,89	6.32	19,89	,1969
19	,97538	—	430,74	524,48	6,26	19,07	,1888
18	,97627	—	454,67	548,47	6,20	18,23	,1805
17	,97718	—	481,42	575,28	6,14	17,38	,1721
16	,97810	—	511,51	605,43	6 08	16,52	,1636
100 + 15	,97905	—	545,61	639,61	6,00	15,63	,1548
14	,98003	—	584,58	678,68	5,90	14,73	,1459
13	,98101	—	629,56	723,73	5,83	13,81	,1368
12	,98203	—	682,02	776,29	5,73	12,88	,1275
11	,98310	—	744,02	838,40	5,62	11,93	,1181
100 + 10	,98412	—	818,42	912,89	5,53	10,95	,1085
9	,98533	—	909,36	1003,96	5,40	9,96	,0986
8	,98651	—	1023,03	1117,76	5,27	8,95	,0886
7	,98775	—	1169,21	1264,03	5.18	7,91	,0783
6	,98902	—	1364,04	1459,02	5,02	6,85	,0679
100 + 5	,99036	—	1630,85	1731,95	4,90	5,77	,0572
4	,99176	—	2046,06	2141,34	4,72	4,67	,0462
3	,99323	—	2728,08	2823,50	4,58	3,54	,0351
2	,99476	—	4092,12	4187,66	4,46	2,39	,0236
1	,99636	—	8184,23	8279.90	4,33	1,21	,0120

TABLE I.   HEAT 78°.

I. Spirit and water by weight.	II. Specific gravity.	III. Spirit by measure.	IV. Water by measure.	V. Bulk of mixture.	VI. Diminution of bulk.	VII. Quantity of spirit per cent.	VIII. Decimal multipliers.	I. Spirit and water by weight.	II. Specific gravity.	III. Spirit by measure.	IV. Water by measure.	V. Bulk of mixture.	VI. Diminution of bulk.	VII. Quantity of spirit per cent.	VIII. Decimal multipliers.
Sp. + W.								Sp. + W.							
100 + 0	,81630	100	—	100,00	—	100,00	,9894	100 + 50	,88875	100	40,90	137,77	3,13	72,59	,7182
1	,81861	—	0,82	100,72	0,10	99,29	,9824	51	,88969	—	41,72	138,55	3,17	72,18	,7142
2	,82086	—	1,64	101,44	0,20	98,59	,9755	52	,89059	—	42,54	139,32	3,22	71,78	,7102
3	,82308	—	2,46	102,16	0,30	97,89	,9686	53	,89148	—	43,35	140,10	3,25	71,38	,7063
4	,82522	—	3,28	102,88	0,40	97,21	,9618	54	,89235	—	44,17	140,88	3,29	70,99	,7024
100 + 5	,82730	—	4,09	103,60	0,49	96,52	,9551	100 + 55	,89321	—	44,99	141,65	3,34	70,60	,6985
6	,82935	—	4,91	104,34	0,57	95,85	,9484	56	,89406	—	45,81	142,43	3,38	70,21	,6947
7	,83134	—	5,73	105,07	0,66	95,18	,9418	57	,89489	—	46,62	143,21	3,41	69,83	,6909
8	,83328	—	6,55	105,81	0,74	94,52	,9352	58	,89572	—	47,44	143,99	3,45	69,45	,6872
9	,83518	—	7,36	106,54	0,82	93,86	,9287	59	,89653	—	48,26	144,77	3,49	69,07	,6835
100 + 10	,83702	—	8,18	107,28	0,90	93,21	,9223	100 + 60	,89733	—	49,08	145,55	3,53	68,70	,6798
11	,83883	—	9,00	108,02	0,98	92,57	,9160	61	,89811	—	49,89	146,33	3,56	68,34	,6762
12	,84059	—	9,82	108,77	1,05	91,94	,9098	62	,89888	—	50,71	147,11	3,60	67,97	,6726
13	,84231	—	10,63	109,51	1,12	91,31	,9035	63	,89964	—	51,53	147,89	3,64	67,61	,6690
14	,84400	—	11,45	110,26	1,19	90,69	,8974	64	,90039	—	52,35	148,68	3,67	67,26	,6655
100 + 15	,84564	—	12,27	111,01	1,26	90,08	,8914	100 + 65	,90113	—	53,16	149,46	3,70	66,91	,6620
16	,84727	—	13,09	111,76	1,33	89,47	,8854	66	,90188	—	53,99	150,24	3,75	66,56	,6586
17	,84886	—	13,91	112,51	1,40	88,88	,8795	67	,90262	—	54,81	151,02	3,79	66,21	,6552
18	,85042	—	14,73	113,26	1,47	88,28	,8736	68	,90336	—	55,62	151,80	3,82	65,87	,6518
19	,85195	—	15,54	114,02	1,52	87,70	,8678	69	,90407	—	56,44	152,58	3,86	65,54	,6484
100 + 20	,85347	—	16,35	114,78	1,57	87,13	,8622	100 + 70	,90477	—	57,26	153,36	3,90	65,20	,6451
21	,85495	—	17,18	115,53	1,65	86,56	,8565	71	,90547	—	58,08	154,16	3,92	64,87	,6419
22	,85641	—	18,00	116,29	1,71	85,99	,8509	72	,90615	—	58,89	154,94	3,95	64,54	,6386
23	,85784	—	18,82	117,04	1,78	85,44	,8454	73	,90683	—	59,71	155,73	3,98	64,22	,6354
24	,85926	—	19,63	117,80	1,83	84,89	,8400	74	,90749	—	60,53	156,51	4,02	63,89	,6322
100 + 25	,86064	—	20,45	118,56	1,89	84,34	,8346	100 + 75	,90816	—	61,35	157,29	4,06	63,57	,6290
26	,86200	—	21,27	119,32	1,95	83,80	,8293	76	,9088c	—	62,17	158,09	4,08	63,26	,6259
27	,86333	—	22,09	120,08	2,01	83,27	,8240	77	,90945	—	62,99	158,87	4,12	62,94	,6228
28	,86465	—	22,90	120,84	2,06	82,75	,8188	78	,91008	—	63,81	159,65	4,16	62,63	,6197
29	,86594	—	23,72	121,60	2,12	82,23	,8137	79	,91071	—	64,62	160,44	4,18	62,33	,6167
100 + 30	,86719	—	24,54	122,37	2,17	81,72	,8086	100 + 80	,91136	—	65,44	161,22	4,22	62,02	,6137
31	,86844	—	25,35	123,14	2,21	81,21	,8036	81	,91195	—	66,26	162,02	4,24	61,72	,6107
32	,86967	—	26,18	123,91	2,27	80,71	,7986	82	,91255	—	67,08	162,81	4,27	61,43	,6078
33	,87089	—	26,99	124,67	2,32	80,22	,7937	83	,91315	—	67,89	163,59	4,30	61,13	,6049
34	,87208	—	27,81	125,44	2,37	79,73	,7889	84	,91374	—	68,71	164,38	4,33	60,83	,6020
100 + 35	,87324	—	28,64	126,20	2,44	79,24	,7841	100 + 85	,91432	—	69,53	165,17	4,36	60,54	,5991
36	,87437	—	29,45	126,97	2,48	78,76	,7793	86	,91490	—	70,35	165,96	4,39	60,26	,5962
37	,87548	—	30,27	127,73	2,54	78,28	,7746	87	,91547	—	71,17	166,74	4,43	59,98	,5934
38	,87658	—	31,09	128,51	2,58	77,81	,7700	88	,91603	—	71,98	167,53	4,45	59,69	,5906
39	,87767	—	31,90	129,28	2,62	77,35	,7654	89	,91659	—	72,80	168,32	4,48	59,41	,5878
100 + 40	,87873	—	32,72	130,06	2,66	76,89	,7609	100 + 90	,91713	—	73,62	169,11	4,51	59,13	,5851
41	,87981	—	33,54	130,83	2,71	76,44	,7564	91	,91767	—	74,44	169,90	4,54	58,86	,5824
42	,88086	—	34,36	131,59	2,77	75,99	,7519	92	,91821	—	75,25	170,69	4,56	58,59	,5797
43	,88191	—	35,17	132,36	2,81	75,55	,7475	93	,91875	—	76,08	171,48	4,60	58,32	,5770
44	,88295	—	35,99	133,13	2,86	75,11	,7432	94	,91928	—	76,90	172,27	4,63	58,05	,5744
100 + 45	,88395	—	36,81	133,90	2,91	74,68	,7389	100 + 95	,91979	—	77,71	173,06	4,65	57,78	,5717
46	,88496	—	37,62	134,67	2,95	74,25	,7347	96	,92030	—	78,52	173,83	4,69	57,52	,5692
47	,88593	—	38,44	135,45	2,99	73,83	,7305	97	,92081	—	79,33	174,63	4,70	57,26	,5666
48	,88689	—	39,26	136,22	3,04	73,41	,7264	98	,92131	—	80,16	175,42	4,74	57,00	,5641
49	,88784	—	40,08	137,00	3,08	73,00	,7223	99	,92182	—	80,97	176,20	4,77	56,75	,5616

TABLE II.   HEAT 78°.

Water and spirit by weight.	Specific gravity.	Spirit by measure.	Water by measure.	Bulk of mixture.	Diminution of bulk.	Quantity of spirit per cent.	Decimal multipliers.
W. + Sp.							
100+100	,92230	100	81,80	177,01	4,79	56,49	,5590
99	,92281	—	82,63	177,81	4,82	56,23	,5564
98	,92331	—	83,48	178,63	4,85	55,97	,5539
97	,92381	—	84,34	179,46	4,88	55,71	,5513
96	,92431	—	85,22	180,31	4,91	55,45	,5488
100+95	,92481	—	86,11	181,17	4,94	55,19	,5461
94	,92531	—	87,02	182,07	4,95	54,93	,5434
93	,92581	—	87,95	182,97	4,98	54,65	,5407
92	,92632	—	88,91	183,90	5,01	54,37	,5380
91	,92683	—	89,89	184,86	5,03	54,09	,5352
100+90	,92734	—	90,89	185,84	5,05	53,81	,5324
89	,92786	—	91,91	186,81	5,10	53,52	,5296
88	,92840	—	92,95	187,84	5,11	53,24	,5267
87	,92894	—	94,02	188,88	5,14	52,95	,5238
86	,92948	—	95,11	189,94	5,17	52,65	,5209
100+85	,93003	—	96,23	191,03	5,20	52,35	,5179
84	,93059	—	97,38	192,14	5,24	52,05	,5149
83	,93115	—	98,55	193,28	5,27	51,74	,5119
82	,93172	—	99,75	194,45	5,30	51,42	,5088
81	,93229	—	101,00	195,65	5,35	51,11	,5057
100+80	,93285	—	102,25	196,87	5,38	50,79	,5025
79	,93343	—	103,54	198,13	5,41	50,46	,4994
78	,93399	—	104,87	199,44	5,43	50,13	,4961
77	,93456	—	106,24	200,78	5,46	49,81	,4928
76	,93513	—	107,64	202,14	5,50	49,47	,4895
100+75	,93570	—	109,07	203,55	5,52	49,12	,4861
74	,93629	—	110,54	205,00	5,54	48,78	,4827
73	,93687	—	112,05	206,48	5,57	48,43	,4792
72	,93743	—	113,61	208,02	5,59	48,08	,4757
71	,93806	—	115,21	209,58	5,63	47,71	,4721
100+70	,93867	—	116,86	211,20	5,66	47,35	,4685
69	,93928	—	118,54	212,86	5,68	46,98	,4649
68	,93991	—	120,29	214,56	5,73	46,60	,4612
67	,94054	—	122,08	216,33	5,75	46,23	,4574
66	,94117	—	123,94	218,14	5,80	45,84	,4536
100+65	,94182	—	125,85	220,01	5,84	45,45	,4497
64	,94246	—	127,81	221,95	5,86	45,05	,4458
63	,94311	—	129,84	223,94	5,90	44,65	,4419
62	,94376	—	131,93	226,00	5,93	44,24	,4378
61	,94442	—	134,10	228,13	5,97	43,83	,4337
100+60	,94508	—	136,33	230,33	6,00	43,42	,4296
59	,94575	—	138,64	232,60	6,04	42,99	,4254
58	,94642	—	141,02	234,96	6,06	42,56	,4211
57	,94709	—	143,50	237,40	6,10	42,12	,4168
56	,94776	—	146,07	239,93	6,14	41,68	,4124
100+55	,94844	—	148,72	242,55	6,17	41,22	,4079
54	,94911	—	151,45	245,26	6,19	40,77	,4034
53	,94979	—	154,30	248,09	6,21	40,31	,3988
52	,95047	—	157,28	251,04	6,24	39,83	,3941
51	,95115	—	160,38	254,13	6,25	39,35	,3894
100+50	,95183	100	163,60	257,28	6,32	38,87	,3846
49	,95253	—	166,94	260,58	6,36	38,37	,3797
48	,95323	—	170,42	264,04	6,38	37,87	,3747
47	,95394	—	174,05	267,64	6,41	37,36	,3697
46	,95465	—	177,83	271,39	6,44	36,84	,3646
100+45	,95535	—	181,78	275,32	6,46	36,32	,3594
44	,95606	—	185,91	279,43	6,48	35,78	,3541
43	,95677	—	190,24	283,74	6,50	35,24	,3488
42	,95748	—	194,77	288,26	6,51	34,69	,3433
41	,95819	—	199,52	292,99	6,53	34,13	,3377
100+40	,95890	—	204,51	297,95	6,56	33,56	,3321
39	,95962	—	209,75	303,18	6,57	32,98	,3264
38	,96034	—	215,27	308,69	6,58	32,39	,3206
37	,96107	—	221,08	314,50	6,58	31,79	,3146
36	,96180	—	227,23	320,64	6,59	31,19	,3086
100+35	,96250	—	233,72	327,11	6,61	30,57	,3025
34	,96326	—	240,59	333,99	6,60	29,94	,2963
33	,96400	—	247,89	341,30	6,59	29,30	,2900
32	,96474	—	255,64	349,05	6,59	28,65	,2835
31	,96549	—	263,88	357,30	6,58	27,99	,2770
100+30	,96624	—	272,67	366,08	6,59	27,32	,2703
29	,96700	—	282,08	375,50	6,58	26,63	,2635
28	,96777	—	292,15	385,60	6,55	25,93	,2566
27	,96855	—	302,97	396,44	6,53	25,22	,2496
26	,96933	—	314,62	408,10	6,52	24,50	,2425
100+25	,97013	—	327,21	420,72	6,49	23,77	,2352
24	,97093	—	340,85	434,38	6,47	23,02	,2278
23	,97175	—	355,66	449,22	6,44	22,26	,2203
22	,97259	—	371,83	466,43	6,40	21,49	,2126
21	,97343	—	389,54	483,17	6,37	20,69	,2048
100+20	,97429	—	409,02	502,70	6,32	19,89	,1968
19	,97516	—	430,54	524,27	6,27	19,07	,1887
18	,97606	—	454,46	548,26	6,20	18,23	,1805
17	,97697	—	481,20	575,05	6,15	17,39	,1721
16	,97790	—	511,27	605,18	6,09	16,53	,1635
100+15	,97885	—	545,35	639,35	6,00	15,64	,1548
14	,97983	—	584,30	678,40	5,90	14,74	,1459
13	,98082	—	629,26	723,42	5,84	13,82	,1368
12	,98185	—	681,70	775,96	5,74	12,88	,1275
11	,98292	—	743,67	838,04	5,63	11,93	,1181
100+10	,98401	—	818,03	912,49	5,54	10,96	,1084
9	,98516	—	908,93	1003,52	5,41	9,97	,0986
8	,98635	—	1022,65	1117,26	5,29	8,95	,0886
7	,98759	—	1168,65	1263,46	5,19	7,91	,0783
6	,98887	—	1363,40	1458,35	5,06	6,86	,0678
100+ 5	,99021	—	1636,08	1731,16	4,92	5,78	,0572
4	,99160	—	2045,10	2140,35	4,75	4,67	,0462
3	,99308	—	2727,80	2822,19	4,61	3,54	,0351
2	,99462	—	4090,20	4185,70	4,50	2,39	,0236
1	,99622	—	8180,39	8276,03	4,36	1,21	,0120

TABLE I.

HEAT 79°.

I. Spirit and water by weight.	II. Specific gravity.	III. Spirit by measure.	IV. Water by measure.	V. Bulk of mixture.	VI. Diminution of bulk.	VII. Quantity of spirit per cent.	VIII. Decimal multipliers.
Sp. + W.							
100 + 0	,81580	100	—	100,00	—	100,00	,9888
1	,81811	—	0,82	100,71	0,11	99,29	,9818
2	,82037	—	1,63	101,43	0,20	98,59	,9749
3	,82258	—	2,45	102,15	0,30	97,89	,9680
4	,82472	—	3,27	102,87	0,40	97,21	,9612
100 + 5	,82681	—	4,09	103,60	0,49	96,52	,9545
6	,82885	—	4,90	104,33	0,57	95,85	,9478
7	,83084	—	5,72	105,06	0,66	95,18	,9412
8	,83278	—	6,54	105,80	0,74	94,52	,9346
9	,83468	—	7,35	106,53	0,82	93,86	,9281
100 + 10	,83652	—	8,18	107,27	0,91	93,22	,9218
11	,83833	—	9,00	108,02	0,98	92,58	,9155
12	,84010	—	9,81	108,76	1,05	91,95	,9093
13	,84182	—	10,63	109,50	1,13	91,32	,9030
14	,84351	—	11,45	110,26	1,19	90,70	,8969
100 + 15	,84516	—	12,27	111,01	1,26	90,09	,8908
16	,84678	—	13,09	111,76	1,33	89,48	,8848
17	,84837	—	13,90	112,51	1,39	88,88	,8790
18	,84993	—	14,72	113,26	1,46	88,29	,8731
19	,85146	—	15,54	114,02	1,52	87,70	,8673
100 + 20	,85298	—	16,35	114,78	1,57	87,13	,8617
21	,85446	—	17,17	115,52	1,65	86,56	,8560
22	,85592	—	17,99	116,28	1,71	86,00	,8504
23	,85735	—	18,81	117,03	1,78	85,45	,8449
24	,85877	—	19,62	117,79	1,83	84,90	,8395
100 + 25	,86015	—	20,44	118,55	1 89	84,35	,8341
26	,86151	—	21,26	119,31	1,95	83,81	,8288
27	,86284	—	22,08	120,08	2,00	83,28	,8235
28	,86416	—	22,89	120,84	2,05	82,76	,8183
29	,86545	—	23,71	121,60	2,11	82,24	,8132
100 + 30	,86670	—	24,53	122,37	2,16	81,73	,8081
31	,86796	—	25,34	123,13	2,21	81,22	,8031
32	,86920	—	26,16	123,90	2,26	80,72	,7982
33	,87041	—	26,98	124,66	2,32	80,22	,7932
34	,87160	—	27,80	125,43	2,37	79,73	,7884
100 + 35	,87276	—	28,02	126,19	2,43	79,24	,7836
36	,87389	—	29,43	126,96	2,47	78,76	,7789
37	,87500	—	30,25	127,73	2,52	78,28	,7742
38	,87610	—	31,07	128,50	2,57	77,82	,7696
39	,87719	—	31,89	129,27	2,62	77,36	,7650
100 + 40	,87824	—	32,71	130,05	2,66	76,90	,7605
41	,87933	—	33,53	130,82	2,71	76,44	,7560
42	,88038	—	34,34	131,58	2,76	76,00	,7515
43	,88143	—	35,16	132,35	2,81	75,56	,7471
44	,88247	—	35,97	133,12	2,85	75,12	,7428
100 + 45	,88348	—	36,80	133,89	2,91	74,69	,7385
46	,88448	—	37,60	134,66	2,94	74,26	,7343
47	,88546	—	38,42	135,44	2,98	73,83	,7301
48	,88642	—	39,24	136,21	3,03	73,41	,7260
49	,88736	—	40,06	136,99	3,07	73,00	,7219
100 + 50	,88829	100	40,88	137,76	3,12	72,59	,7178
51	,88921	—	41,70	138,54	3,16	72,18	,7138
52	,89011	—	42,52	139,31	3,21	71,78	,7098
53	,89100	—	43,33	140,09	3,24	71,39	,7059
54	,89187	—	44,15	140,87	3,28	71,00	,7020
100 + 55	,89273	—	44,97	141,64	3,33	70,60	,6981
56	,89358	—	45,79	142,42	3,37	70,21	,6943
57	,89442	—	46,60	143,20	3,40	69,83	,6905
58	,89525	—	47,42	143,98	3,44	69,45	,6868
59	,89606	—	48,24	144,76	3,48	69,07	,6831
100 + 60	,89686	—	49,06	145,54	3,52	68,70	,6794
61	,89765	—	49,87	146,32	3,55	68,34	,6759
62	,89842	—	50,69	147,10	3,59	67,97	,6723
63	,89918	—	51,51	147,88	3,63	67,61	,6687
64	,89993	—	52,32	148,66	3,66	67,26	,6651
100 + 65	,90067	—	53,14	149,45	3,69	66,91	,6617
66	,90142	—	53,96	150,23	3,73	66,56	,6583
67	,90216	—	54,78	151,01	3,77	66,21	,6549
68	,90289	—	55,59	151,79	3,80	65,87	,6515
69	,90361	—	56,41	152,57	3,84	65,54	,6482
100 + 70	,90431	—	57,23	153,35	3,88	65,21	,6448
71	,90501	—	58,05	154,13	3,90	64,87	,6416
72	,90569	—	58,86	154,92	3,94	64,54	,6383
73	,90637	—	59,68	155,71	3,97	64,22	,6351
74	,90703	—	60,50	156,49	4,01	63,90	,6319
100 + 75	,90770	—	61,32	157,28	4,04	63,58	,6287
76	,90834	—	62,14	158,07	4,07	63,26	,6256
77	,90899	—	62,96	158,85	4,11	62,95	,6225
78	,90963	—	63,78	159,63	4,15	62,64	,6194
79	,91026	—	64,59	160,42	4,17	62,33	,6164
100 + 80	,91091	—	65,41	161,20	4,21	62,03	,6134
81	,91149	—	66,23	162,00	4,23	61,73	,6104
82	,91209	—	67,05	162,79	4,26	61,43	,6075
83	,91269	—	67,86	163,57	4,29	61,14	,6046
84	,91328	—	68,68	164,36	4,32	60,84	,6017
100 + 85	,91386	—	69,50	165,15	4,35	60,55	,5988
86	,91444	—	70,31	165,94	4,37	60,26	,5959
87	,91501	—	71,13	166,72	4,41	59,98	,5931
88	,91557	—	71,95	167,51	4,44	59,69	,5903
89	,91613	—	72,77	168,30	4,47	59,41	,5875
100 + 90	,91668	—	73,59	169,09	4,50	59,14	,5848
91	,91722	—	74,40	169,88	4,52	58,86	,5821
92	,91776	—	75,22	170,67	4,55	58,59	,5794
93	,91830	—	76,04	171,46	4,58	58,32	,5767
94	,91883	—	76,86	172,24	4,62	58,05	,5741
100 + 95	,91935	—	77,68	173,03	4,65	57,79	,5715
96	,91986	—	78,48	173,81	4,67	57,53	,5689
97	,92037	—	79,29	174,60	4,68	57,27	,5663
98	,92087	—	80,12	175,40	4,72	57 01	,5638
99	,92138	—	80,93	176,18	4,75	56,75	,5613

TABLE II.   HEAT 79°.

I. Water and spirit by weight.	II. Specific gravity.	III. Spirit by measure.	IV. Water by measure.	V. Bulk of mixture.	VI. Diminution of bulk.	VII. Quantity of spirit per cent.	VIII. Decimal multipliers.
W. + Sp.							
100+100	,92186	100	81,76	176,98	4,78	56,50	,5587
99	,92237	—	82,59	177,79	4,80	56,24	,5562
98	,92286	—	83,44	178,61	4,83	55,98	,5536
97	,92337	—	84,30	179,44	4,86	55,72	,5511
96	,92387	—	85,18	180,29	4,89	55,46	,5485
100+95	,92437	—	86,07	181,15	4,92	55,20	,5459
94	,92487	—	86,98	182,04	4,94	54,93	,5432
93	,92537	—	87,91	182,95	4,96	54,66	,5405
92	,92588	—	88,87	183,88	4,99	54,38	,5378
91	,92639	—	89,85	184,84	5,01	54,10	,5350
100+90	,92690	—	90,85	185,81	5,04	53,82	,5322
89	,92743	—	91,87	186,79	5,08	53,53	,5294
88	,92797	—	92,91	187,81	5,10	53,25	,5265
87	,92852	—	93,98	188,85	5,13	52,95	,5236
86	,92905	—	95,07	189,91	5,16	52,65	,5207
100+85	,92960	—	96,19	191,00	5,19	52,35	,5177
84	,93016	—	97,33	192,11	5,22	52,05	,5147
83	,93072	—	98,50	193,25	5,25	51,75	,5117
82	,93129	—	99,70	194,42	5,28	51,43	,5086
81	,93187	—	100,95	195,62	5,33	51,12	,5055
100+80	,93243	—	102,20	196,84	5,36	50,80	,5023
79	,93301	—	103,49	198,10	5,39	50,47	,4992
78	,93357	—	104,82	199,40	5,42	50,14	,4959
77	,93414	—	106,19	200,74	5,45	49,81	,4926
76	,93472	—	107,59	202,11	5,48	49,47	,4893
100+75	,93529	—	109,02	203,51	5,51	49,13	,4859
74	,93588	—	110,49	204,96	5,53	48,79	,4825
73	,93646	—	112,00	206,45	5,55	48,44	,4790
72	,93702	—	113,55	207,98	5,57	48,08	,4755
71	,93765	—	115,15	209,54	5,61	47,72	,4719
100+70	,93826	—	116,80	211,16	5,64	47,36	,4683
69	,93888	—	118,49	212,82	5,67	46,99	,4647
68	,93951	—	120,23	214,52	5,71	46,61	,4610
67	,94014	—	122,02	216,29	5,73	46,23	,4572
66	,94077	—	123,88	218,10	5,78	45,85	,4534
100+65	,94142	—	125,79	219,97	5,82	45,46	,4495
64	,94207	—	127,75	221,91	5,84	45,06	,4456
63	,94272	—	129,78	223,90	5,88	44,66	,4417
62	,94337	—	131,87	225,96	5,91	44,25	,4376
61	,94403	—	134,04	228,09	5,95	43,84	,4336
100+60	,94469	—	136,27	230,28	5,99	43,43	,4294
59	,94537	—	138,57	232,55	6,02	43,00	,4252
58	,94604	—	140,95	234,91	6,04	42,57	,4209
57	,94671	—	143,43	237,35	6,08	42,13	,4166
56	,94738	—	146,00	239,88	6,12	41,69	,4122
100+55	,94806	—	148,65	242,50	6,15	41,23	,4078
54	,94874	—	151,38	245,21	6,17	40,78	,4033
53	,94942	—	154,23	248,04	6,19	40,32	,3987
52	,95010	—	157,21	250,98	6,23	39,84	,3940
51	,95079	—	160,31	254,06	6,25	39,36	,3893

I. Water and spirit by weight.	II. Specific gravity.	III. Spirit by measure.	IV. Water by measure.	V. Bulk of mixture.	VI. Diminution of bulk.	VII. Quantity of spirit per cent.	VIII. Decimal multipliers.
W. + Sp.							
100+50	,95147	100	163,52	257,22	6,30	38,88	,3845
49	,95217	—	166,86	260,52	6,34	38,38	,3796
48	,95288	—	170,34	263,97	6,37	37,88	,3746
47	,95359	—	173,97	267,57	6,40	37,37	,3696
46	,95430	—	177,74	271,32	6,42	36,85	,3645
100+45	,95501	—	181,69	275,25	6,44	36,33	,3593
44	,95572	—	185,82	279,36	6,46	35,79	,3540
43	,95644	—	190,15	283,67	6,48	35,25	,3487
42	,95715	—	194,68	288,19	6,49	34,70	,3432
41	,95787	—	199,43	292,91	6,52	34,14	,3376
100+40	,95858	—	204,41	297,87	6,54	33,57	,3320
39	,95930	—	209,65	303,10	6,55	32,99	,3263
38	,96003	—	215,17	308,60	6,57	32,40	,3205
37	,96076	—	220,98	314,42	6,56	31,80	,3145
36	,96149	—	227,13	320,56	6,57	31,20	,3085
100+35	,96220	—	233,61	327,01	6,60	30,58	,3024
34	,96296	—	240,48	333,89	6,59	29,95	,2962
33	,96371	—	247,77	341,20	6,57	29,31	,2899
32	,96445	—	255,52	348,94	6,58	28,66	,2834
31	,96521	—	263,76	357,18	6,58	28,00	,2769
100+30	,96596	—	272,54	365,96	6,58	27,33	,2702
29	,96673	—	281,95	375,38	6,57	26,64	,2635
28	,96750	—	292,01	385,47	6,54	25,94	,2566
27	,96829	—	302,83	396,30	6,53	25,23	,2496
26	,96908	—	314,47	407,95	6,52	24,51	,2424
100+25	,96988	—	327,06	420,57	6,49	23,78	,2351
24	,97069	—	340,69	434,22	6,47	23,03	,2278
23	,97151	—	355,49	449,06	6,43	22,27	,2203
22	,97236	—	371,66	466,26	6,40	21,49	,2125
21	,97321	—	389,36	482,99	6,37	20,70	,2047
100+20	,97407	—	408,83	502,51	6,32	19,90	,1968
19	,97494	—	430,34	524,07	6,27	19,08	,1887
18	,97585	—	454,25	548,04	6,21	18,24	,1804
17	,97676	—	480,97	574,82	6,15	17,39	,1720
16	,97770	—	511,03	604,94	6,09	16,53	,1635
100+15	,97865	—	545,10	639,09	6,01	15,65	,1547
14	,97963	—	584,03	678,12	5,91	14,75	,1458
13	,98063	—	628,97	723,10	5,85	13,83	,1367
12	,98167	—	681,38	775,63	5,75	12,89	,1275
11	,98274	—	743,32	837,68	5,64	11,94	,1180
100+10	,98384	—	817,65	912,10	5,55	10,96	,1084
9	,98499	—	908,51	1003,08	5,43	9,97	,0986
8	,98619	—	1022,07	1116,76	5,31	8,95	,0886
7	,98743	—	1168,09	1262,89	5,20	7,92	,0783
6	,98872	—	1362,76	1457,69	5,07	6,86	,0678
100+5	,99006	—	1635,31	1730,37	4,94	5,78	,0571
4	,99146	—	2044,14	2139,35	4,79	4,67	,0462
3	,99293	—	2725,51	2820,86	4,65	3,54	,0351
2	,99447	—	4088,22	4183,74	4,53	2,39	,0236
1	,99607	—	8176,55	8272,15	4,40	1,21	,0120

Left half

I. Spirit and water by weight.	II. Specific gravity.	III. Spirit by measure.	IV. Water by measure.	V. Bulk of mixture.	VI. Diminution of bulk.	VII. Quantity of spirit per cent.	VIII. Decimal multipliers.
Sp. + W.							
100 + 0	,81530	100	—	100,00	—	100,00	,9882
1	,81761	—	0,82	100,71	0,11	99,29	,9812
2	,81987	—	1,63	101,43	0,20	98,59	,9743
3	,82207	—	2,45	102,15	0,30	97,89	,9674
4	,82422	—	3,27	102,87	0,40	97,21	,9606
100 + 5	,82631	—	4,09	103,60	0,49	96,52	,9539
6	,82835	—	4,90	104,33	0,57	95,85	,9472
7	,83034	—	5,72	105,06	0,66	95,18	,9406
8	,83228	—	6,54	105,80	0,74	94,52	,9340
9	,83418	—	7,35	106,53	0,82	93,86	,9275
100 + 10	,83603	—	8,17	107,27	0,90	93,22	,9212
11	,83783	—	8,99	108,01	0,98	92,58	,9149
12	,83960	—	9,81	108,76	1,05	91,95	,9087
13	,84133	—	10,62	109,50	1,12	91,32	,9025
14	,84302	—	11,44	110,25	1,19	90,70	,8964
100 + 15	,84467	—	12,26	111,00	1,26	90,09	,8903
16	,84629	—	13,08	111,75	1,33	89,48	,8843
17	,84788	—	13,89	112,50	1,39	88,88	,8784
18	,84944	—	14,71	113,26	1,45	88,29	,8726
19	,85097	—	15,53	114,01	1,52	87,71	,8668
100 + 20	,85248	—	16,34	114,77	1,57	87,13	,8611
21	,85397	—	17,16	115,52	1,64	86,56	,8555
22	,85543	—	17,98	116,28	1,70	86,00	,8499
23	,85686	—	18,80	117,03	1,77	85,45	,8444
24	,85827	—	19,61	117,79	1,82	84,90	,8390
100 + 25	,85966	—	20,43	118,55	1,88	84,35	,8336
26	,86102	—	21,25	119,31	1,94	83,81	,8283
27	,86235	—	22,07	120,07	2,00	83,28	,8231
28	,86307	—	22,88	120,83	2,05	82,76	,8179
29	,86496	—	23,70	121,59	2,11	82,24	,8128
100 + 30	,86622	—	24,52	122,36	2,16	81,73	,8077
31	,86748	—	25,33	123,12	2,21	81,22	,8027
32	,86872	—	26,15	123,89	2,26	80,72	,7977
33	,86993	—	26,97	124,65	2,32	80,23	,7928
34	,87112	—	27,79	125,42	2,37	79,74	,7880
100 + 35	,87228	—	28,60	126,18	2,42	79,25	,7832
36	,87341	—	29,42	126,95	2,47	78,77	,7784
37	,87452	—	30,24	127,72	2,52	78,29	,7737
38	,87562	—	31,06	128,49	2,57	77,82	,7691
39	,87670	—	31,87	129,26	2,61	77,36	,7645
100 + 40	,87776	—	32,69	130,04	2,65	76,90	,7600
41	,87884	—	33,51	130,81	2,70	76,45	,7555
42	,87990	—	34,32	131,57	2,75	76,00	,7511
43	,88095	—	35,14	132,34	2,80	75,56	,7467
44	,88199	—	35,96	133,11	2,85	75,12	,7424
100 + 45	,88301	—	36,78	133,88	2,90	74,69	,7381
46	,88400	—	37,59	134,65	2,94	74,26	,7339
47	,88498	—	38,41	135,43	2,98	73,84	,7297
48	,88594	—	39,23	136,20	3,03	73,42	,7256
49	,88688	—	40,04	136,98	3,06	73,01	,7215

Right half

I. Spirit and water by weight.	II. Specific gravity.	III. Spirit by measure.	IV. Water by measure.	V. Bulk of mixture.	VI. Diminution of bulk.	VII. Quantity of spirit per cent.	VIII. Decimal multipliers.
Sp. + W.							
100 + 50	,88781	100	40,86	137,75	3,11	72,60	,7174
51	,88873	—	41,68	138,53	3,15	72,19	,7134
52	,88963	—	42,50	139,30	3,20	71,79	,7094
53	,89052	—	43,31	140,08	3,23	71,40	,7055
54	,89139	—	44,13	140,86	3,27	71,01	,7016
100 + 55	,89225	—	44,95	141,63	3,32	70,61	,6977
56	,89310	—	45,77	142,41	3,36	70,22	,6939
57	,89394	—	46,58	143,19	3,39	69,84	,6901
58	,89477	—	47,40	143,97	3,43	69,46	,6864
59	,89559	—	48,22	144,75	3,47	69,08	,6827
100 + 60	,89639	—	49,04	145,53	3,51	68,71	,6791
61	,89718	—	49,85	146,31	3,54	68,35	,6755
62	,89795	—	50,67	147,09	3,58	67,98	,6719
63	,89871	—	51,49	147,87	3,62	67,62	,6683
64	,89947	—	52,30	148,65	3,65	67,27	,6648
100 + 65	,90021	—	53,12	149,44	3,68	66,92	,6613
66	,90095	—	53,94	150,22	3,72	66,57	,6579
67	,90169	—	54,76	151,00	3,76	66,22	,6545
68	,90242	—	55,57	151,78	3,79	65,88	,6511
69	,90314	—	56,39	152,56	3,83	65,55	,6478
100 + 70	,90385	—	57,21	153,34	3,87	65,21	,6445
71	,90454	—	58,03	154,13	3,90	64,88	,6412
72	,90522	—	58,84	154,91	3,93	64,55	,6379
73	,90590	—	59,66	155,70	3,96	64,23	,6347
74	,90657	—	60,48	156,48	4,00	63,91	,6315
100 + 75	,90723	—	61,29	157,27	4,02	63,59	,6284
76	,90788	—	62,11	158,05	4,06	63,27	,6253
77	,90853	—	62,93	158,84	4,09	62,96	,6222
78	,90917	—	63,75	159,62	4,13	62,65	,6191
79	,90980	—	64,56	160,41	4,15	62,34	,6161
100 + 80	,91046	—	65,38	161,19	4,19	62,04	,6131
81	,91103	—	66,20	161,98	4,22	61,74	,6101
82	,91163	—	67,02	162,77	4,25	61,44	,6071
83	,91223	—	67,83	163,55	4,28	61,14	,6042
84	,91282	—	68,65	164,34	4,31	60,85	,6013
100 + 85	,91340	—	69,47	165,13	4,34	60,56	,5985
86	,91398	—	70,28	165,92	4,36	60,27	,5956
87	,91455	—	71,10	166,71	4,39	59,99	,5928
88	,91511	—	71,92	167,49	4,43	59,70	,5900
89	,91567	—	72,74	168,28	4,46	59,42	,5872
100 + 90	,91622	—	73,55	169,07	4,48	59,15	,5845
91	,91677	—	74,37	169,86	4,51	58,87	,5818
92	,91721	—	75,19	170,65	4,54	58,60	,5791
93	,91785	—	76,00	171,44	4,56	58,33	,5764
94	,91838	—	76,82	172,22	4,60	58,06	,5738
100 + 95	,91891	—	77,64	173,01	4,63	57,80	,5712
96	,91942	—	78,45	173,79	4,66	57,54	,5686
97	,91993	—	79,26	174,58	4,68	57,28	,5661
98	,92043	—	80,08	175,37	4,71	57,02	,5635
99	,92093	—	80,90	176,16	4,74	56,76	,5610

TABLE II.　　　　　　　　HEAT 80°.

I. Water and spirit by weight.	II. Specific gravity.	III. Spirit by measure.	IV. Water by measure.	V. Bulk of mixture.	VI. Diminution of bulk.	VII. Quantity of spirit per cent.	VIII. Decimal multipliers.
W. + Sp.							
100 + 100	,92142	100	81,73	176,96	4,77	56,51	,5584
99	,92192	—	82,56	177,77	4,79	56,25	,5559
98	,92242	—	83,40	178,59	4,81	55,99	,5533
97	,92293	—	84,26	179,42	4,84	55,73	,5508
96	,92343	—	85,14	180,27	4,87	55,47	,5482
100 + 95	,92393	—	86,03	181,13	4,90	55,21	,5456
94	,92443	—	86,94	182,02	4,92	54,94	,5429
93	,92493	—	87,87	182,93	4,94	54,67	,5402
92	,92544	—	88,83	183,86	4,97	54,39	,5375
91	,92595	—	89,81	184,81	5,00	54,11	,5347
100 + 90	,92646	—	90,81	185,78	5,03	53,83	,5319
89	,92699	—	91,83	186,77	5,06	53,54	,5291
88	,92753	—	92,87	187,78	5,09	53,26	,5262
87	,92807	—	93,94	188,82	5,12	52,96	,5233
86	,92862	—	95,03	189,88	5,15	52,66	,5204
100 + 85	,92917	—	96,15	190,96	5,19	52,36	,5175
84	,92973	—	97,29	192,07	5,22	52,06	,5144
83	,93029	—	98,46	193,21	5,25	51,75	,5114
82	,93086	—	99,66	194,38	5,28	51,44	,5083
81	,93144	—	100,90	195,58	5,32	51,13	,5052
100 + 80	,93201	—	102,16	196,80	5,36	50,81	,5021
79	,93259	—	103,45	198,07	5,38	50,48	,4989
78	,93315	—	104,77	199,37	5,40	50,15	,4957
77	,93372	—	106,13	200,70	5,43	49,82	,4924
76	,93430	—	107,53	202,07	5,46	49,48	,4890
100 + 75	,93488	—	108,97	203,48	5,49	49,14	,4856
74	,93546	—	110,44	204,93	5,51	48,80	,4822
73	,93605	—	111,95	206,41	5,54	48,45	,4788
72	,93664	—	113,50	207,94	5,56	48,09	,4753
71	,93724	—	115,10	209,51	5,59	47,73	,4717
100 + 70	,93785	—	116,75	211,12	5,63	47,37	,4681
69	,93845	—	118,44	212,78	5,66	47,00	,4645
68	,93910	—	120,18	214,48	5,70	46,62	,4608
67	,93973	—	121,97	216,25	5,72	46,24	,4570
66	,94037	—	123,82	218,06	5,76	45,86	,4532
100 + 65	,94102	—	125,73	219,93	5,80	45,47	,4493
64	,94167	—	127,69	221,86	5,83	45,07	,4454
63	,94232	—	129,72	223,85	5,87	44,67	,4415
62	,94298	—	131,81	225,91	5,90	44,26	,4375
61	,94364	—	133,98	228,04	5,94	43,85	,4334
100 + 60	,94431	—	136,21	230,23	5,98	43,43	,4292
59	,94498	—	138,51	232,50	6,01	43,01	,4250
58	,94565	—	140,89	234,86	6,03	42,58	,4207
57	,94632	—	143,37	237,30	6,07	42,14	,4164
56	,94700	—	145,94	239,83	6,11	41,69	,4120
100 + 55	,94768	—	148,58	242,45	6,13	41,24	,4076
54	,94836	—	151,31	245,16	6,15	40,79	,4031
53	,94904	—	154,16	247,99	6,17	40,33	,3985
52	,94973	—	157,13	250,93	6,20	39,85	,3938
51	,95042	—	160,23	253,99	6,24	39,37	,3891
W. + Sp.							
100 + 50	,95111	100	163,45	257,16	6,29	38,89	,3843
49	,95181	—	166,79	260,46	6,33	38,39	,3794
48	,95252	—	170,26	263,91	6,35	37,89	,3744
47	,95324	—	173,89	267,50	6,39	37,38	,3694
46	,95395	—	177,66	271,25	6,41	36,86	,3643
100 + 45	,95467	—	181,61	275,18	6,43	36,34	,3591
44	,95538	—	185,74	279,29	6,45	35,81	,3538
43	,95610	—	190,06	283,60	6,46	35,26	,3485
42	,95682	—	194,59	288,11	6,48	34,71	,3431
41	,95754	—	199,33	292,82	6,50	34,15	,3375
100 + 40	,95826	—	204,32	297,79	6,53	33,58	,3319
39	,95898	—	209,56	303,02	6,54	33,00	,3261
38	,95971	—	215,07	308,53	6,54	32,41	,3203
37	,96044	—	220,88	314,34	6,54	31,81	,3144
36	,96118	—	227,02	320,47	6,55	31,20	,3084
100 + 35	,96192	—	233,51	326,92	6,59	30,59	,3023
34	,96266	—	240,37	333,79	6,58	29,96	,2961
33	,96341	—	247,66	341,09	6,57	29,32	,2898
32	,96416	—	255,40	348,83	6,57	28,67	,2833
31	,96492	—	263,64	357,07	6,57	28,01	,2768
100 + 30	,96568	—	272,42	365,85	6,57	27,33	,2701
29	,96645	—	281,82	375,26	6,56	26,65	,2634
28	,96723	—	291,88	385,34	6,54	25,95	,2565
27	,96802	—	302,69	396,16	6,52	25,24	,2495
26	,96882	—	314,33	407,81	6,52	24,52	,2424
100 + 25	,96963	—	326,91	420,42	6,49	23,79	,2351
24	,97045	—	340,53	434,07	6,46	23,04	,2277
23	,97128	—	355,33	448,90	6,43	22,27	,2202
22	,97212	—	371,49	466,09	6,40	21,50	,2125
21	,97298	—	389,18	482,81	6,37	20,71	,2047
100 + 20	,97385	—	408,64	502,32	6,32	19,91	,1967
19	,97473	—	430,14	523,87	6,27	19,09	,1886
18	,97563	—	454,04	547,83	6,21	18,25	,1804
17	,97655	—	480,75	574,59	6,16	17,40	,1720
16	,97749	—	510,79	604,70	6,09	16,54	,1635
100 + 15	,97845	—	544,85	638,83	6,02	15,65	,1547
14	,97943	—	583,76	677,84	5,92	14,75	,1458
13	,98044	—	628,67	722,82	5,85	13,83	,1367
12	,98148	—	681,06	775,30	5,76	12,89	,1275
11	,98256	—	742,97	837,32	5,65	11,94	,1180
100 + 10	,98367	—	817,27	911,70	5,57	10,97	,1084
9	,98482	—	908,08	1002,64	5,44	9,97	,0986
8	,98602	—	1021,59	1116,26	5,33	8,96	,0885
7	,98727	—	1167,53	1262,32	5,21	7,92	,0783
6	,98856	—	1362,12	1457,03	5,09	6,86	,0678
100 + 5	,98991	—	1634,54	1729,58	4,96	5,78	,0571
4	,99131	—	2043,18	2138,18	4,82	4,68	,0462
3	,99278	—	2724,23	2819,54	4,69	3,55	,0350
2	,99432	—	4086,35	4181,79	4,56	2,39	,0236
1	,99592	—	8172,70	8268,20	4,44	1,21	,0120

Specific gravity of water at the different degrees of heat.

Heat.	Specific gravity.	Heat.	Specific gravity.	Heat.	Specific gravity.	Heat.	Specific gravity.	Heat.	Specific gravity.	Heat.	Specific gravity.	Heat.	Specific gravity.	Heat.	Specific gravity.	Heat.	Specific gravity.	Heat.	Specific gravity.
30	1,00074	35	1,00090	40	1,00094	45	1,00086	50	1,00068	55	1,00038	60	1,00000	65	,99950	70	,99894	75	,99830
31	1,00078	36	1,00092	41	1,00093	46	1,00083	51	1,00063	56	1,00031	61	,99991	66	,99939	71	,99882	76	,99816
32	1,00082	37	1,00093	42	1,00092	47	1,00080	52	1,00057	57	1,00024	62	,99981	67	,99928	72	,99869	77	,99802
33	1,00085	38	1,00094	43	1,00090	48	1,00076	53	1,00051	58	1,00016	63	,99971	68	,99917	73	,99856	78	,99788
34	1,00088	39	1,00094	44	1,00088	49	1,00072	54	1,00045	59	1,00008	64	,99961	69	,99926	74	,99843	79	,99774
																		80	,99759

Although the titles of the Columns in the preceding Tables, with what has been said in the introductory discourse, may render the Tables sufficiently obvious to the generality of readers, yet to some perhaps an example may be necessary; I shall therefore shew in what manner the quantity by measure of pure Spirit of ,825 specific gravity at 60° of heat, may readily be found by the help of these Tables.—In order to which, we must know the heat, the specific gravity, and quantity of spirit which the vessel contains.

EXAMPLE. Suppose the heat to be 35°, the specific gravity ,909, and the quantity of spirit 138,99 measures.

Under 35° of heat, and in Column II. of specific gravity, find ,909 ; and in the same horizontal line, take out from Column VIII, the decimal multiplier ,7297, by which multiply 138,99, cutting off as many figures to the right as there are decimals in both factors ; then we shall have 101,421003 for the measures of pure spirit, of the specific gravity ,825 at 60° of heat.

Now a mere inspection of the Columns in the Tables will shew that a spirit of that strength was obtained, by adding 51 parts of water by weight to 100 parts of spirit, as in Column I. which produced the specific gravity found in the same horizontal line, Column II. It will as readily be seen, that the same specific gravity results from adding together their equivalents in measure, Columns III. and IV. and that Column V. contains the quantity, which the two quantities really measure after the mixture has been made.

ERRATA.

Page 281, column IV. line 34, *for* 126,12, *read* 125,12.
283, ———— I. — 17, *for* 100 + 54, *read* 100 + 34.
Ibid. ———— IV. — 34, *for* 126,05, *read* 125,05.
Ibid. ———— V. — 2 from bottom, *for* 4886,96, *read* 4286,96.
285, ———— VIII. — 23, *for* ,5046, *read* ,5056.
286, ———— IV. — 8 from bottom, *for* 77,99, *read* 76,99.
287, ———— VIII. — 23, *for* ,5044, *read* ,5054.
289, ———— II. — 28, *for* ,95485, *read* ,95385.
290, ———— IV. — 1 *dele* 0,84.

www.ingramcontent.com/pod-product-compliance
Lightning Source LLC
Chambersburg PA
CBHW021826190326
41518CB00007B/760